Welding Essentials:
Questions & Answers
Second Edition

_____

William L. Galvery, Jr.

Professor of Welding Technology
American Welding Society QC-1, CWI, CWE
Orange Coast College
Costa Mesa, California

and

Frank M. Marlow, PE

_____

Illustrations by

Pamela Tallman
&
Lawrence Smith

Industrial Press Inc.

STATEMENT OF NON-LIABILITY

The process of welding is inherently hazardous. Welding is often used to assemble buildings, structures, vehicles, and devices whose failure can lead to embarrassment, property damage, injury, or death. The authors have carefully checked the information in this book, have had others with professional welding experience and credentials review it as well and believe that it is correct and in agreement with welding industry practices and standards. However, the authors cannot provide for every possibility, contingency, and the actions of others. If you decide to apply the information, procedures, and advice in this book, the authors refuse to be held responsible for your actions. If you decide to perform welding, use common sense, have a competent person check your designs, the welding processes, and the completed welds themselves. Seek the help of people with professional welding inspection credentials to check critical welds until you gain experience. Do not attempt life-critical welds without competent on-the-spot assistance.

ISBN 978-0-8311-3301-6

Manufactured in the U.S.A.

Visit us at our website:
www.industrialpress.com

10   9   8   7   6   5   4   3   2

About the Authors

William L. Galvery graduated from California State University, Long Beach with a Bachelor of Vocational Education, has more than 30 years of industrial welding experience, and is an American Welding Society (AWS) Certified Welding Inspector and a Certified Welding Educator. In 2003 the AWS presented him with its prestigious national teaching award, the Howard E. Adkins Memorial Instructor Award, and for the second time he was chosen AWS District 21 Educator of the Year. In the same year he was also presented with an Excellence in Education Award by the University of Texas, Austin. Bill has served as an officer for the AWS Long Beach/Orange County Section and is currently Professor of Welding Technology and Welding Department coordinator at Orange Coast College in Costa Mesa, California.

Frank M. Marlow is a Registered Professional Engineer and holds a BA and BSEE from Lehigh University, an MSEE from Northeastern University and an MBA from the University of Arizona. With a background in electronic circuit design, industrial power supplies, and electrical safety, he has worked for Avco, Boeing, Raytheon, DuPont, and Emerson Electric. Frank has served as both Secretary and Treasurer of the AWS Long Beach/Orange County Section and is author of *Welding Fabrication and Repair: Questions and Answers* (Industrial Press) and *Machine Shop Essentials: Questions and Answers* (Metal Arts Press).

Contents

Acknowledgements

The authors would like to express their thanks and appreciation to the following organizations which provided invaluable illustration source materials for this book.

The American Welding Society granted us permission to use illustrations from several publications, including:

- *Welding Handbook, Volume 1, Welding Technology*
- *Welding Handbook, Volume 2, Welding Processes*
- *Introductory Welding Metallurgy*
- *Welding Inspection Technology*
- *Standard Welding Terms and Definitions*
- *Brazing Handbook*
- *Structural Welding Code — Steel, ANSI/AWS D1.1*

These publications are the classics of the welding industry and they greatly helped us illustrate the text. AWS official terminology is used throughout the book.

ESAB Welding and Cutting Products graciously gave permission to use illustrations from their publication *The Oxyacetylene Handbook.*

Lincoln Electric Company kindly provided product technical information on SMAW electrodes and permission to reproduce illustrations from their publication *Weldirectory—Stick Electrodes for Carbon and Low Alloy Steel.*

Thermadyne/Victor provided technical information and illustrations on oxyacetylene torches, check valves and flame arrestors.

The authors want to thank the reviewers of this book: John Brady, Gene Lawson, Bob Saddler, David Randal, Professor Richard Hutchison and his wife Catherine.

Thanks also to Lawrence Smith and Joan Cordova for their assistance in completing changes for the second edition.

Introduction

The authors had the following objectives in mind when writing this book:

- Develop an easy-to-read, concise book with all the technical and safety information a beginning welder needs.
- Use a question-and-answer format to divide this information into small, easily understandable blocks.
- Provide illustrations to clarify and detail every explanation.
- Eliminate non-essential items to prevent overloading the reader.
- In a single chapter cover each major, commercially important process: oxyacetylene welding, oxyfuel cutting, brazing & soldering, SMAW, GMAW/FCAW, and GTAW/PAW.
- Describe less important welding processes in a survey chapter.
- Explain the effects of welding heat on metals.
- Present welding power supply and electrical safety issues.
- Introduce the beginning welder to weld inspection, welding symbols and qualification and certification issues.

Because this book initially did not provide much practical how-to-do-it advice in this expanded first edition we have added a chapter on fabrication and repair tips: assembling angle iron frames, repairing cracked heavy truck C-channels, and soldering copper tubing. These are relatively common tasks the welder will see and we want him to know how to do them.

We have not included torch manipulation instructions. This is a hands-on skill that is best learned from classroom demonstrations, personal instruction.

Although this book is primarily based on English measurement units, we have usually included metric units also.

WILLIAM L. GALVERY, JR.　　　　FRANK M. MARLOW

Costa Mesa, California　　　　Huntington Beach, California

Chapter 1

Oxyacetylene Welding

The chapter of knowledge is a very short one, but the chapter of accidents is a very long one.

Philip Dormer Stanhope

Introduction

Oxyacetylene welding was first used industrially in the early years of the twentieth century. Although this process makes excellent welds in steel, it is little used for welding today except for a few specialties (light aircraft, race car frames and American Petroleum Institute natural gas distribution), since there are other more efficient welding processes available. However, oxyacetylene has many other important uses: cutting, hardening, tempering, bending, forming, preheating, postheating, brazing, and braze welding. Because of the precise control the welder has over heat input and its high-temperature flame, together with its low equipment cost, portability, and versatility, it remains an essential tool. No industrial shop is complete without an oxyacetylene outfit. As with all effective tools, using oxyacetylene carries risk. We will cover the theory and use of oxyacetylene equipment so you can use them with confidence and safety. It will also prepare you for the next chapter on oxyfuel cutting, because many components and issues are common to both processes.

Process Name

What is the name covering all welding processes using oxygen and a fuel gas?
Oxygen fuel (oxyfuel) welding.

What is the American Welding Society (AWS) abbreviation for oxyfuel welding?
The abbreviation for all oxyfuel welding processes; those using oxygen and any fuel gas is *OFW*.

A particularly important member of the OFW process family is oxy-acetylene welding. What is the AWS abbreviation for this process?
The abbreviation for oxyacetylene welding is *OAW*. Note that OAW is just one member of the OFW family.

Equipment

Figure 1-1 shows and labels the components of the basic oxyacetylene welding equipment (outfit) showing how they are connected.

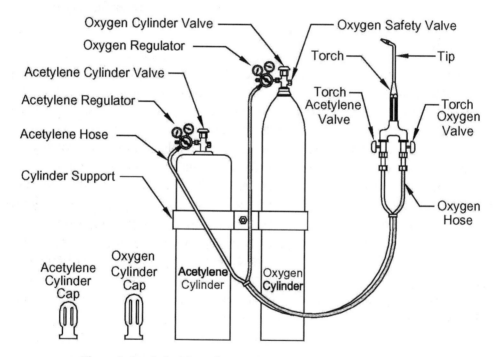

Figure 1–1   A drawing of oxygen acetylene welding equipment

Process

How does oxyacetylene equipment perform the welding process?
Oxygen and acetylene gases when combined, in the proper proportions, in the torch handle's mixing chamber, produce an approximately 5600°F (3100°C) flame at the torch tip. This flame melts the edges of the base metals to be joined into a common pool. Sometimes additional filler metal is added to the molten

pool from a welding rod. When this common pool cools and the metal freezes solid, the joined metals are fused together and the weld complete.

What are some advantages of the OAW process?
- Low cost
- Readily portable
- Excellent control of heat input and puddle viscosity
- No external power required
- Good control of bead size and shape
- Fuel mixture is hot enough to melt steel

Setup (and related safety)

List the steps for installation of oxygen and acetylene pressure regulators on full cylinders. Be sure to include all safety precautions.
- Put on your welding safety equipment: tinted safety goggles (or tinted face shield), cotton or wool shirt and pants, high-top shoes, and welding gloves at a minimum.
- Make sure the valves on previously used or empty cylinders are fully closed and their valve protection covers are securely screwed in place. Then remove the empty cylinders from the work area and secure them against tipping during the wait for a refill shipment. Secure the newly replaced or full cylinders to a welding cart, building column, or other solid anchor to prevent the cylinders from tipping over during storage or use.
- Momentarily open each cylinder valve to the atmosphere and reclose the valve quickly purging the valve; this is known as cracking a valve. Cracking serves to blow out dust and grit from the valve port and to prevent debris from entering the regulators and torch.
- With a clean, oil-free cloth, wipe off the cylinder valve-to-regulator fittings on both cylinders to remove dirt and grit from the fittings' connection faces and from the fittings' threads. Do the same to both regulators' threads and faces. Remember, never use oil on high-pressure gas fittings. Oxygen at high pressures can accelerate combustion of oil into an explosion.
- Make sure reverse-flow check valves are installed on the torch or the regulators.
- Check to see that both the oxygen and acetylene regulator pressure adjustment screws are unscrewed, followed by threading each regulator to its respective cylinder. Snug up the connections with a wrench. Caution: Oxygen cylinder-to-regulator threads are right-handed; so are oxygen hose-

to-torch screw fittings. Acetylene cylinder-to-regulator fittings and acetylene hose-to-torch fittings threads are left-handed. This arrangement prevents putting the wrong gas into a regulator or torch connection.

- Stand so the cylinders are between you and the regulators, S-L-O-W-L-Y open the oxygen cylinder valves. Open the oxygen cylinder valve until it hits the upper valve stop and will turn no further. Also standing so the cylinders are between you and the regulator, open the acetylene cylinder valve gradually and not more than 1 1/2 turns. If there is an old-style removable wrench on the acetylene cylinder, keep it on the valve in case you must close it in an emergency.

- Look at the high-pressure—cylinder side—pressure gauges to indicate about 225 psi (15.5 bar) in the acetylene cylinder and 2250 psi (155 bar) in the oxygen cylinder. Note: 1 bar = 1 atmosphere = 14.5 psi = 0.1 MPa. Cylinder pressures vary with ambient temperature. The pressures given above are for full cylinders at 70°F (21°C).

- Purge each torch hose of air separately: Open the oxygen valve on the torch about three-quarters of a turn, then screw in the pressure control screw on the oxygen regulator to your initial pressure setting—about 6 psi (0.4 bar). After several seconds, close the torch valve. Do the same for the acetylene hose. Comment: We do this for two reasons, (1) to make sure we are lighting the torch on just oxygen and acetylene, not air, and (2) to set the regulators for the correct pressure while the gas is flowing through them.

- Caution: never adjust the acetylene regulator pressure above 15 psi (1 bar) as an explosive disassociation of the acetylene could occur.

- Recheck the low-pressure gauge pressures to make sure the working pressures are not rising. If the working pressure rises, it means the regulator is leaking. Immediately shut down the cylinders at the cylinder valves as continued leaking could lead to a regulator diaphragm rupture and a serious accident. Replace and repair the defective regulator.

- Test the system for leaks at the cylinder-to-regulator fittings and all hose fittings with special leak detection solutions; bubbles indicate leaks.

If you are using a small to medium torch tip on a job for the first time, what regulator pressures should be set as a starting point?
Set both the acetylene and the oxygen regulator pressures to 6psi (0.4 bar).

What are the steps for adjusting the torch to a neutral flame?
- Open the acetylene valve no more than 1/16 turn and use a spark lighter to ignite the gas coming out of the tip. A smoky orange flame will be the result, Figure 1–2 (A).

- Continue to open the acetylene valve until the flame stops smoking (releasing soot). Another way to judge the proper amount of acetylene is to open the acetylene valve until the flame jumps away from the torch tip, leaving about 1/16 inch gap (1.6 mm), Figure 1–2 (B). Then close the valve until flame touches the torch tip.
- Open the oxygen valve slowly. As the oxygen is increased, the orange acetylene flame turns purple and a smaller, white inner cone will begin to form. With the further addition of oxygen, the inner cone goes from having ragged edges, Figure 1–2 (C), to sharp, clearly defined ones. The flame is now neutral and adding oxygen will make an oxidizing flame, Figure 1–2 (D).
- If a larger flame is needed while keeping the same tip size, the acetylene may be increased and the oxygen further increased to keep the inner cone's edges sharp. This process of increasing the acetylene, then the oxygen is usually done in several cycles before the maximum flame available from a given tip is achieved. Adjusting the flame below the minimum flow rate for the tip orifice permits the flame to ignite *inside* the nozzle. This is *flashback* and makes a popping sound. If you need a smaller flame, use a smaller torch tip. See the section on flashback.

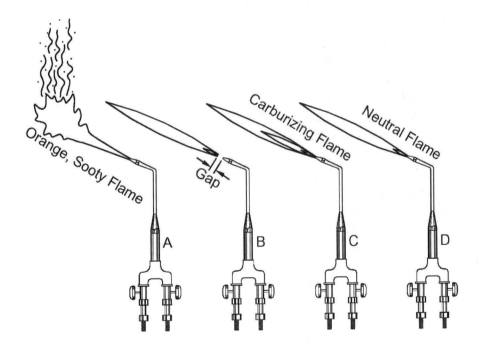

Figure 1–2   Shows flame adjustments from carburizing to a neutral flame

What it the hottest part of a neutral flame?

The tip of the inner cone is the hottest part of the flame. The inner cone is where the optimum mixture of oxygen and acetylene burn. The outer envelope where any unburned acetylene burns with oxygen from the atmosphere. A neutral flame is when enough oxygen is present in the flame to be burning all of the acetylene gas and is used for most welding processes. See Figure 1–3.

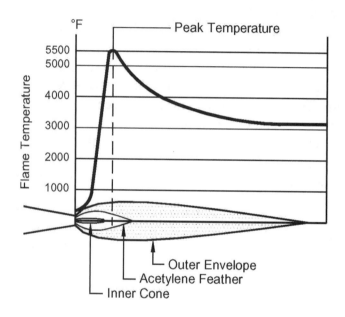

Figure 1–3   Graph of an oxyacetylene flame temperature profile

What effect do oxidizing and carburizing flames have on molten metal in the weld pool?

An oxidizing flame contains more oxygen than the flame can burn and this oxygen combines (or burns) with carbon in the steel to carbon dioxide gas. The result is the weld metal has a change in carbon content and in its properties. Strength is always degraded and brittleness increased.

A carburizing flame contains more acetylene than the flame can burn and the carbon in the acetylene adds to the carbon in the weld pool causing gas bubbles in the weld. When the weld freezes these gas bubbles create porosity holes.

What are the proper steps to shut down an oxyacetylene torch and its cylinders?

- First turn off the oxygen and then the acetylene with the torch handle valves. Turning off the acetylene first can cause a flashback.
- Turn off the oxygen and acetylene cylinder valves at the upstream side of the regulators.
- Separately, open and reclose the oxygen and acetylene valves on the torch handle to bleed the remaining gas in the hoses and regulator into the atmosphere. Verify that both the high-pressure and low-pressure gauges on both regulators indicate zero.
- Unscrew the regulator pressure adjustment screws on both cylinders in preparation for the next use of the equipment. The regulator screws should be loose but not about to fall from their threads.

Gases

What is acetylene gas?

It is a clear gas having a specific gravity slightly lighter than air at 0.906 (air = 1.000). Acetylene's chemical formula, C_2H_2, indicates that each molecule of this hydrocarbon compound contains two carbon atoms and two hydrogen atoms.

What is the odor of acetylene gas?

It has a distinctive garlic odor. Because the liquid acetone in the acetylene cylinder also has an odor, this acetone odor is frequently mistaken for that of acetylene, when in fact it is the odor of the mixture of *both* acetylene and acetone.

How is acetylene made?

Acetylene results from dissolving calcium carbide in water and capturing the resulting gas. One pound of calcium carbide generates about 10 cubic feet of acetylene (1 kg calcium carbide generates about 618 liters of acetylene).

Where does calcium carbide come from?

Calcium carbide results from an industrial process where lime and coke are smelted in an electric furnace. A gray, hard solid, it is supplied in a variety of forms: bricks, powders, pellets, or granules.

What is the chemical equation of combustion of acetylene in a neutral flame?

$$C_2H_2 + 2.5O_2 \rightarrow 2CO_2 + H_2O$$

What does this equation tell us?
One part acetylene and two and a half parts oxygen combine to produce a neutral flame (a neutral flame has just the right amounts of fuel and oxygen so there is neither an excess of oxygen or fuel after combustion). What the equation does not tell us is that equal volumes of acetylene and oxygen from the compressed gas cylinders combine with another one and a half parts of oxygen from the atmosphere to make the flame.

How is oxygen made for welding?
Atmospheric air is repeatedly cooled and compressed until it becomes a very cold liquid. This liquid is gradually warmed, and as each component gas of the liquid air reaches its vaporization temperature, it comes out of the liquid air, and separates itself. This is the fractional distillation of liquid air. Other gases important in welding—nitrogen, carbon dioxide, and argon—are also made using this process. Oxygen can also be made by electrolysis of water, but this is not a cost-effective process to make industrial quantities.

What are the two main ways of supplying welding shops with oxygen?
Compressed gas cylinders are used in smaller shops; liquid oxygen cylinders in larger shops. The liquid oxygen flows from its cylinder into a radiator that warms the liquid oxygen, and converts it into gaseous oxygen.

How are large welding shops supplied with acetylene?
Multiple cylinders are manifolded together and their output piped around the plant to each welding or cutting station.

What is peculiar about the filling and draining of acetylene cylinders?
Because the acetylene is dissolved in acetone, not just pumped into the pressure vessel, the filling process takes seven hours as the absorption process occurs. Similarly, an acetylene cylinder can only deliver one-seventh of its capacity per hour as the acetylene will not come out of solution in the acetone faster. More acetylene capacity will require cylinders manifolded together. This can become an important issue when using large multi-flame heating tips (in the industry called a rosebud tip) which consume many times more gas than a welding tip.

Why is acetylene potentially so dangerous?

Acetylene will form explosive mixtures with air at *all* concentrations between 2.5 and 80%. This is the widest range of any common gas and almost insures an explosion if leaking gas is ignited.

Can other fuel gases be used in place of acetylene?

Certainly, but their maximum heat potential is below that required for welding steel. Acetylene is the best gas for welding because it:

- Has the highest temperature of all fuel gases.
- Acetylene delivers a higher concentration of heat than other fuel gases.
- Has the lowest chemical interaction with the weld pool's molten metal than all other gases.

However, other gases such as natural gas, methylacetylene-propradene stabilized (also called MPS or MAPP® gas), propane, hydrogen, and proprietary gases based on mixtures of these are frequently used for other non-welding processes for cost reasons. They work well for soldering, brazing, preheating, and oxygen cutting, and are seldom used for welding. Small changes, like different torch tips, may be necessary to accommodate alternate fuel gases. Table 1–1 shows the maximum temperature achievable with different fuel gases. Where even lower temperatures are needed (sweating copper tubing and many small soldering tasks) a single cylinder of fuel gas using only atmospheric oxygen is effective and economical.

Fuel Gas	Oxygen-to-Fuel Gas Combustion Ratio	Neutral Flame Temperature with Oxygen		Total Heat Btu/ft^3	Specific Gravity (Air = 1.0)
		°F	°C		
Acetylene	2.5	5589	3087	1470	0.906
Propane	5.0	4579	2526	2498	1.52
Methylacetylene-Propadiene (MPS, MAPP®)	4.0	5301	2927	2460	1.48
Propylene	4.5	5250	2900	2400	1.48
Natural Gas	2.0	4600	2538	1000	0.60
Hydrogen	0.5	4820	2660	325	0.07

Table 1–1   Combustion properties of fuel gases

Compressed Gas Cylinders

What is the difference between acetylene cylinder and oxygen cylinder construction?
Oxygen cylinders are seamless vessels of special high-strength alloy steel. They are made from a single billet by a draw-forming process and they contain no welds. Acetylene cylinders are fabricated and contain welds.

What materials other than acetylene are found inside acetylene cylinders?
Under certain conditions above 15 psi (1 bar), acetylene may spontaneously disassociate into its components of carbon and hydrogen. Acetylene cylinders are packed with an inert porous monolithic filler to prevent this dangerous disassociation. Acetylene cylinders are also contains acetone that can dissolve 25 times its own volume of acetylene per atmosphere of pressure. This greatly increases the cylinder's acetylene capacity.

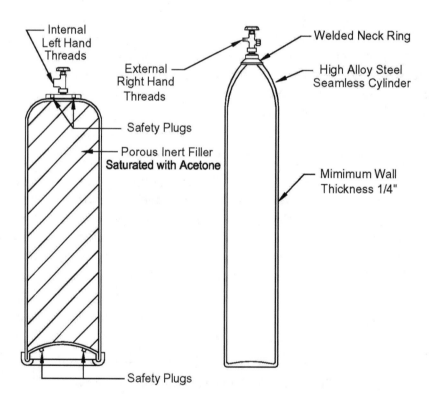

Figure 1–4 Oxygen and acetylene cylinder cross sections

What is the purpose of safety valves and plugs in oxygen and acetylene cylinders?

Their purpose is to prevent the cylinder bursting from overpressure when it is heated. Oxygen cylinders have a small metal diaphragm in a section of the valve which ruptures, releasing cylinder pressure to the atmosphere and preventing a cylinder burst. Disk rupture occurs above 3360psi (232 bar), the cylinder test pressure.

Acetylene cylinders contain one to four fusible safety plugs depending on their capacity. These fusible plugs, made of a special metal alloy, melt at 212°F (100°C).

They also release the cylinder contents to atmosphere to prevent rupturing (and then exploding) when the cylinder is exposed to excessive temperatures, usually from a fire. Acetylene cylinders may have the plugs on the top, or top and bottom.

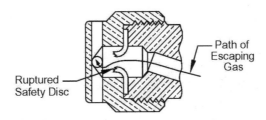

Figure 1–5   Detail of pressure safety relief on oxygen valve

Why should the welder open the oxygen cylinder valve all the way, but open the acetylene cylinder valve just one turn?

Because the oxygen cylinder is filled to such a high pressure (2250 psi or 155 bar) to prevent leakage around the valve stem, oxygen and all other high-pressure cylinders have a second valve seat to make a solid seal around the valve stem when the valve is open. See Figure 1–6. Because the acetylene cylinder valve sees a relatively low pressure (225 psi or 15.5 bar), leakage around the valve stem in use is small and a single seat is used. Since the acetylene valve can deliver adequate volume with one turn open, opening the valve more just increases the closing time in an emergency. For similar reasons the welder must never remove the removable wrench from the valve of old-style acetylene cylinders while the cylinder is in use.

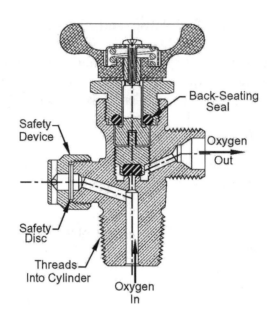

Figure 1–6   Details a cross section of oxygen valve

Why should the welder position the cylinders between himself and the regulators when opening the cylinder valves?
If a regulator fails internally, releasing high-pressure gas from a cylinder into the regulator's low-pressure side, the regulator housing and gauges may explode. Fatalities have resulted from such malfunctions.

If an acetylene cylinder has been incorrectly transported on its side, why should the welder avoid immediate use?
The acetylene gas and the acetone in which it is dissolved may become mixed in the area just below the valve, resulting in both gaseous acetylene and liquid acetone at the top of the cylinder. This is where acetylene exits the cylinder and goes through the valve to enter the regulator. Both acetylene gas and liquid acetone will be drawn into the regulator possibly ruining the rubber components of the regulator and torch and creating a safety hazard. The weld metallurgy may also be contaminated.

What should the welder do knowing that a newly delivered acetylene cylinder has been incorrectly transported on its side?
Upright the cylinder and wait at least one-half hour before connecting and using the cylinder to allow the liquid phase of the acetone to separate from the

acetylene gas in the upper portion of the cylinder. That way no acetone will be drawn into the regulator possibly damaging its seals. Also, acetone in the weld flame will contaminate the weld pool and spoil the weld.

How can one readily distinguish between the oxygen fitting swivel nut and one for acetylene?
Acetylene, like most other fuel gas handling equipment, has a notch or groove cut in the middle of the edges of the hexagonal faces of the swivel nut. This is a flag for a left-handed thread. See Figure 1–7.

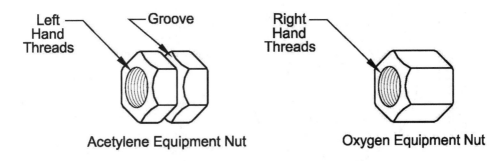

Figure 1–7   Compares connector nuts used on oxygen and acetylene equipment

Are all oxygen cylinders painted green?
Frequently, oxygen cylinders are painted green or have a green band, but the only sure way to determine the contents of a compressed gas cylinder is to *read* the adhesive label on it. This label is required by law and should not be removed. Do not go by its color as there is no color code. Unlike civilian industry, the US armed forces *do* color code their cylinders.

What pressures should full oxygen cylinder and full acetylene cylinder gauges show at 70°F (21°C)?
The acetylene should show 225 psi (15.5 bar) and the oxygen 2250 psi (155 bar). Note that these pressures will fluctuate with ambient temperature.

What do the letters and numbers stamped on the neck of high-pressure cylinders indicate?
The stampings indicate which US Department of Transportation specifications the cylinder meets, what type steel was used, who fabricated it, and when.
- Steel stamp markings such as "DOT-3A-2400" indicate the cylinder was made to US Government Department of Transportation (DOT) specifica-

tions, the "3A" denotes chrome manganese steel (or "AA" for molybdenum steel), and the "2400" the maximum filling pressure in psi.

- The oldest date indicates the month and year of manufacture. Subsequent dates, usually at five year intervals, indicate when mandatory hydrostatic pressure testing was performed and by whom. See Figures 1–8 and 1–9.

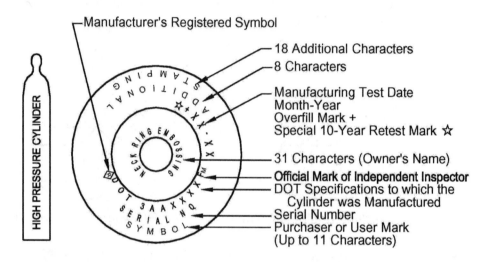

Figure 1–8   High-pressure cylinder markings

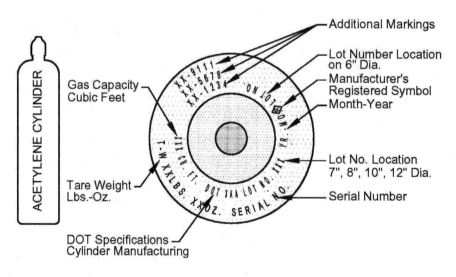

Figure 1–9   Acetylene cylinder markings

What are common oxygen cylinder sizes?
Figure 1–10 shows high-pressure cylinder sizes. Many gases in addition to oxygen, like nitrogen, carbon dioxide, and argon come in high-pressure cylinders.

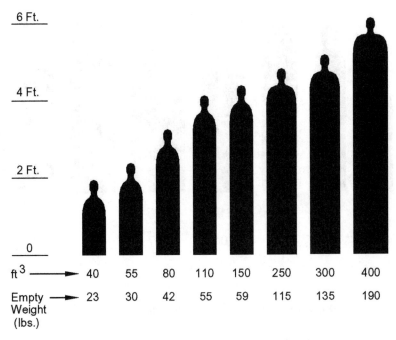

Figure 1–10   Oxygen cylinder sizes

What factors govern the choice of oxygen and acetylene cylinder size?
Smaller cylinders are suitable for refrigeration repairman who must climb ladders with OAW equipment, but they are impractical for most work. For example, a 55 ft^3 (1557 liter) oxygen cylinder would last under two hours cutting 1/8 inch (3 mm) steel plate. For the larger cylinders, their size and weight can be major drawbacks where forklifts and loading docks are not available. In general the mid-sized cylinders offer the best compromise of economy and convenience.

What are the common acetylene cylinder sizes?
See Figure 1–11.

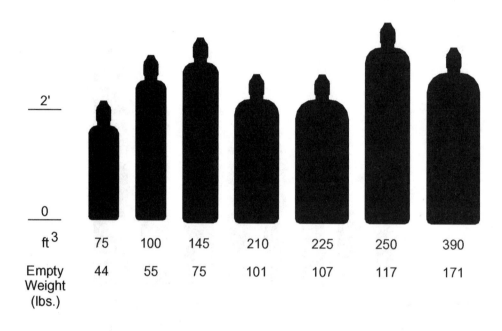

Figure 1–11   Acetylene cylinder sizes

Why are pressure gauges on acetylene cylinders poor indicators of remaining gas quantity?
Gas pressure remains nearly constant at a given temperature as acetylene gas is withdrawn from solution in acetone until very little gas remains. The best way to determine a cylinder's remaining contents is to compare its current weight with its empty weight. Note that the weight of the empty cylinder is stamped on its top. There are 14.6 ft^3 of acetylene for every pound of cylinder weight over its empty weight, or one liter of acetylene for every 1.1 grams of cylinder weight over the empty weight. Drawing acetylene from cylinders at pressures below 25 psi (1.7 bar) can cause acetone to be withdrawn from the cylinder.

Under what ownership arrangement may compressed gas cylinders be offered to users by welding equipment suppliers?
•   Outright sale of the cylinder with the right to exchange it for a filled one of equal size by paying for refill is most economical in the long run. Usually one gas supplier will accept the cylinders you obtained from another at no

additional charge. There will be a problem swapping cylinders if an embossed owner's name appears on the neck ring.

- Cylinders may be rented by the month or year. Excellent when you don't have a long-term need for them, just an immediate one.
- Some distributors lease cylinders for a year or more; some for 99 years.
- There is a very silly practice that is worth mentioning. Some gas distributors try to sell you a brand new cylinder for more money than one with previous use—an "old" one, but since most cylinders cannot be refilled immediately, chances are that you will exchange your "new" one for an "old" one when you get your first refill. The "old" ones usually cost less. And since cylinders can last well over 30 years and the cylinders due for pressure hydro testing are tested at the gas supplier's expense, there is little point in paying extra to get your own "new" cylinder.

Regulators

What is the purpose of pressure regulators?

Regulators reduce the pressures of welding gases from the very high cylinder pressures to the low pressures needed by the torch to function properly. Also, as the cylinder pressure falls with gas consumption, the regulator maintains the constant pressure needed by the torch, even though the cylinder supply pressure drops greatly. For example, an oxygen cylinder may contain oxygen at 2250 psi (155 bar) and the torch requires about 6 psi (0.4 bar) to operate. Similarly, a full acetylene tank may contain gas at 225 psi (15.5 bar) and the torch needs fuel gas at 6 psi (0.4 bar).

How does a single-stage pressure regulator work?

There are two designs for single-stage regulators, the stem-type and the nozzle-type. In the stem-type, the balance of forces on each side of the diaphragm and attached stem perform pressure regulation. There are four forces acting on the diaphragm and stem. In Figure 1–12 the combined forces of the large upper spring and atmospheric pressure act to *open* the regulator valve and admit gas into the regulator and hoses; in the opposite direction the combined forces of the high-pressure gas on the lower side of the diaphragm and the small stem spring act to *close* the regulator. When the adjusting screw is unscrewed (or in the up position in the diagram), there is little pressure exerted downward on the diaphragm by the large spring and the regulator stays closed. When the adjusting screw is tightened downward to increase regulator pressure, the increased pressure on the attached spring exerts more pressure on the diaphragm and opens the valve, admitting high-pressure gas to the lower chamber and hose. As gas continues to enter this chamber, chamber pressure rises. When it rises above the pressure called for, the high-pressure gas in the lower chamber partially or fully closes the valve to maintain the desired pressure.

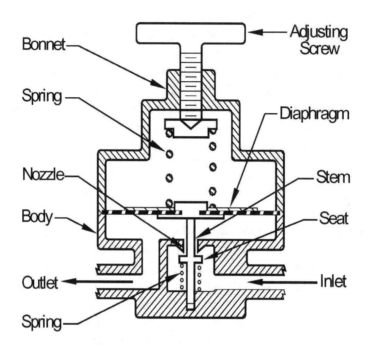

Figure 1–12   Single-stage stem-type regulator

The nozzle-type regulator is very similar to the stem-type regulator, but instead of the valve being closed by inlet or cylinder pressure as in the stem-type, the inlet pressure works to open the valve. The result is the same: a balance of pressure across the diaphragm accurately controls pressure to the torch.

How does a two-stage regulator work?

A two-stage regulator is basically two single-stage regulators connected in series inside the same housing with the total pressure drop being split across the two regulator stages. The first stage pressure is factory-set; the second stage pressure is user-set. See Figure 1–14.

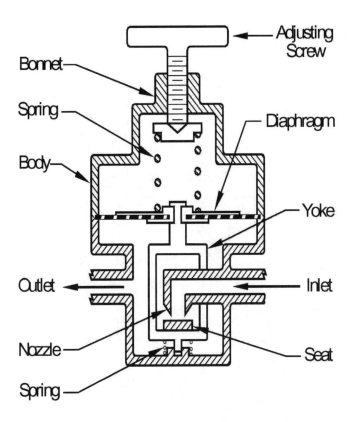

Figure 1–13   Single-stage nozzle-type regulator

What are the advantages and disadvantages of a two-stage pressure regulator over a single-stage one?

The two-stage regulator's advantage is that a higher volume of gas may be withdrawn from the cylinder with less pressure fluctuation than produced by a single-stage regulator. The combination of two regulators working together in series maintains a very constant torch pressure over wide cylinder pressure changes. Its disadvantage is cost. They are only needed when large gas volumes are needed as with multiple stations or rosebud tips.

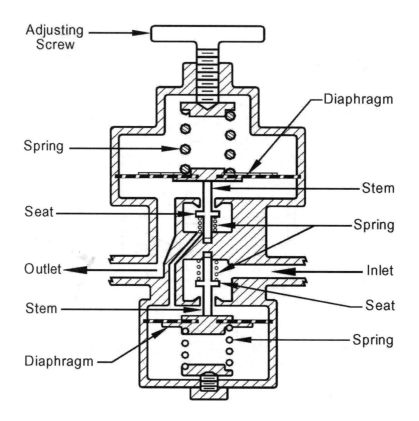

Figure 1–14   Two-stage regulator

Torches, Tips, and Hoses

What are the major parts of an oxyacetylene torch?

Shown below is the most common oxyacetylene torch design. Other designs are available. Some have very small flames for jewelry and instrument work, while others take no accessories and are much lighter in weight than standard torch designs to reduce operator fatigue.

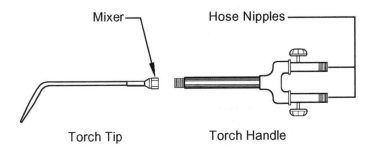

Figure 1–15   Oxyacetylene torch and tip

Besides a selection of tip sizes for different sized jobs, what other devices can be put on the torch handle and what are they used for?
- *Cutting heads* also called *cutting attachments* (see Chapter 2).
- *Multi-flames* for heating metals prior to bending, brazing, or heat-treating.

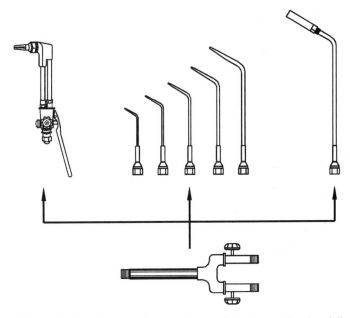

Figure 1–16   Oxyacetylene torch attachments: cutting head (left),
welding tips (center) and multi-flame tip (right)

Why are there different size torch tips?
Matching the size of the flame and the resulting volume of gas to the thickness
of the metal in the weld is important. Too much flame and the base metal
around the weld may be damaged, too little and there is inadequate heat to melt
metal for full penetration.

How are torch tip sizes designated?
There is no industry standard; each torch manufacturer has its own numbering
system. Cross-reference tables compare each manufacturer's tip sizes with
numbered drill sizes.

The American Welding Society (AWS) has been urging tip manufacturers to
stamp tips with the material thickness size to eliminate the confusion of tip size
numbers. The AWS C4.5M Uniform Designation System for Oxy-Fuel Nozzles
calls for tips to be stamped with the name of the manufacturer, a symbol to
identify the fuel gas, the maximum material thickness, and a code or part number to reference the manufacturer's operating data; many manufacturers are not
in compliance. Most companies making welding tips do provide information
booklets available to cross reference their tip sizes to tip drill sizes. See Table
1-2 tip drill size to material thickness.

Material Thickness Range (in.)	Filler Rod Diameter (in.)	Tip Drill Size	Orifice Size (in.)	Approximate Flame Cone Length
22-16 gauge	1/16	69	0.029	3/16
1/16-1/8	3/32	64	0.036	1/4
1/8-3/16	1/8	57	0.043	5/16
3/16-5/16	1/8	55	0.052	3/8
5/16-7/16	5/32	52	0.064	7/16
7/16-1/2	3/16	49	0.073	1/2
1/2-3/4	3/16	45	0.082	1/2
3/4-1	1/4	42	0.094	9/16
Over 1	1/4	36	0.107	5/8
Heavy Duty	1/4	28	0.140	3/4

Table 1-2   Matching welding tip size to weld material thickness

How can the drill size of a tip be determined?
Using a tip cleaner find the round file which fits into the tip easily but snuggly

then check the drill size of that file listed on the body of the tip cleaner cover.

When should the torch tip be cleaned and how is it performed?

When sparks from the weld puddle deposit carbon *inside* the nozzle and on the tip face. These act as spark plugs and cause premature ignition of the gas mixture. Torch tips should be cleaned at the start of each day's welding and whenever flashback occurs, the flame splits, or when the sharp inner cone no longer exists. To clean, select the largest torch tip cleaning wire file that fits easily into the nozzle and use the serrated portion to remove any foreign material. Be careful not to bend the tip cleaner file into the tip which can cause the cleaning file to break inside the tip; if the file breaks inside the tip it is nearly impossible to remove. Also be sure not to enlarge the existing hole. Then touch up the face of the tip with a file or emery cloth to remove any adhering dirt. Use compressed air or oxygen to blow out the tip. Never use a twist drill to clean the tip; it will cause bell-mouthing.

Why is it important to purge each gas hose separately and not simultaneously?

All possibility of permitting gas to enter the wrong hose and regulator must be prevented as it can lead to a deadly explosion.

How are gas hoses color coded?

Hoses for oxygen and acetylene welding and cutting are coded red for acetylene and green or black for oxygen.

What is *flashback* and what hazards does it present?

Flashback occurs when a mixture of fuel and oxygen burns inside the mixing chamber in the torch handle and reaches the hoses to the regulators or cylinders. *Such burning in the hoses, regulators, or cylinders is likely to cause an accident with burns, a major fire, explosion, shrapnel injuries, and fatalities.* If either through operator horseplay (like turning on both the acetylene and the oxygen with the torch tip blocked), or through regulator failure, an explosive mixture of acetylene and oxygen is forced back toward the cylinders. This explosive mixture may enter:

- One hose, or
- One hose and one regulator, or
- One hose, one regulator, and one cylinder.

The stage has been set for a catastrophic explosion. When the torch is lit, this explosive mixture will go off. See Figure 1–17.

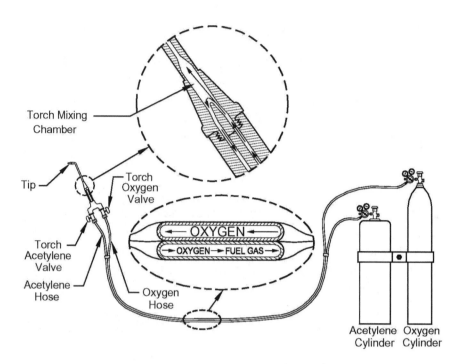

Figure 1–17   How flashback can occur

How can flashback be prevented?

Flashback is easily prevented by installation of flashback arrestors consisting of both a check valve and a flame arrestor. One flashback arrestor fits between each hose and the torch handle hose fitting. The check valves prevent the gas from one hose from entering the torch handle and then crossing to the other gas hose inside the back of the mixing chamber. Without the mixing of gases into an explosive mixture in the hoses, there can be no explosion in the hoses, regulator, or cylinder. The flame arrestor consists of a compressed stainless steel or sintered metal cylinder. The flame arrestor cylinder tends to stop fire from passing through it by both lowering the temperature of the flame front by absorbing its heat and by forcing the flame through small passages.

These devices are about the diameter of the gas hoses and about 1 3/4 inches long. Some newer torch designs incorporate check valves and flashback arrestors into the torch handle itself. Some arrestors fit between the regulator and the hose. See Figure 1–18. The best arrestors include a thermally-activated, spring-loaded shut-off valve which closes on sensing a fire.

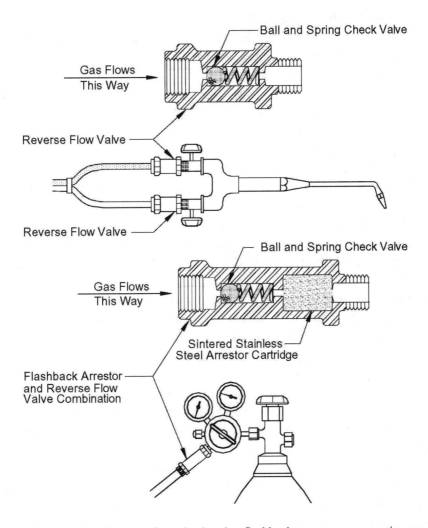

Figure 1–18   Reverse-flow check valve flashback arrestor cross section

What is *backfire* and what hazards does it present?

A backfire is a small explosion of the flame at the torch tip. The biggest hazard is that the detonation from the tip may blow molten weld metal five to ten feet from the weld and injure someone. Also, a series of repeated, sustained back-fires, which can sound like a machine gun, may overheat the tip or torch, per-manently damaging them.

How can backfire be prevented?

The most frequent cause of backfire is pre-ignition of the mixed acetylene and oxygen. Here are the most common causes of pre-ignition and their solutions:

- The mixed welding gases are flowing out through the tip more slowly than the flame front burns and the flame front ignites the gas in the tip and/or mixing chamber causing a pop. Solution: Slightly increase both the oxygen and acetylene pressures and if this results in too large a flame for the job, reduce the torch tip size.
- The tip may be overheated from being held too close to the weld or from working in a confined area like a corner. Solution: Let the tip cool off and try again holding the tip farther from the weld pool.
- Carbon deposits or metal particles inside the tip act like spark plugs prematurely igniting the mixed gases. Solution: Let the tip cool, then clean it thoroughly with your tip cleaning kit.

You are using a multi-flame (rosebud) tip that has a large flame for heating metal prior to welding, bending, or brazing. Soon after the torch is lit, it starts to pop (either once, or in a series of pops), or begins to squeal. What is the most likely problem and how is it best corrected?

This is flashback. Most likely low acetylene gas pressure is not pushing the oxygen/fuel mixture out of the tip faster than the flame can burn back on itself inside the tip. This allows the flame to burn inside the torch either in a single pop, a series of pops, or in a rapid series of pops that sounds like a squeal. Not only can one ruin a tip by allowing this to continue unchecked, but if the flame burning inside the tip reaches back into the hoses, these can explode and/or burn off and leave the welder holding a burning rubber hose, a very serious condition. To avoid this hazard: Immediately turn off the torch, oxygen first, then the acetylene. Allow the torch to cool down for several minutes, increase the acetylene regulator pressure setting to 15 psi (1 bar), reignite the torch and open the acetylene valve to obtain full flow, followed by adjusting the oxygen.

What is the proper way to set-up, light, adjust and use a multi-flame?

When using a multi-flame tip you first set the acetylene pressure at or just below 15psi (1 bar) and the oxygen pressure at 30psi (2 bar); open the acetylene torch valve far enough to light the acetylene flame once the flame is ignited open the acetylene valve until you have full flow of gas; now you can open the oxygen torch valve and adjust the flame to slightly carburizing. You may now use the multi-flame (rosebud) to heat materials but keep the sharp inner cone flame away from the material and only touch the carburizing flame to the material being heated. A heat sensing device such as a pyrometer or temperature sensing stick can be applied to the material to indicate the temperature of the material being heated.

Flames

What are the three types of flames that different ratios of oxygen and acetylene can produce and what are the characteristics of these flames?
- *Oxidizing flames* result when there is an excess of oxygen over acetylene. This flame will change the metallurgy of the weld pool metal by lowering the carbon content as it is converted to carbon dioxide.
- *Neutral flames* result when there is just enough oxygen to burn all the acetylene present. This flame has the least effect on weld pool metal as only carbon monoxide and hydrogen combustion products result and is most frequently used in welding common materials.
- *Carburizing flames* result when there is an excess of acetylene gas over the amount that can be burned by the oxygen present. The opposite of an oxidizing flame, it adds carbon to the weld pool and can change its metallurgy, usually adversely.

Which of the three types of flames, oxidizing, neutral, and carburizing produces the hottest flame?
An oxidizing flame is significantly hotter than the other two flames, but is less useful as it will introduce more contaminants into the weld pool.

For what applications is an oxidizing flame used?
An oxidizing flame is often used in braze welding or in fusion welding of heavy, thick parts with brass or bronze rod. In these applications, we are not concerned with weld pool contamination by carbon. An oxidizing flame is required for oxygen-fuel cutting.

Applications

For what type jobs is OAW best suited?
- Repair and maintenance where one type of equipment can perform many different repairs
- Welding of thin sheet, tubing, and small diameter pipe
- In field operations for natural gas distribution systems up to four inch diameter schedule forty pipes

For what type jobs are OAW definitely not a good choice?
OAW welds of thick sections are not economical when compared with shielded metal arc welding, flux-cored arc welding, or gas metal arc welding.

What disadvantages does OAW have over other welding processes?
In general most other processes are faster: they can apply more weld filler metal in a given time.

What metals can the OAW process readily weld?

- Copper
- Bronze
- Lead
- Low alloy steels
- Wrought Iron
- Cast steel

What materials can be welded by the OAW process if additional steps are taken?
Aluminum and stainless steel may be welded, provided one or more of the following steps are taken—preheat, postheat, use of fluxes, or special welding techniques.

What are three major problems associated with welding aluminum?

- Does not change color prior to melting, so it requires extra welder skill to control heat input.
- Has *hot shortness*—lacks strength at high temperatures.
- Exposed aluminum has a very thin oxide layer that requires the use of flux and also the oxide surface does not let the welder see a wet-looking molten weld pool.

Besides welding, what other processes can an oxyacetylene welding assembly perform?
With minor additional equipment, it may perform:

- Brazing and soldering
- Case hardening
- Descaling
- Post-heating
- Pre-heating
- Stress relieving
- Oxyacetylene cutting
- Flame hardening
- Flame straightening
- Shrink-to-fit parts assembly
- Surface treatment
- Forging
- Heating for bending and forming
- Tempering and annealing

Why is the carbon content of steel important to the welder?
Carbon content determines its weldability and controls the steel's tendency to harden upon rapid cooling. The greater the carbon content, the harder it may become.

Why should clothes hangers not be used as welding rod?
Both safety and quality suggest clothes hangars should not be used for welding. Hangers are usually painted and may release toxic fumes, as they burn; they may also be plated, also possibly toxic. From a quality standpoint, their metal content is unknown, variable and unlikely to provide a good weld.

What factors are important in the selection of filler metal (welding rod)?
Usually the filler metal is a close match to the base metal. Sometimes the filler metal will have deoxidizers added which will improve the weld more than just a base metal match. Rod diameters vary from $1/16$ to $3/8$ inch diameter. The prefix R in the description of the oxy-acetylene welding wire means *rod* which is followed by two or three numbers designating the ultimate tensile strength of the as welded filler material in thousands of pounds per square inch (psi). See Table 1–3.

AWS Classification	Minimum Tensile Strength (ksi)	Elongation in 1 inch (minimum %)
R45	—	—
R60	60	20
R65	65	16
R100	100	14

Table 1–3   Oxyacetylene steel welding rods

What procedures should be followed in welding common metals and what welding rods would make a good starting point?

For all metals, begin by removing all surface dirt, scale, oxide, grease, and paint. Refer to Table 1–4 for technique, flux, flame, and suggested method.

Metal Welded	Technique and Potential Problems	Flux Used	Flame Type	Suggested Rod
Aluminum	Al does not show color change before melting and has poor hot strength. Tack joint before welding. Remove all flux after welding.	Al flux	SR	Match base metal
Brass	Braze	Borax	SO	Navy brass
Bronze	Braze	Borax	SO	Copper-tin
Copper	Braze	—	N	Copper
Iron, Grey Cast	Pre-heat to avoid cracking. Weld at dull red heat. Flux applied to rod by dipping hot rod into flux. Allow joint to cool slowly or it will crack.	Borax	N	Cast iron
Iron, Malleable Cast	Welds to poor strength. Better to braze weld using bronze rods.	Borax		Bronze
Iron, Wrought	Weld or braze	—	N	Steel
Steel, Low-Carbon	Weld or braze	—	N	Steel
Steel, Medium-Carbon	Weld or braze	—	SR	Steel
Steel, High-Carbon	Weld or braze	—	R	Steel
Steel, Low-Alloy	Weld or braze	—	SR	Steel
Steel, Stainless	Weld or braze	SS flux	SR	Match base metal

SR = Slightly Reducing   SO = Slightly Oxidizing   N = Neutral

Table 1–4   Information for welding various metals

Weld Preparation

What joint preparation is used for OAW butt welds?
See Figure 1–19.

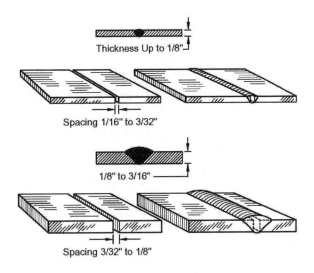

Thickness Up to 1/8"

Spacing 1/16" to 3/32"

1/8" to 3/16"

Spacing 3/32" to 1/8"

Figure 1–19   Preparation for OAW butt welds

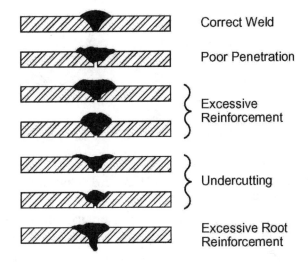

Correct Weld

Poor Penetration

Excessive
Reinforcement

Undercutting

Excessive Root
Reinforcement

Figure 1–20   Correct and defective butt weld profiles

Weld Profiles

On a butt weld, what do the following weld profiles look like: correct weld, poor penetration, excessive reinforcement, undercutting, and excessive root reinforcement?
See Figure 1–20.

Safety

What essential pieces of safety equipment are needed to begin OAW?
See Figure 1–21.
- Non-synthetic fabric (cotton or wool) long-sleeved shirt buttoned to the top to prevent sparks from entering.
- Tinted welding goggles with minimum of number 5 shade lenses.
- Leather gloves.
- Spark igniter.
- Pliers for moving hot metal.

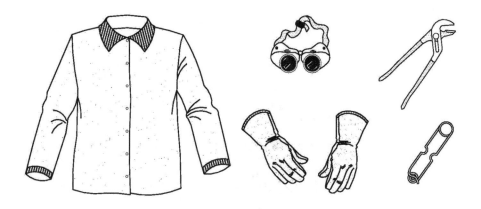

Figure 1–21   Oxyacetylene safety equipment

What are the main hazards of OFW and what safety equipment can prevent these injuries?
- External eye injuries from welding or grinding sparks are prevented by using welding goggles, safety glasses, or safety shields.
- Internal (retinal) eye damage from viewing hot metal and the radiation being emitted during welding and while cooling (until the metal is no longer red), prevented by using a number 5 tinted lens.

- Burns from weld sparks and hot metal prevented by leather or heavy cotton welding gloves, fire retardant clothing, leathers or specially treated welding jacket or cape-sleeves and bibs when working overhead, cuffless pants, high-top leather shoes.
- Fume hazards from the vapors of metals and flux, must be avoided by proper ventilation, fume filters, and welder air supplies to the welding hood.
- Fires from the welding process prevented by moving flammables away from the weld zone and having water or fire extinguishers close at hand.

What are 15 important welding safety practices whose violations can lead to serious accidents?
- Never use oxygen in place of compressed air.
- Never use oxygen for starting engines or cleaning clothing.
- Store and use acetylene and propane cylinders valve end up.
- Secure cylinders to prevent them from being knocked over in use.
- Use valve protection caps on cylinders while moving them.
- Never leave a lighted torch unattended.
- When a cylinder is empty, close the valve and mark it *EMPTY (MT)*.
- Do not attempt repair of cylinder valves or regulators; send them to a qualified repair shop.
- Never use compressed gas cylinders as rollers.
- Never attempt welding on a compressed gas cylinder.
- Keep power and welding cables away from compressed gas cylinders.
- Prevent sparks from falling on other persons, combustible materials, or falling through cracks in the floor.
- On old-style acetylene cylinders with a removable valve wrench, always leave the wrench in place when using the equipment, so it can be shut off quickly in an emergency.
- When transporting compressed gas cylinders by vehicle have the cylinder caps in place and secure the cylinders so they will not move around as the vehicle starts and stops. Never transport cylinders with the regulators in place.
- Never carry compressed gas cylinders inside a car or car trunk.

What is the best way to weld on a sealed cylinder, tank, vessel or container?
Never weld on a sealed container regardless of its size. Even if the vessel is clean and empty, penetration of the shell could release hot gases from the interior. They could also drive the torch flame back towards the welder. If the cylinder is empty and contains no residual vapors, vent it to atmosphere by opening a valve, hatch, bung, or by drilling a hole. An even more dangerous

situation results when the cylinder contains residual flammable vapors whether it is vented to atmosphere or not. This will almost certainly result in an explosion. Clean or purge the cylinder with an inert gas, then have it checked for lack of explosive vapors by a qualified person. Vent it to the atmosphere and begin welding. In some cases filling the vessel with water, or other liquid and welding below the liquid is acceptable, but this is an area for experienced, knowledgeable welders.

Why use a striker to light an oxyfuel torch and not a match?
The striker keeps your fingers away from the flame that can ignite into a large flame. The use of a butane cigarette lighter for torch ignition can cause a large fire or explosion with the potential power of a half-stick of dynamite.

Why should the welding area be well ventilated to draw the weld fumes away from the welder?
Many fumes from the welding process are poisonous and must be avoided. Welding fumes from cadmium plating, galvanized sheet metal, lead, brass (which contains zinc), and many fluxes (especially those containing fluorine) are poisonous. They can have both immediate and long-term adverse health effects. Welding supply companies, welding equipment manufacturers, and materials suppliers will provide MSDSs (Material Safety Data Sheets) on request. Often they are available for downloading via the Internet from the manufacturer. They detail the hazards of materials and equipment and show how to deal with them safely. They are particularly helpful in understanding the fume hazards of fluxes, solders, and brazing materials.

Chapter 2

Thermal Cutting

Practice is the best of all instruction.
Aristotle

Introduction

Oxygen-fuel cutting is an important industrial process. Much more acetylene is used for cutting metals than for welding them. For many cutting applications, there is no more effective and efficient process. Used in construction, manufacturing, and repair operations, cutting equipment is inexpensive, portable, and easy to use. In some applications, propane or natural gas may be more cost-effective and replaces acetylene as the fuel gas. We will explain how the oxygen-fuel cutting process works, its capabilities and limitations. We will also cover cutting torches, troubleshooting, operating tips, safety, and then present some helpful accessories. Finally we will discuss other thermal cutting processes important in today's industry.

Process Name

What term does the AWS use for any cutting process using oxygen and any fuel gas?
Oxy-fuel cutting.

What is the AWS abbreviation for oxy-fuel cutting?
The abbreviation for all oxy-fuel cutting is *OFC.*

What is the AWS abbreviation for the oxyacetylene cutting process?
The abbreviation for oxygen-acetylene cutting is *OAC.* OAC is just one member of the OFC family.

Equipment

How do you convert oxyacetylene welding equipment into an oxygen-acetylene cutting equipment?

Conventional OAW equipment (outfit) is readily converted to perform light to heavy OAC by exchanging the welding nozzle on the torch handle to a cutting accessory head fitting into the handle, Figure 2–1.

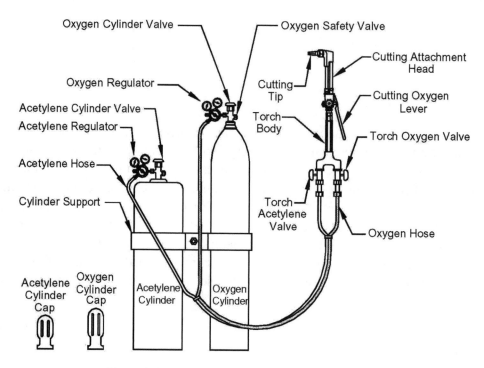

Figure 2–1   Oxygen-acetylene cutting equipment

How does an OAC cutting torch or accessory cutting assembly differ from an OAW torch?

The OAC cutting head still contains a means of mixing oxygen and acetylene to produce an approximate temperature of 6300°F (3100°C). But it has added means to deliver a stream of pure oxygen to the cutting point. An oxygen lever opens this pure oxygen stream when the welder, fitter or cutting-operator depresses it. See Figure 2–2.

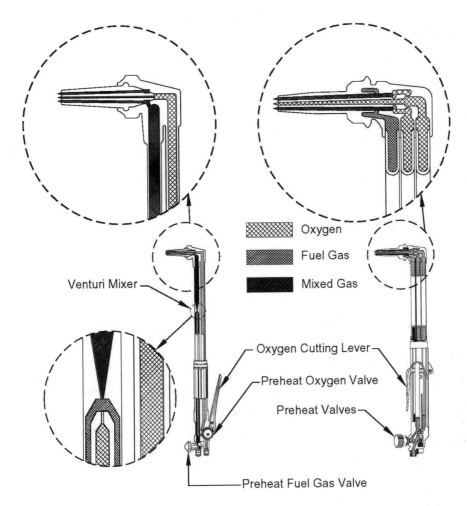

Oxygen

Fuel Gas

Mixed Gas

Venturi Mixer

Oxygen Cutting Lever

Preheat Oxygen Valve

Preheat Valves

Preheat Fuel Gas Valve

Figure 2–2   Oxy-fuel cutting torches. An injector cutting torch (left) and a mixing chamber positive pressure cutting torch (right)

What are the advantages of a cutting accessory head over a regular cutting torch?

A cutting accessory head is less expensive than a one-piece cutting torch; it is quicker and easier to change back and forth between the cutting and welding functions than between a welding torch and cutting torch. With its greater length, a one-piece cutting head puts more distance between the cutting action and the welder and usually can handle greater oxygen flows for large jobs. Some cutting torches have cutting heads at a particular angle for a given task to relieve operator fatigue. The position of the cutting handle is a matter of preference and varies by manufacturer.

The two cutting torches in Figure 2–2 have different designs. How do they differ and what are their advantages?

The torch on the left side of Figure 2–2 uses an injection chamber or venturi to draw the fuel gas into the oxygen stream and operates with fuel pressures 6 oz/in^2, such as those supplied by an acetylene generator or a regulated natural gas system delivering in water column inches or about 1/3 pound. The torch depicted on the right uses a mixing chamber to bring the gases together and is also known as balanced-pressure, positive-pressure, or medium-pressure torch. The advantage of the mixing chamber design torch is that it operates at higher fuel gas pressures and can supply more heat than the venturi design: the venturi design when adjusted properly creates a near perfect cutting flame which uses the fuel gas more efficiently. See detail in Figure 2–2.

What changes are needed to cut steel thicker than one inch?

Because cutting thick steel requires more oxygen than thin steel, a special oxygen regulator with the capacity of delivering more oxygen volume at higher than welding pressures may be needed. Larger diameter hoses may also be required. The welding acetylene regulator is fine for cutting. Also since there is much higher oxygen consumption and more rapid cylinder depletion than in welding operations, the typical cutting regulator is a two-stage regulator to maintain a constant working pressure as the cylinder gas dwindles. Oxygen regulators specifically for cutting usually have low-pressure gauges (on the output or torch side of the regulators) with higher pressure calibrations than welding regulators. OFC operations on extremely thick metals can require 100 to 150 psi (6.8 to 10 bar) oxygen pressures.

Plate Thickness in inches	Oxygen PSIG	Fuel Gas PSIG	Oxygen bore drill size
1/8	20-25	3-5	0.031
1/8-1/4	20-25	3-5	0.036
¼-1/2	25-35	3-5	0.040
½-3/4	30-35	3-5	0.046
3/4-1 1/2	35-45	3-7	0.059
1 ½-2/12	40-50	4-10	0.067
2 1/12-3	45-55	5-10	0.093
3 1/2-5	45-55	5-10	0.110
5-8	45-60	7-10	0.120

Table 2–1   Optimum pressure and gas flow settings for cutting various metal thicknesses

What important facts should be remembered regarding cutting tips?
- They made are of copper and can easily be damaged if dropped. Tips from one torch maker cannot, in general, be used in another manufacturer's torch.
- If you have removed the tip nut that retains the torch tip, and the torch tip is stuck in the torch body, a gentle tap on the back of the torch head with a plastic hammer will release the tip.
- Care should be taken when cleaning the tip to avoid breaking off the tip cleaner inside the torch tip.

How do high-speed cutting tips differ from standard ones?
Regular cutting tips have a straight-bore oxygen channel and operate from 30 to 60 psi (2 to 4 bar). High-speed tips have a diverging taper and permit operation at oxygen pressures from 60 to 100 psi (4 to 7 bar). This permits a 20% increase in cutting speed. They are used only on cutting machines. See Figure 2–3.

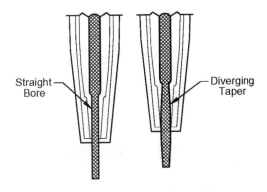

Straight Bore — Diverging Taper

Figure 2–3   Standard cutting tip (left) and high-speed cutting tip (right)

What fuels other than acetylene are used for preheating in the OFC process? These are often called *alternative* fuels.
- Propane
- Natural gas
- Propylene
- Methyl acetylene-propadiene stabilized (known as MPS or MAPP®)

What changes must be made to OAC equipment to properly utilize alternative fuels?
Torch tip designs are frequently different because alternative fuels may be supplied at lower pressures, have different ratios of fuel to oxygen, and different

flame and burn rate characteristics. Several manufacturers offer alternative fuel torches. Figures 2-4 and 2-5 shows alternative fuels cutting tips.

Figure 2–4   Alternative fuel gas cutting tip

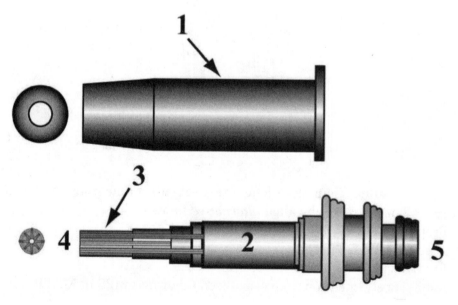

Figure 2–5   Alternative fuel gas tips are two piece tips consisting of [1] an outer shell, [2] an inner member, [3] grooves for preheating flames, [4] extremities of grooves and [5] a cutting oxygen bore

Why are these alternative fuels used when acetylene always produces a higher pre-heat temperature?

Alternative fuels are far more stable than acetylene therefore much safer to handle. An alternative fuel may also offer significant cost savings. Fuel selection is a complex matter involving material thickness, cutting speeds, preheat time, fuel performance on straight lines, curves and bevels, and their impact on the total cost. The availability of fuel, labor, the cost of preheat oxygen, and the suitability of the alternative fuel to perform related processes like welding, heating, and brazing also influence total production cost. While fuels other than acetylene do not produce the high flame temperature of acetylene, some can produce a greater *volume* of heat output throughout the outer flame envelope. This gives an advantage to some alternative fuels in cutting thick steels.

What other differences are there between cutting with OAC tips and OFC tips?

The main difference is the distance the cutting tip is held above the metal. The acetylene cutting tip is held so the pre-heat inner- cones are just above the metal; the alternative fuel cutting tip should be held outside the skirt which further away from the metal. See Figure 2–5A.

Why is the alternative fuel cutting tip held further away from the metal?

The heat characteristics of acetylene are different from other fuels; in Chapter 1, Figure 1–3 shows the hottest part of an acetylene flame is at the inner-cone. Alternative fuel gases characteristics are different and the hottest part of the flame with these gases is outside the skirt. See Figure 2–5A.

Is there a difference in the way alternative fuel gas tip is lit and adjusted?

The acetylene tips are adjusted by opening the fuel valve until no soot, at the end of the carburizing flame, is visible at the end of the acetylene flame then adjust the flame to neutral by adding oxygen. Once the heat cone flames are at neutral depress the oxygen lever and look at the flame if it appears to have a feather or carburizing flame adjust the oxygen with the oxygen stream lever depressed until the flame again looks neutral; now cutting may proceed.

Alternative fuel tips are adjusted by opening the fuel valve enough so the gas can be ignited followed by adding a small amount of oxygen reducing the flame enough to see the inner-cones; then alternate between opening the fuel valve a small portion at a time followed by opening the oxygen until you see the heat cones sticking out of the end of the tip approximately 3/16" then increase the oxygen flow until you hear the tip whistle and see the skirt.

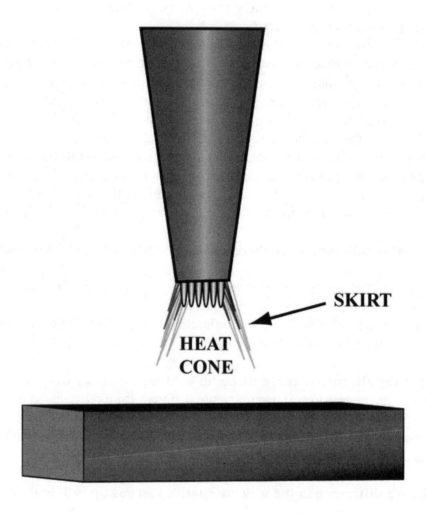

Figure 2–5A   Appearance of the OFC flame inner-cones and skirt

What other OAC tip designs are available?
A wide variety of tips are available. See Figure 2–6.

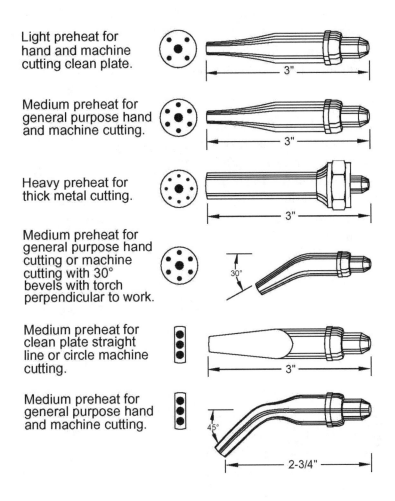

Light preheat for hand and machine cutting clean plate.

Medium preheat for general purpose hand and machine cutting.

Heavy preheat for thick metal cutting.

Medium preheat for general purpose hand cutting or machine cutting with 30° bevels with torch perpendicular to work.

Medium preheat for clean plate straight line or circle machine cutting.

Medium preheat for general purpose hand and machine cutting.

Figure 2–6   Special purpose oxy-fuel cutting tips

Process

How does oxyacetylene cutting equipment perform the cutting process?
The oxyacetylene flame brings the steel at the beginning of the cut up to kindling temperature of 1600°F (871°C). At the kindling temperature, steel will readily burn in the presence of oxygen. When the oxygen lever is turned on, the pure oxygen stream with the steel at kindling temperature burns, this combination causes a chemical reaction called oxidation. The mixture of oxides of iron is called slag. This slag has a melting point much lower than the melting point of steel itself that is 2600°F (1427°C) and readily runs out of the cut or kerf. The force of the oxygen stream provides additional help to clear the kerf of

molten oxides. In addition to the oxyacetylene preheat flame, the burning of the iron in the oxygen stream releases large amounts of heat. This aids cutting action particularly when cutting thick steel. Moving the torch across the work produces continuous cutting action; straight, curved or beveled cuts are readily made.

What is the *kerf* of a cut?
When cutting is performed material is removed, the width of the cut is the *kerf*; when flame cutting the oxidation of the metal along the line of the cut removes a thin strip of metal or *kerf* which is the thickness of the cut which is also the bore size of the cutting tip. In steel under two inches in thickness, it is possible to hold the kerf to about $1/64$ inch (0.4 mm). In making patterns to fabricate parts by flame cutting them out of flat stock, allowance must be made for the *kerf*. See Figure 2–7 and Table 1–1.

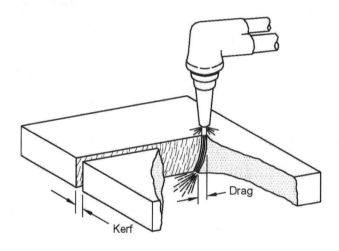

Figure 2–7   Kerf and drag in an oxy-fuel cut

What factors determine kerf size?
Kerf size depends on the following:
- Torch oxygen bore (orifice) size
- Torch tip design
- Oxygen pressure and flow rate
- Preheat flame size
- Cutting speed

What is the proper size cutting tip to use for various material thicknesses?
Cutting tip bore drill sizing, like welding tip orifice drill sizing, numbering system is not standardized in the welding industry. The drill sizing is standard but the manufacturer's numbers placed on the tips are not standard. One company

may identify a number one tip for cutting one inch steel while the same bore drill size from another company may call for a number two tip. The American Welding Society (AWS) has been urging manufacturers to stamp cutting tips with material thickness size to eliminate confusion with the publication AWS C4.5 Uniform Designation System for Oxy-Fuel Nozzles. Compliance is not mandatory therefore manufacturers have not followed through with using this standard.

What does the bore drill size indicate?
Cutting drill bore size indicates cutting orifice size and material thickness which can be cut. See Table 2–2.

Bore Size for Oxy-Fuel Cutting

Plate Thickness inches (mm)	Bore Drill Size inches (mm)
1/4-1/2 (6.35–12.7)	68-53 DR 0.031-0.059 (0.794-1.51)
3/4 (19.05)	62-53 DR 0.038-0.059 (0.965-1.51)
1 (25.1)	56-53 DR 0.046-0.059 (1.18-1.51)
1 ½-2 (38.1-50.8	51-46 DR 0.067-0.081 (1.70-2.06)
3-5 (76.2-127.0)	46-44 DR 0.081-0.086 (2.06-2.18)
6-8 (152.4-203.2)	40-39 DR 0.098-0.010 (2.49-2.53)
10 (254)	39-35 DR 0.010-0.011 (2.53-2.94)

Table 2–2   Material thickness to bore size for cutting tips

Why does kerf width grow larger with increasing steel thickness?
Cutting thicker steel requires more oxygen, which requires a larger oxygen orifice size, greater oxygen flow rates and a larger oxygen stream. These lead to a wider kerf.

How can the bore size be determined?
Using a tip cleaner find the round file which will fit snuggly into the bore then determine the bore size by the chart list on the tip cleaner's container.

What is *drag*?
The distance between the cutting action at the top and bottom of the kerf is called *drag*. When the oxygen stream enters the top of the kerf and exits the bottom of the kerf directly below, the drag is said to be zero. If the cutting speed is increased (or the oxygen flow decreased), oxygen in the lower portion of the kerf decreases and the kinetic energy of the oxygen stream drops, slowing cutting action in the bottom of the cut. This causes the cutting action at the bottom

of the kerf to lag behind the cutting action at the top. Drag may also be expressed as a percentage of the thickness of the cut. See Figure 2–6.

What are the effects of excessive drag?

Excessive drag can cause loss of cutting action in thick cuts and restarting the cutting action may cause the loss of a part being flame cut.

When can reverse drag occur?

Excessive oxygen flow, too slow a cut, or damaged orifices may cause reverse drag leading to rough cut edges and excessive slag adhesion.

What is the chemistry of the OAC process?

There are three principal reactions producing three different iron oxides. Notice that the second reaction releases the most heat that helps sustain the cutting action. The equations show the ratios of oxygen to fuel (iron) needed. To a chemist these equations indicate that about 104 ft^3 of oxygen will oxidize 2.2 lb. of steel to Fe_3O_4.

$$Fe + O_2 \rightarrow FeO + \text{heat of 267 Kj (Kilojoules)}$$
$$3Fe + 2O_2 \rightarrow Fe_3O_4 + \text{heat of 1120 Kj}$$
$$2Fe + 1.5O_2 \rightarrow Fe_2O_3 + \text{heat of 825 Kj}$$

What are advantages of the OAC process?

- Low cost compared with machine tool cutting equipment.
- No external power required.
- Readily portable.
- Steels usually cut faster than by conventional machining process.
- Cutting direction may be changed easily.
- OAC is an economical method of plate edge preparation for groove and bevel weld joints.
- Large plates may be cut in place.
- Parts with unusual shapes and thickness variations hard to produce with conventional machinery are easily produced with OAC.
- Can be automated using tracks, patterns, or computers to guide the torch.

What are disadvantages of the OAC process?

- Dimensional tolerance of OAC is dramatically poorer than machine tool based cutting.
- OAC process is commercially limited to steel and cast steel.
- Both the preheat flame and the stream of molten slag present fire and burn hazards to plant and personnel.

- Proper fume control is required.
- Hardenable steels may need pre-heat, post-heat, or both to control the metallurgy and properties of the steel adjacent to the cut.
- High-alloy steels and cast iron need additional process modifications.

What is the maximum steel thickness that may be cut with OAC?
OAC has no practical limit. Steel seven feet thick is routinely cut in heavy industry, and fourteen-foot cuts are not uncommon.

What is the minimum mild steel thickness that may be cut with OAC?
OAC's lower limit is 20 gauge (0.035 inch or 0.88 mm) steel. Below this thickness the cut becomes irregular with uncontrollable melting, but it can be cut with a large tip-to-plate angle and fast travel speed. Thinner steel sheets are best cut with laser or plasma cutters.

Setup (and Related Safety)

How can the welder determine what cutting tip size and what oxygen and acetylene pressures to use on a given thickness of material?
Given a material and thickness, use a torch manufacturer's table to convert metal thickness to tip size, starting oxygen pressure and acetylene pressure. Remember these are suggested starting pressure ranges. Fine-tuning of the pressures may be needed to get the best combination of speed and quality.

What steps are required to set up a cutting torch to cut 1/4 inch carbon steel? Be sure to include all safety precautions.
- Inspect and clean the torch using the cleaning kit.
- Put on your welding safety equipment: goggles with filter lens (or tinted face shield), cap, high-top shoes, fire retardant coat, cape sleeves and bib or cotton or wool long-sleeved shirt, and pants and welding gloves.
- Avoid wearing trousers with cuffs when cutting as they tend to catch hot sparks and can easily catch your pants on fire. *Wear no synthetics*. If you will be doing overhead cutting, leather skins, fire retardant coats, cape sleeves and bid or aprons are necessary to protect your clothing from falling sparks. Goggles and face shields should be of number 5 shade.
- Firmly secure the oxygen and acetylene cylinders to a welding cart, building column, or other solid anchor to prevent tipping during storage or use. Non-flammable material must be used to secure the cylinders. Remove the safety caps.

- Verify the cutting torch has flashback arrestors installed.
- Check to make sure there are no nearby sources of ignition and then momentarily open each cylinder's valve to the atmosphere and re-close the valve quickly to purge the valve; this is known as *cracking* a valve. Cracking serves to blow out dust and grit from the valve port and to prevent debris from entering the regulators and torch. Stand on the opposite side of the cylinder from the valve port when cracking.
- With a clean, oil-free cloth, wipe the valve-to-regulator fittings on both cylinders to remove dirt and grit from the fittings' connection faces and threads. Cleanse to both regulators' threads and faces. *Remember*, to never use any oil on high-pressure gas fittings. Oxygen at high pressures can accelerate combustion of oil into an explosion.
- Check to see that both the oxygen and acetylene regulator pressure adjustment screws are loosened (but not falling out of their threads), then screw each regulator to its respective cylinders. Snug up the connections with a wrench. Caution: Oxygen cylinder-to-regulator threads are *right-handed*; so are oxygen hose-to-torch screw fittings. Acetylene cylinder-to-regulator fittings and acetylene hose-to-torch fittings are *left-handed* threads. This arrangement prevents putting the wrong gas into a regulator or torch connection.
- Stand so the cylinders are between you and the regulators. S-L-O-W-L-Y open the oxygen cylinders valves. Be sure to open the oxygen cylinder valve until it hits the upper valve stop and will turn no further.
- With the cylinders between you and the regulators, open the acetylene cylinder valve gradually and not more than one and a half turns. If there is an old style removable wrench on the cylinders, make sure to keep it handy in case you must close the cylinder valve immediately in an emergency.
- Look for the high-pressure—or cylinders side—gauges to indicate about 225 psi (15.5 bar) in the acetylene cylinders and 2250 psi (155 bar) on the oxygen cylinders. These pressures at 70°F (21°C) will indicate the cylinders are fully charged. Note that these pressures will vary with ambient temperature of the cylinders. The pressures given above are for full cylinders at 70°F (21°C), but the actual pressure will vary with cylinder temperature.
- Install the cutting torch on the hoses, or if using a combination welding and cutting handle, install the cutting accessory on the torch handle.
- First, check the area for ignition sources, other than your torch igniter. Then purge each torch hose of air separately: Open the oxygen valve on the torch about three-quarters of a turn, then screw in the pressure control screw on the oxygen regulator to your initial pressure setting. After several seconds, close the torch valve. Do the same for the acetylene hose. Comment: We do

this for two reasons, (1) to make sure we are lighting the torch on just oxygen and acetylene, not air, and (2) to get the regulators set for the correct pressure while the gas is flowing through them. If the gas hoses are more than 50 feet (15 m) long, a higher regulator setting will be needed to compensate for the pressure drop in the hoses.
- Test the system for leaks at the cylinder-to-regulator fittings and all hose fittings with soapy water. Bubbles indicate leaks.
- Proceed to light and adjust the cutting torch as detailed below.

What are the steps for lighting and adjusting the cutting torch to cut ¼ inch thick mild steel? Include all safety precautions.
- Follow the steps of securing the cylinders, installing the regulators, hoses, and torch, purging the hoses of air, and setting the regulator pressures from cutting reference tables for ½ inch steel: acetylene at 6 psi (0.4 bar) and oxygen at 30psi (2 bar).
- *Never* adjust the acetylene regulator pressure above 15 psi (1 bar) as an explosive disassociation of the acetylene could occur.
- Open the oxygen valve on the back end of the torch all the way.
- Recheck the low-pressure gauge pressures to make sure the working pressures are not rising. If the working pressure should rise, it means that the regulator is leaking. The cylinders must be immediately shut down at the cylinder valves as continued leaking could lead to regulator diaphragm rupture and a serious accident.
- Light the torch by opening the acetylene valve on the torch handle about ¹⁄₁₆ turn and light the acetylene using your flint igniter. A large, smoky, orange flame will result. Also, you must have your tinted welding facemask over safety glasses (or your welding goggles with a number 5 lens shade) on prior to lighting the flame.
- Increase the flame size by slowly opening the acetylene valve until most of the smoke disappears.
- Open the oxygen preheat valve on middle of the torch and adjust for a neutral flame.
- Actuate the cutting oxygen lever and examine the preheat flame. Further adjustment of the preheat oxygen valve may be needed to keep the preheat flame large enough when the cutting oxygen is used. This is because cutting oxygen use may cause the hose pressure to drop so much the oxygen to the preheat flame must be increased to keep a proper preheat flame.
- You are ready to begin cutting.

What is the best way to flame cut ¼ inch (6 mm) through ½ inch (12.7mm) thick metal?
See Figure 2–8.
- The cutting torch tip is held perpendicular to the metal.
- Start the cut at the edge of the stock by preheating the edge of the stock. In thicker material, the torch may be angled away from the direction of travel so the preheat flame strikes down the edge of the material. When the stock becomes a dull cherry red, begin cutting by squeezing the oxygen cutting lever. Remember to hold the torch tip perpendicular to the surface of the stock when cutting action has begun.
- Move the torch along the cut line in a steady motion. For right-handed welders, cutting from right to left allows the welder to see the marks of the cutting line more easily. Left-handers will usually prefer cutting left to right.

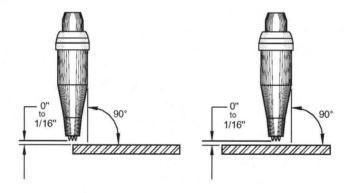

Figure 2–8   Position of cutting torch tip on 1/4 inch and thicker plate, starting (left) and cutting (right)

What is the best way to cut thin metal (10 guage- ⅛ inch or thinner)?
- Utilize the smallest cutting tip available with two preheat flames.
- Hold the torch at a 20 degree to 40 degree angle to the metal surface to increase the kerf thickness.
- Adjust the flame to the smallest preheat flame that will permit cutting.
- Set oxygen pressure at 15psi (1 bar). See Figure 2–9.

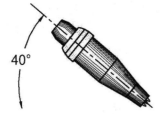 

Figure 2–9   Torch position for cutting thin sheet metal, starting (left) and cutting (right)

When cutting thin-gauge sheet metal, what step can be taken if slag accumulates on the underside of the good part?

Tipping the torch away from the side you will use allows the slag to form on the scrap side of the kerf will keep slag off the good part.

What are the proper steps to shut down an oxyacetylene torch and its cylinders?

- Turn off the oxygen and then the acetylene with the torch handle valves.
- Turn off the oxygen and acetylene cylinders valves on the cylinders.
- One at a time, open and reclose the oxygen and acetylene valves on the torch handle to bleed the remaining gas in the lines and regulator to atmosphere. Verify that both the high-pressure and low-pressure gauges on both gases indicate zero pressure. Bleed off the oxygen first to eliminate the possibility of providing oxygen to the remaining acetylene.
- Unscrew the regulator pressure adjustment screws on both regulators in preparation for the next use of the equipment.

Applications

What metals can readily be cut using the OFC process?

- Cutting of new steel plate, beams, and pipes to size both in mills, in fabrication plants, and on construction sites.
- Cutting risers, gates, and defects from cast iron and steel castings in steel mills.
- Cutting up old steel and cast iron equipment for removal and salvage.
- Removing rivets from old equipment without damaging the surrounding steel.
- Removing damaged parts prior to welding new ones in place.

- Gouging the surface of steel plate edges in preparation for welding.
- Manufacturing steel parts by cutting them out of flat stock instead of machining them.
- Removing backing bars from a weld.

What is *stack cutting*?
In order to make multiple parts in a single cutting pass, multiple sheets or plates are stacked and clamped or welded together. Then the cutting proceeds manually, or more likely by machine. When the cutting is complete, the stack comes apart, leaving multiples of flame cut parts. Stack cutting is also useful in cutting stock so thin it could not be cut in a single layer. It can be used in place of shearing or die stamping when the production run does not justify making dies. There may also be substantial savings of fuel and oxygen in stack cutting as gas consumption is not directly proportional to total cut thickness. Generally the maximum thickness of plates in stack cutting must not exceed 0.5 inch (12.7 mm). Note that high-quality stack-cut parts require clean, flat plates (or sheets) securely clamped in position with no air gaps between the plate layers. If there is air between the plates, the cutting action will be extinguished and the parts will be ruined.

What metal would OFC definitely not a good choice?
Here are some examples:

- Aluminum
- Brass
- Copper
- Lead
- Magnesium
- Stainless steel
- Zinc

The OAC process can readily cut what metals?
- Mild steel (steel with a carbon content of less than 0.3% carbon)
- Low-alloy steels
- Cast iron (though not readily)
- Titanium

What materials can be cut by the OAC process if additional steps are taken?
- Stainless steel
- High-alloy steel (must be preheated)

Why do high-alloys of steel resist OFC?
As the number and percentage of alloying elements increase, OFC becomes less

effective. Oxides of the alloying elements have a higher melting point than the alloying elements themselves and are refractory in nature. (Remember that oxides of iron have melting points *lower* than the melting point of iron so they become fluid and they readily leave the kerf as molten slag.) Unlike iron oxides, an alloy's oxides do not readily run out of the kerf to expose new iron to oxygen to keep the burning process going, and cutting becomes more difficult.

By what means can OFC be extended to metals and alloys not readily cut?
- Torch Oscillation
- Waster Plate
- Wire Feed
- Metal Powder Cutting
- Flux Cutting

How does *torch oscillation* work?
By torch manipulation the entire starting edge of the cut is brought to a bright red color before beginning the cut. This technique is usually used in conjunction with one of the other four cutting enhancement methods on low-alloy stainless steel up to 4 inches thick and on resistant cast iron.

How does a *waster* plate work?
A low-carbon steel *waster* plate is secured to the top of the stainless steel to be cut, and the cutting action begun on the waster plate. The heat released from the waster plate's burning provides additional heat to the cutting action in the stainless below. Hot slag from the waster plate tends to flush the kerf of the stainless steel's refractory oxides. Waster plate disadvantages are the extra cost of the waster plate, additional set-up time, slow cutting speeds, and rough cut.

How does *wire feed* cutting work?
A small diameter carbon steel wire is fed into the torch preheat flame just ahead of the cut and melts onto the surface of the alloy steel. The additional carbon steel works just like a waster plate to enhance cutting action. A motor feed and wire guides are needed as accessories.

How does *metal powder* cutting work?
Powder metal cutting (AWS abbreviation is *POC*) uses iron-rich powder that is dropped into the kerf or injected into the cutting oxygen stream to add heat. Some powders also chemically combine with the alloying oxides to increase their fluidity and increase the ability of the oxygen jet to wash them out of the kerf. Frequently, cutting speed of POC in high-alloy steels can match OFC in mild steel of the same thickness.

How does *flux* cutting work?
Flux cutting uses a granular flux introduced into the oxygen cutting stream. The flux combines with the alloying metals' oxides to lower their melting temperatures to near those of iron oxides and get them to flow out of the kerf. Flux cutting can eliminate torch oscillation and can increase cutting speeds in stainless to that of carbon steel of the same thickness.

Tips, Techniques, and Helpful Accessories

Why is it important to keep the torch tip face clean and flat and to clean out the orifices with tip cleaners regularly?
A damaged tip face or plugged tip orifice can cause an unsymmetrical flame. Such a flame will produce irregular rough edges, and slow cutting action. Dirt inside the torch tip may cause flashbacks.

What are the two best ways of marking the line of cut?
• With welder's soapstone.
• By a series of center punch marks along the line of cut.

What readily available material may be used to improve the quality of some cutting tasks?
Angle-iron may be used as a straight edge guide, or as a bevel guide. See Figure 2–10.

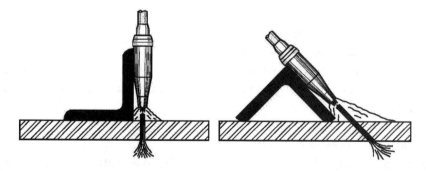

Figure 2–10   Angle iron used as a straight edge (left) and bevel guide (right)

What are two mechanical aids that can be attached to a cutting torch?
• Wheels to keep the torch tip at a constant height above the work and reduce operator fatigue. See Figure 2–11.

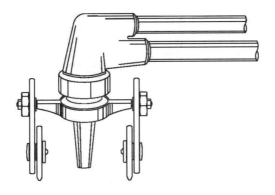

Figure 2–11   Torch wheels

- Compass attachment to make nearly perfect circles easily. See Figure 2–12.

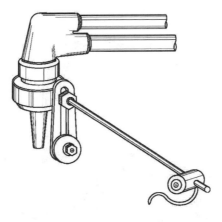

Figure 2–12   Compass attachment

Name the four main types of electrical or electronic aids to guide a cutting torch to increase the accuracy and quality of cutting.

- The motorized cutting head is the most primitive improvement over a hand-held torch. Its wheel is motor-driven to maintain optimum cutting speed; the wheel also keeps the torch-to-work distance constant. However, the welder must still guide the motorized head manually. The small wheel in the rear of the unit is useful for cutting accurate bevels. See Figure 2–13.

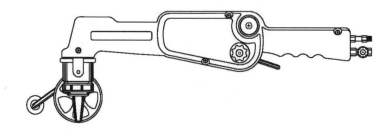

Figure 2–13   A motorized cutting head

- A portable track cutting machine travels along a pair of steel rails driven by a 120 VAC motor. It can make 90° bevel angle and chamfer cuts. It also cuts circles from 4 to 96 inches (0.1 to 2.4 m) diameter. With a second torch, it makes two parallel cuts simultaneously to produce a strip of metal with two parallel edges. It is especially useful in cutting accurate bevels and chamfers for proper fit up. See Figure 2–14 and 2–15.

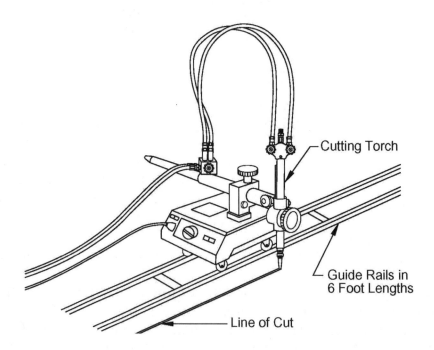

Figure 2–14   Portable track cutting machine

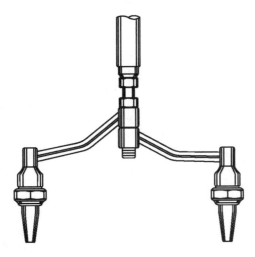

Figure 2–15 Two-torch strip cutting attachment for portable cutting machine

- A pattern tracer is the next level of improvement. A stylus follows the edge of a metal pattern, or a photoelectric eye follows the lines on paper and a torch (or could be multiple torches) is directly connected to the pattern tracing mechanism, permitting the torch to reproduce the pattern shape in steel. These systems require heavy cutting operator involvement and supervision.
- Computer driven cutting machines produce the most accurate and best quality cuts. These machines store the path of the cutting torch in their memories. Advanced machines control torch-to-work distance, adjust the torch speed on curves and around corners, adjust the pre-heat and cutting flame to starting and ending cuts. The most sophisticated machines require only loading raw stock and removing scrap and finished parts and can pierce holes to make inside cuts without operator assistance.

What degree of dimensional accuracy can be maintained in cutting machines?
About 1/32 inch (0.8 mm) can be achieved.

How should you adjust the torch tip preheat orifices in a normal cut?
If there are two preheat orifices, the tip should be rotated in the torch so that a line drawn between orifices will be perpendicular to the cut line. If more orifices, two should fall on the cut line and the rest divided equally on each side

of the cut line. This symmetrical preheating improves the quality of the cut. See Figure 2–16.

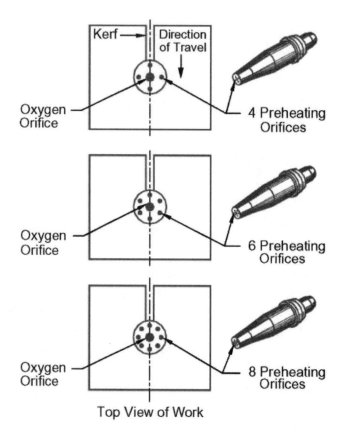

Figure 2–16   Location of preheat orifices in relation to kerf for a normal cut

For making a bevel cut how should you adjust the preheat orifices?
See Figure 2–17.

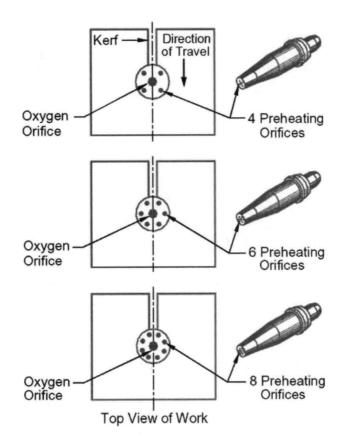

Figure 2–17   Location of preheat orifices in relation to kerf for a bevel cut

What is the best torch technique to start a heavy cut?
See Figure 2–18.

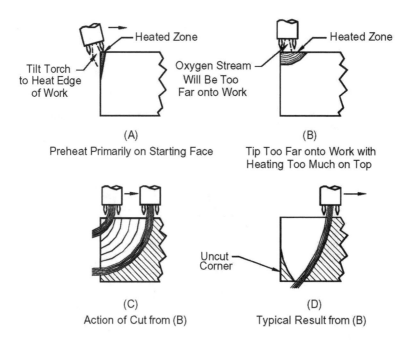

Figure 2–18   How to start a cut on heavy steel

What are the proper terminating conditions when making a heavy cut so as not to let the effect of drag permit the cutting action to skip a small triangular area at the bottom of the end of the cut?
As the end of the cut nears, tilt the torch away from the direction of travel. This permits the bottom of the cutting action to proceed ahead of the top cutting action and eliminates premature breakout of the flame which leaves a triangle at the end of the cut. See Figure 2–19.

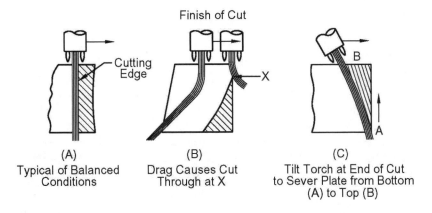

Finish of Cut

(A)
Typical of Balanced
Conditions

(B)
Drag Causes Cut
Through at X

(C)
Tilt Torch at End of Cut
to Sever Plate from Bottom
(A) to Top (B)

Figure 2–19   How to complete a cut on heavy steel

What is the easiest way to remove a rivet with a cutting torch?
Follow the steps in the Figure 2–20.

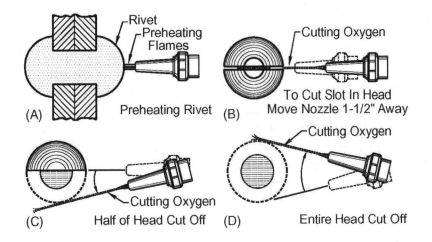

(A)   Preheating Rivet

(B) To Cut Slot In Head
Move Nozzle 1-1/2" Away

(C)   Half of Head Cut Off

(D)   Entire Head Cut Off

Figure 2–20   Steps to remove rivet

How can a countersunk rivet be removed?

See Figure 2–21.

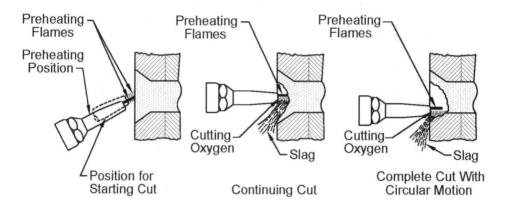

Figure 2–21   Steps to remove a countersunk rivet

What are two ways to pierce steel with OFC?

- Begin by preheating the material in the pierce location to a dull red color, kindling temperature, with the torch perpendicular to the metal. When metal becomes dull red, slightly raise the torch from the surface and angle the tip away from perpendicular. This prevents the slag blown back from the surface from landing on or in the torch tip. Then squeeze the oxygen lever to start cutting action. As soon as the material is completely pierced, restore the tip to perpendicular and the preheat flame to just above the surface. Complete cutting the opening wanted.

- If a small hole is wanted and the surrounding material is to be protected from cutting action, drill a 1/4 inch (6 mm) hole at the starting point. Begin the cutting action through the hole. See Figure 2–22.

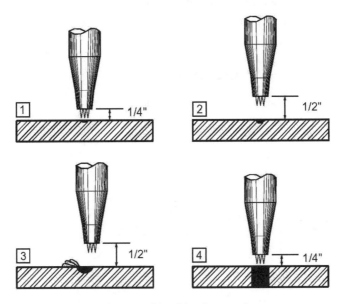

Figure 2–22   Piercing steel

What is the best way to cut out a circle?

Pierce the material *inside* the circle and away from the finished edge. When cutting action is established, extend the cut into a spiral and begin cutting the circle itself, Figure 2–23. With small circles to avoid damaging the finished edge, drill a 1/4 inch (6 mm) hole in the center of the circle and begin the cut through the inside of the hole, then spiral out to the edge.

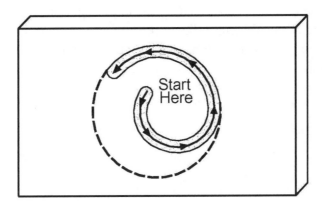

Figure 2–23   Cutting a circle

What is the easiest way to sever an I-beam?
Follow the numbered steps in Figure 2–24.

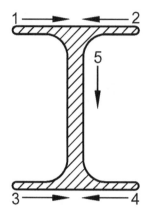

Figure 2–24   Severing an "I" beam

What is the AWS designation for oxygen lance cutting?
The abbreviation for oxygen lance cutting is *LOC*.

What is an *oxygen cutting lance* and how is it used?
An oxygen lance is a length of steel pipe connected to a source of oxygen. An oxyacetylene welding or cutting torch is used to bring a spot on the work up to ignition temperature. The torch is then removed and the lance pipe end placed over the heated spot and the oxygen supply opened. The lance cuts through the work like a large cutting torch. The steel lance pipe is usually 1/8 to 1/4 inch (3 to 6 mm) diameter and is consumed by the cutting action. It has the advantage of being able to poke holes into the work several feet deep. Lances are used to cut large steel or cast iron sections and to cut through reinforced concrete as well. See Figure 2–25.

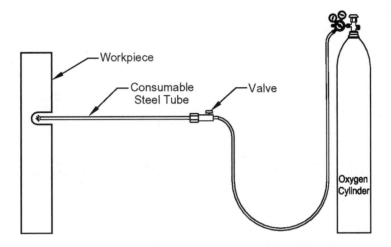

Figure 2–25   Oxygen lance process

Trouble Shooting

What causes a *bell-mouthed* kerf?
Excessive oxygen pressure, see Figure 2–26.

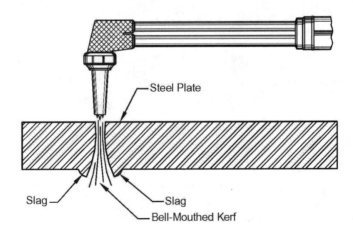

Figure 2–26   Bell-mouthed cut

When the cut is not smooth, how can you determine what corrective action to take?

Compare the defective edge with the drawings in Figure 2–27 to diagnose the problem.

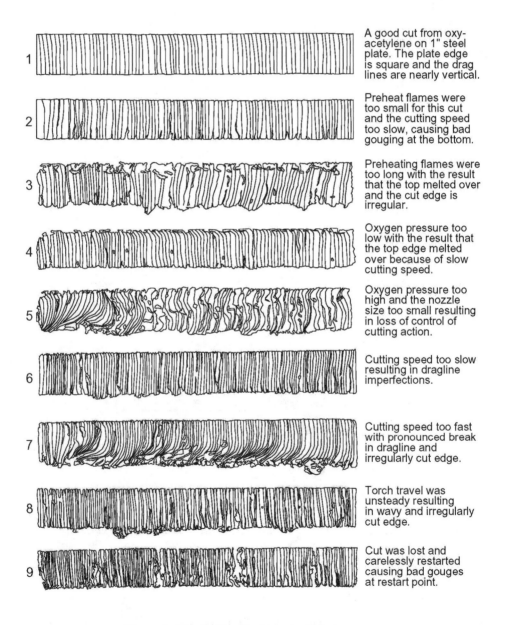

1. A good cut from oxy-acetylene on 1" steel plate. The plate edge is square and the drag lines are nearly vertical.

2. Preheat flames were too small for this cut and the cutting speed too slow, causing bad gouging at the bottom.

3. Preheating flames were too long with the result that the top melted over and the cut edge is irregular.

4. Oxygen pressure too low with the result that the top edge melted over because of slow cutting speed.

5. Oxygen pressure too high and the nozzle size too small resulting in loss of control of cutting action.

6. Cutting speed too slow resulting in dragline imperfections.

7. Cutting speed too fast with pronounced break in dragline and irregularly cut edge.

8. Torch travel was unsteady resulting in wavy and irregularly cut edge.

9. Cut was lost and carelessly restarted causing bad gouges at restart point.

Figure 2–27   Cutting Problems and their causes

Air Carbon Arc Cutting

What is the AWS abbreviation for Air Carbon Arc cutting?
The American Welding Society acronym is CAC-A

How does the CAC-A cutting process work?
This process uses an electric arc to melt the metal which is blown away by a high-velocity jet of compress air. See a CAC-A torch in Figure 2–28.

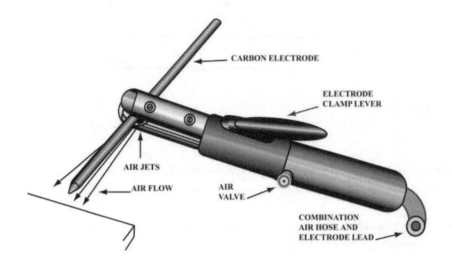

Figure 2–28   Air Carbon Arc Torch

Why was this process developed?
Myron D. Stepath originated the process during WW II, while working as a welding engineer with the U. S. Navy, where he conceived the idea to solve the problem of removing defective stainless steel welds in armor plate on warships; the conventional methods, at the time, were chipping and grinding, which had proved infeasible due to time and cost factors.

How is the CAC-A process used?
Today this process is used to rapidly remove defects in welds and base metal.

What are the electrodes made of?
The electrodes are rods made from a mixture of graphite and carbon and most are coated with copper to increase their current-carrying capacity. Manufacturers make both DC and AC rods for this process. See Figure 2–29.

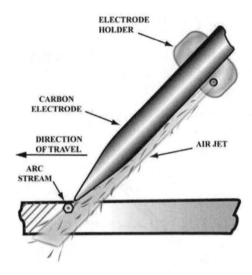

Figure 2–29   Carbon electrode air stream and travel direction

What type of power source is required for CAC-A?
Standard constant current welding power sources are used to provide current.

What is the required source of air for this process?
A jet of ordinary compressed shop air is all that is required.

What are the amperage requirements for this cutting or gouging process?
Depending on the electrode diameter and job requirements the amperage ranges can be as low as 60 amperes or as high as 2200 amperes. See Table 2–3.

Electrode Size Inch mm	Amperage Min.	Amperage Max.	Electrode Size Inch mm	Amperage Min.	Amperage Max.
1/8 3	60	90	3/8 10	450	600
5/32 4	90	150	1/2 13	800	1000
3/16 5	200	250	5/8 16	1000	1250
1/4 6	300	400	3/4 19	1250	1600
5/16 8	350	450	1 25	1600	2200

Table 2–3   Matching electrode diameter to current

Can we calculate electrode consumption?
When used correctly every inch of carbon consumed by the user will get approximately eight inches of groove when making a gouge that is equal in depth to the diameter of the carbon electrode; the gouge should be 1/8" wider than the diameter of the electrode. Never burn the electrode closer than three inches from the electrode holder because the heat from the electrode will damage the torch.

What are the advantages of the CAC-A process?
• The primary advantage of this process is the rapid removal of defects so repairs may be made in a timely manner.
• The CAC-A torches are relatively inexpensive.

Are there disadvantages to the use of this process?
• Operator can leave carbon deposits in the area that will be re-welded.
• Carbon deposits must be ground or brushed away before re-welding.
• This process requires compressed air.

Safety for CAC-A

What considerations should be made when using this equipment?
All of the electrical safety considerations covered in Chapter 13 should be followed. The minimum shade lens requirements are the same as those found in Chapter 5, Table 5–7. All of the clothing requirements covered in Chapter 4, Figure 4–26.

Safety

What precautions in handling oxygen and fuel gas cylinders and related equipment apply to OFC?
In addition to the safety precaution covered in this chapter beginning on the next page, all precautions listed in the Safety section of Oxyacetylene Welding, Chapter 1 must be followed.

What are the main hazards of OFC and what safety equipment can prevent these injuries?

- External eye injuries from cutting sparks prevented by safety glasses, or safety shields.
- Internal eye (retinal) damage from viewing hot metal and the radiation coming off it prevented by using a number 5 tinted lens while cutting or oxyfuel welding.
- Burns from weld sparks and hot metal prevented by leather gloves, non-flammable clothing, leather *skins* when working overhead, cuffless pants, pocketless shirts, a welding cap, and high-top shoes.
- Fires from the welding process are prevented by moving flammables away from the weld zone and having water or fire extinguishers close at hand.
- Fumes from paint or plating vaporized by the cutting process prevented by good ventilation and keeping out of the cutting plume.

What fire safety considerations are important in OFC?

- When cutting near materials that will burn, make sure that flame, sparks, hot slag, and hot metal do not reach them. Cutting creates more sparks than OAW.
- If the work to be cut can be moved, bring it to a safe location before cutting it.
- When flammable materials cannot be moved, use sheet metal shields or guards to keep the sparks away from burnables.
- Prevent sparks from falling into holes or cracks in wooden buildings.
- Do not use tarps or fabric covers to protect other materials from sparks as they will catch fire.
- If cutting on a wooden floor, first sweep it clean, then wet it down before starting cutting.
- Avoid using excessive oxygen pressure while cutting as this will propel sparks farther and make more of them.
- Plan ahead where hot metal will fall when cut; be especially careful to avoid your legs, feet, gas hoses, cylinders, and regulators.
- Have fire extinguishers, buckets of sand, or water on hand should a fire start.
- Make sure jacketed or hollow parts are vented before beginning cutting operations.

What is best way to cut into a sealed tank or container?

Never cut into a sealed container regardless of its size. Even if the container is clean and empty, penetration of the shell could release hot gases or send the cutting flame back toward the welder. If the container is empty *and* contains no residual vapors, vent it to atmosphere by opening a valve, hatch, bung or drilling a hole, then proceed to cut or weld. See Figure 2–30.

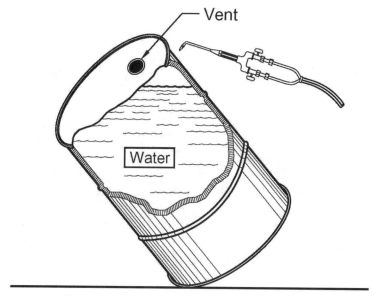

Figure 2–30   How to cut into a clean container

How can cutting or welding be done on a tank or container which has contained flammable materials?

An even more dangerous situation results when the vessel contains residual flammable vapors, whether it is vented to atmosphere or not. This will almost certainly result in an explosion. Flood the vessel with water to just below the cutting or welding point. Get the container cleaned usually by boiling with a caustic if the container is small or purged with a non-flammable gas like nitrogen, carbon dioxide or steam. Have the vessel checked for lack of explosive vapors by a qualified person. Then begin cutting or welding. See Figure 2–31.

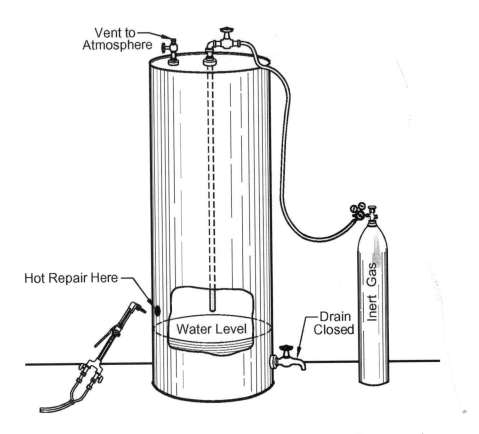

Figure 2–31   Using nitrogen or carbon dioxide to purge oxygen from a container
which has held flammable materials

Plasma Arc Cutting

How does *plasma arc cutting* work and what are its applications?
Plasma arc cutting, AWS designation *PAC*, is an arc cutting process that uses a
constricted arc and removes molten metal with a high-velocity jet of ionized
gas issuing from a constricting orifice. There are two variations:
- The first variations are in low-current plasma systems which use the nitrogen in compressed air for the plasma and are usually manual.
- The second variations are in high-current plasma systems which use pure
  nitrogen for the plasma and are usually automatic.

The PAC torch works very much like the plasma arc welding torch performing keyhole welding, except that the keyhole is not allowed to close. Plasma heat input is very high and melts the work metal. Then the plasma jet blows away the molten material completing the cut. Some PAC systems inject water into the plasma to reduce fumes and smoke; others perform the cutting under water to reduce noise and airborne metal vapor.

What are the capabilities of PAC?

High-current PAC systems cut ⅛ inch (3 mm) thick metal with a 100 inch/minute (2.5 m/minute) travel speed, 0.050 inch (1.25 mm) thick metal with a 200 inch/minute (5 m/minute) travel speed. The smaller, hand-held torches are used in sheet metal and auto body work. Attachments are available to convert PAW torches for PAC.

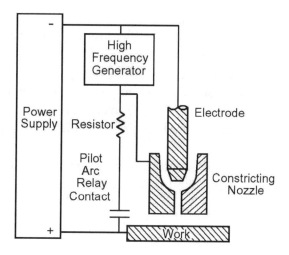

Figure 2–32   PAC schematic.

What are PAC's advantages?
- PAC cuts all metals.
- Cutting action is so rapid that despite the high heat input, there is a smaller heat affected zone than in most other processes.
- PAC can pierce metals cleanly without the starting hole needed by OAC.
- PAC is ideal for cutting parts under computer-driven control.
- With its 30,000°F (16,600°C) plasma, it cuts materials with melting points too high for OAC.

- All positions can be used.
- Surface smoothness of the cut edges is equal to or better than OAC.

What are the drawbacks to PAC?
- Equipment may be expensive. Small units today are more cost effective than in the past.
- Metal vapor produced from the cutting must be captured.
- Thick cuts are normally done under water so the metal vapor can be captured; the water container must then be periodically cleaned usually requiring a HAZ MAT crew.

What safety rules should be followed?
All of the safety rules suggested in OAC cutting should be applied to PAC including:
- This process used electricity with voltage ranges from 150 to 400 volts of direct current; this equipment must be properly grounded to avoid electric shock.
- Keep electrical circuits dry.
- Keep all mechanical electrical connection tight; this includes the work lead. Poor electrical connection can cause over heating and fires.
- Proper ventilation is required to prevent inhalation of hazardous metal vapors and gases.
- When securing the equipment always be sure the power supply has been properly shut down and the torch placed in back on it's proper insulated storage position.

Laser Beam Cutting

How does *laser beam cutting* work and what are its applications?
Laser beam cutting, AWS designation *LBC*, is a thermal process using laser beam energy to cut materials by melting or vaporizing. A gas is sometimes used to assist in the removal of melted or vaporized material. Cutting and drilling power densities in the range of 6.5×10^6 to 6.5×10^8 W/in^2 (10^4 to 10^6 W/mm^2) are achieved. Lasers can also drill holes using higher power densities and shorter dwell times than when cutting. Hole dimensions range from 0.0001 to 0.060 inches (0.0025 to 1.5 mm). Although a high-power CO_2 laser cuts carbon steel up to 1 inch thick (25 mm), good quality cuts are made on material 0.375 inch (9.5 mm) and less in thickness. The depth of focus limits the quality of thick cuts.

What are the advantages of LBC?
- Narrow kerf widths
- High cutting speeds
- High quality edge surfaces
- Low heat input/minimum distortion
- Cuts most materials
- Easily automated
- Repeatable, precision dimensions
- Multiple layers of material may be cut at the same time

What are the drawbacks to LBC?
- Equipment is expensive.
- Replacement lens' and consumables are expensive.

See Figure 2–33.

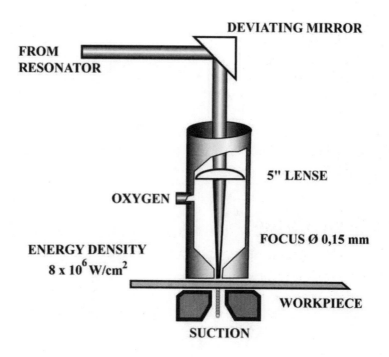

Figure 2–33   Schematic of a LBC

Chapter 3

Brazing & Soldering

It is possible to fly without motors, but not without knowledge and skill.
Wilbur Wright

Introduction

Soldering has been used for thousands of years, but a solid theoretical understanding has come only in the last one hundred. Brazing is similar to soldering but performed at higher temperatures. It has come into use more recently as higher temperature torches became available. Although soldering and brazing do not make joints as strong as welded ones, they are widely used in making and repairing a wide range of products from airplanes to computers to household plumbing to jewelry. Soldering and brazing processes can be as simple as using a soldering iron, a propane torch or as complex as using radio frequency energy in a vacuum to join parts. We will discuss the theory, materials, fluxes, and common industrial processes. These processes have some advantages over welding and we will present them. There are safety issues too. Because soldering copper water pipe is so useful and something every welder should know how to do, we cover step-by-step instructions in Chapter 15.

Process Names

What is the AWS designation for *brazing*?
The AWS designation for brazing is *B*.

What common names are sometimes used for *brazing*?
Silver soldering, silfloss, and *hard soldering* are non-preferred names for brazing. Silver brazing filler metals are not solders; they have melting points above 840°F (450°C).

How are brazing processes classified and what are the commercially significant ones?

Brazing processes are classified by heat-source type. Commercially important ones are:

- Dip Brazing (DB)
- Furnace Brazing (FB)
- Induction Brazing (IB)
- Infrared Brazing (IRB)
- Resistance Brazing (RB)
- Torch Brazing (TB)

What is the AWS designation for *soldering*?

The AWS designation for soldering is *S*.

What are the commercially important soldering processes?

- Dip Soldering (DS)
- Furnace Soldering (FS)
- Induction Soldering (IS)
- Infrared Soldering (IRS)
- Iron Soldering (INS)
- Resistance Soldering (RS)
- Torch Soldering (TS)
- Wave Soldering (WS)

Note that with the exception of iron soldering and wave soldering both brazing and soldering share most common processes.

Process

In general terms how does the brazing process work?

Brazing joins materials that have been heated to the brazing temperature followed by adding a brazing filler metal having a melting point above 840°F (450°C). This temperature will be below the melting point of the base metals joined by this joining process. The non-ferrous filler metal is drawn into and fills the closely fitted mating or *faying* surfaces of the joint by capillary attraction. The filler material, usually aided by fluxes, *wets* the base metal surfaces allowing the brazing material to flow or capillary more readily through and between the two surfaces. When the filler metal cools and solidifies, the base materials are joined. See Figure 3–1.

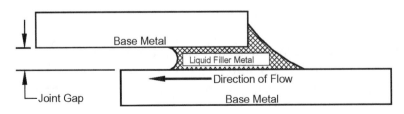

Figure 3–1   Capillary forces draw molten filler metal into brazed joint

How does *soldering* differ from *brazing*?

Brazing takes place *above* 840°F (450°C) and soldering *below* 840°F (450°C), otherwise, the processes are quite similar. They both depend on capillary attraction to draw filler metal or solder into the joint. In general brazed joints are stronger than soldered ones because of the strength of the alloys used.

What is the difference between *brazing* and *braze welding*?

Brazing depends on capillary attraction to draw the filler metal into the mating joint, while in braze welding the filler metal is deposited in grooves or fillets at the points where needed for joint strength. Capillary attraction is *not* a factor in distributing the filler metal, as the joints are open to the welder. Braze welding is *not* a brazing process, but welding with brazing filler metal. Braze welding is frequently used to repair cracked or broken cast iron parts. Joint design is similar to those for OAW: V-groove butt joints, lap joints, T-joints, fillets, and plug joints.

What advantage does braze welding have over welding?

Because braze welding does not melt the base metal and is performed at a lower temperature than welding, there is less distortion of the part. Also, the process of braze welding is performed only on a small area of the part, so the entire part does not need to be brought up to braze temperature all at once.

How does a brazed joint hold the base metals together?

Filler metal atoms have a stronger attraction to the base metal's atoms than to their own. This force between two surfaces is called *adhesion*. This preference for atoms other than its own causes the wetting action of the base metal. When wetting action takes place in a small diameter tube or between closely spaced parallel plates as in a joint, capillary action occurs. Capillary attraction is so strong it readily opposes gravity and works to the welder's advantage by bringing filler material into the joint and distributing it evenly. A soldered joint's strength comes from the same forces. See Figure 3–2.

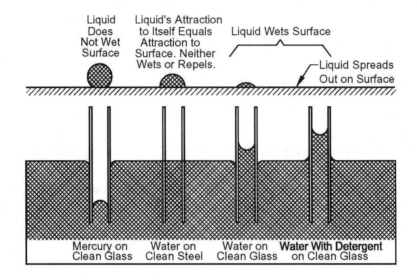

Figure 3–2   Capillary attraction

On what metals can brazing be performed?

Most common metals can be brazed or soldered including:

- Aluminum
- Bronze
- Brass
- Cast iron
- Copper
- Stainless Steel
- Steel
- Titanium
- Tool steels (some)
- Tungsten carbide (tool bits)

What metals can be soldered?

Most metals may be soldered.

What are the advantages of the brazing and soldering processes over other joining methods?

- Ability to join dissimilar metals—Steel is easily joined to copper, cast iron to stainless steel, and brass to aluminum. Many combinations of metals are readily joined.
- Ability to join nonmetals to metals—Ceramics are easily joined to metals, or each other.
- Ability to join parts of widely different thicknesses—either thin-to-thin or thin-to-thick parts may be joined without burn-through or overheating.

- Excellent stress distribution—Many of the distortion problems of fusion welding are eliminated because of the lower process temperature and the even distribution of heat with more gradual temperature changes.
- Low temperature process—the components being joined like semi-conductors are less likely to be damaged, since the base metals are not subjected to melting temperatures.
- Economical for complex assemblies—many parts can be joined in a single process step.
- Joins precision parts well—with proper jigs and fixtures, parts may be very accurately positioned.
- Parts may be temporarily joined, subjected to other manufacturing processes, and then separated without damage.
- Parts can be assembled rapidly—processes are readily adaptable to batch and automatic assembly operations.
- Mistakes are readily fixed—a misaligned part can be repositioned without damage.
- Ability to make leak-proof and vacuum-tight joints—many tanks are soldered or brazed; high-power radio transmitter vacuum tubes and integrated circuits with metal to ceramic joints are brazed.
- Joints require little or no finishing—with proper process design the brazed or soldered joint can be nearly invisible.
- Combined brazing and heat treatment cycles—when protective atmosphere brazing is used, the brazing process may be incorporated into the heat treatment cycle.

What are some disadvantages of the brazing process?
- While brazing processes can produce high-strength joints, they are rarely as strong as a fusion-welded joint.
- The brazed parts and the filler metal may lack a color match.

Important Processes Detailed

How do each of the commercially important brazing and soldering processes work and what are their advantages and applications?

Dip Brazing

There are two types. Molten-metal bath dip brazing uses a pot of molten filler metal, usually temperature controlled and heated by electricity, oil, or gas. The cleaned, fluxed parts are immersed into the molten filler metal that enters the joints by capillary attraction. When the assembly is withdrawn and cooled, the brazing is complete. The parts are usually self-jigging. A layer of flux usually covers the molten metal to retard oxidation.

The other method is molten chemical bath dip brazing. Here the parts are cleaned, filler metal is placed between the joints, the parts to be joined are then assembled with filler in place, preheated, then dipped into a pot of molten chemicals serving as a flux.

The advantage of dip brazing over torch brazing is that even heating of the part reduces distortion. Dip brazing may be manual or automated; it is used on small to medium parts. See Figure 3–3.

Dip soldering very much resembles molten metal dip brazing using a molten metal bath but at a lower temperature.

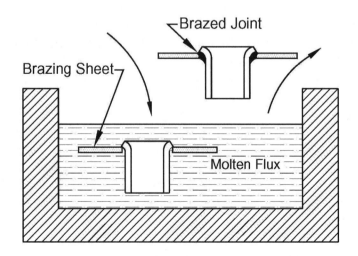

Figure 3–3   Chemical bath dip brazing or soldering

Furnace Brazing

The parts are cleaned, brazing filler metal is placed inside the joints, and the parts assembled using fixtures to hold them in proper position. The pre-placed brazing filler metal can be in the form of filings, foil, paste, powder, tape, or special shapes called *preforms* that fit the joint. Both batch and continuous conveyor furnaces are used. Furnaces may have multiple heat zones for preheat, brazing, and cool-down.

Flux is used in furnaces with an air atmosphere, but air can be eliminated by using a special atmosphere (argon or helium) or a vacuum. If flux is not used in the brazing process, it will not have to be removed in a later step—a significant advantage. Furnace brazing offers lower distortion than torch brazing, and may also perform heat-treating. See Figure 3–4.

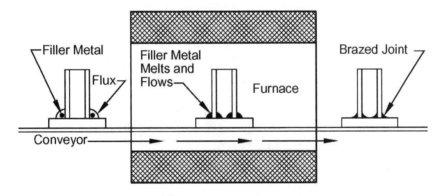

Figure 3–4   Furnace brazing

Furnace soldering is similar to furnace brazing, but at a lower temperature and normally in an air atmosphere.

Induction Brazing

This process depends on inducing an alternating current in the part. As this induced current flows around inside the part, it generates heat from the resistance of the part itself, and brings it up to brazing temperature. A solid-state, or vacuum-tube oscillator generates alternating current from 10 to 500 kHz. This current is fed to a coil of copper tubing that usually surrounds the part. This copper coil acts as the primary of a transformer and the parts themselves act as the secondary. The copper tubing coil itself is kept from melting by cooling water flowing through the tubing's interior. Power from one to several hundred kilowatts is used. Induction coils are designed in shapes that maximize the transfer of current to the assembly being brazed and take many shapes and sizes. See Figures 3–5 and 3–6.

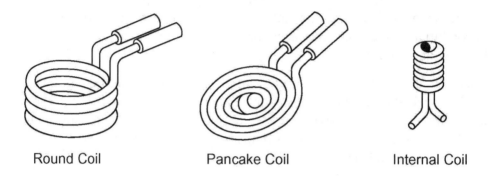

Round Coil   Pancake Coil   Internal Coil

Figure 3–5   Different shapes of induction brazing coils

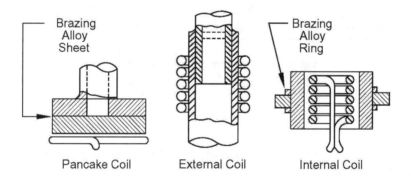

Figure 3–6   Coils in place on brazing work

Cleaning the parts, inserting brazing filler metal and flux into the joints, assembling the parts and heating them with an induction coil, perform induction brazing. The process is very fast, usually measured in seconds, and used to make consumer and military parts. It is often automated. Some processes utilize a vacuum and use no flux. Induction Soldering is similar to induction brazing, but at a lower temperature.

Infrared Brazing

Infrared brazing is a form of furnace brazing. High-intensity quartz lamps supply long-wave heat of up to 5 kW each. Concentrating reflectors focus heat on the parts. This process is sometimes performed in a vacuum and is usually employed in a conveyor-fed production process. Infrared Soldering is similar to infrared brazing, but at a lower temperature.

Resistance Brazing

Electric current flowing through the joint to be brazed provides heat for this process. The joint is cleaned, fluxed, and braze filler material is placed inside the joint in the form of wire, washers, shims, powder, or paste. Then the joint is placed between two electrodes, squeezed together, and electricity is applied. The source of electricity is usually a step-down transformer providing from 2 to 25 volts. Current runs from 50 amperes for small jobs to thousands of amperes for large ones. The electrodes are high-resistance electrical conductors like carbon or graphite blocks, or tungsten or molybdenum rods. Most of the heat is produced in the electrodes raising them to incandescence. This heat flows into the joint completing the braze joint. The resistance of the joint alone is not usually an adequate heat source. The cycle time varies from one second to several minutes depending on part size. See Figure 3–7.

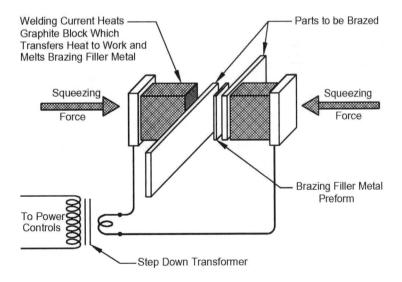

Figure 3–7   Resistance brazing or soldering method

Torch Brazing

Heating is done with one or more gas torches. Depending on the size of the parts and the melting point of the filler metal, a variety of torch fuels (acetylene, propane, methylacetylene-propadiene stabilized, or natural gas) may be burned in oxygen, compressed air, or atmospheric air. A neutral or oxidizing flame will usually produce excellent results.

Flux is required with most braze filler metals and can usually be applied to the joint ahead of brazing. The filler metal can be preplaced in the joint or face fed. Manual torch brazing is successfully used on assemblies involving components of unequal mass. It is frequently used in the repair of castings in the field and is often automated for high-production of small and medium-sized parts. It is a versatile process and probably the most popular brazing technique.

Torch Soldering

This is very much like torch brazing but at a lower temperature. Usually propane, methylacetylene-propadiene gas, or natural gas burning in air supplies the heat.

The joint is cleaned to shiny metal with emery cloth, wire brushes, steel wool, or commercial abrasive pads. The flux (used for wetting the joining surfaces) is usually applied in liquid or paste form, or may be alloyed inside the solder wire.

While widely used in manufacturing and maintenance, it is also used in plumbing to join copper tubing for potable water. See Chapter 15 for a detailed procedure for the torch soldering of copper tubing.

Iron Soldering

Traditional soldering irons contain a copper tip on a heat-resistant handle. They are heated electrically or in a gas, oil, or coke furnace. The copper tip stores and carries heat to the solder joint. This transfer is made possible by heat being transferred from the heated tip of the iron to the part to be soldered; when the joint is raised to soldering temperature, solder is applied to the joint itself, and wets the entire joint. Flux core solder is used for electronic work. This solder has been formed concentrically around a core of one or more strands of flux. In sheet metal and other non-electrical work, the flux may also be in the solder core, or applied as a paste or a liquid.

Today most soldering irons are heated electrically and are available from just a few watts for electronic work to 1250 watts for roofing and heavy sheet- metal work. Many irons for electronic work have temperature-controlled tips to avoid damage to the sensitive components. See Figure 3–8.

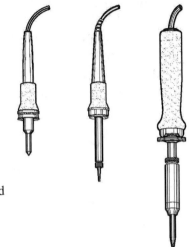

Figure 3–8   Electrically-heated
             soldering irons

Wave Soldering

This process is used to solder electronic components onto printed circuit boards. A conveyor belt draws a printed circuit board with components over a fountain (or *wave*) of solder. This solders all the components in place in a single step. Wave soldering machines are available which can flux, dry, preheat, solder, and clean the flux off a finished board on a single conveyor line. See Figure 3–9.

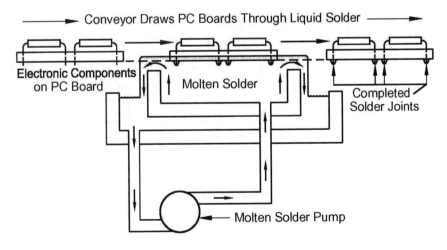

Figure 3–9   Wave soldering

Joint Design for Brazing

Why is joint design an important part of brazing process design?

Well-designed braze joints start with fundamental butt and lap joints. Figures 3–10A shows a butt joint angled to provide more surface area for the brazed or soldered material to bond; this angles joint is called a scarf joint. Figure 3-10B shows both a lap joint and a square edged butt joint. Good joint design insures a reliable, repeatable production brazing process that will provide a strong joint.

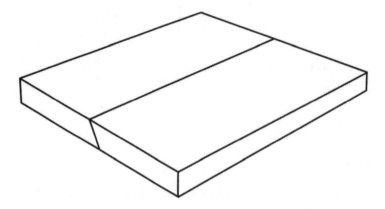

Figure 3–10A   Butt joint prepared at an angle and called a scarf joint

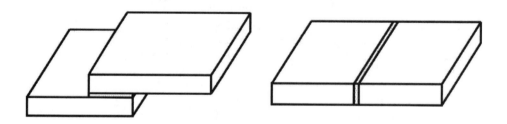

Figure 3–10B   Lap and butt joints

What design changes can be made in butt and lap joints to increase their strength?

See Figures 3–11 through 3–13.

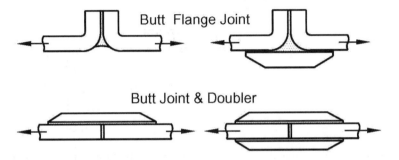

Figure 3–11   Progressive design changes to increase butt joint strength

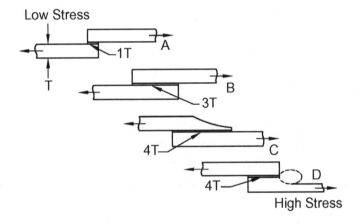

Figure 3–12   Progressive design changes to increase lap joint strength

In Figure 3–12, the maximum strength of a simple lap joint appears in B, an overlap of 3 times the thickness of the base metal (3T). More overlap without additional refinements will not improve strength.

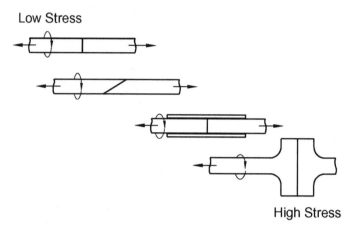

Figure 3–13   Progressive design changes to increase butt joint strength against torsional (twisting) forces

What factors influence joint good design?

- Base metal selection—differences in base metal thermal expansion coefficients may lead to poor joint fit at brazing temperature with too much or too little clearance.
- Effect of flux on clearance—flux must enter the joint *ahead* of the braze filler metal and then be displaced by it, but when joint clearance is too small the flux may be held in the joint by capillary forces. This prevents proper braze filler entry and leaves voids.
- Effect of base metal to filler metal combinations on clearance.
- Effect of brazing metal filler on clearance.
- Effect of joint length and geometry on clearance.
- Dissimilar base metals form a cell that leads to electrolytic corrosion.

What are typical clearances used in brazed joints?

Joint clearances range from 0.002 to 0.010 inches (0.05 to 0.25 mm). Assuming that the joint clearance is adequate to admit braze filler material, the lower the joint clearance, the stronger the joint. Too much clearance will reduce joint strength, too little will permit voids in the joint.

Joint Preparation

How parts to be joined usually cleaned for brazing?

There are two classes of cleaning: chemical and mechanical. There are many different processes in use. Some common chemical ones are:

- Solvent cleaning with petroleum solvents or chlorinated hydrocarbons.
- Vapor degreasing with perchlorethylene or other solvents.
- Alkaline cleaning with mixtures of phosphates, silicates, carbonates, detergents, hydroxides, and wetting agents.
- Electrolytic cleaning.
- Salt baths.
- Ultrasonic cleaning.
- Mechanical cleaning processes include:
- Grinding
- Filing
- Machining
- Blasting
- Wire brushing

Wire brushes must be free of contaminating materials and selected so none of the wire wheel material is transferred to the part being cleaned. A stainless steel wire brush is a good choice for most materials.

Blasting media must be chosen so it does not embed in the base metal and is easily removed after blasting. For this reason, blasting media like alumina, zirconia, and silicon-carbide should be avoided. These processes are used to remove all dirt, paint and grease so the flux and braze filler metal can readily and completely wet the base metal surface.

How are joints prepared for soldering?

Many of the same processes used in brazing are used for soldering. However, in a non-production situation, mechanical cleaning especially with emery cloth, steel wool, or commercial abrasive pads or by filing will be effective. Getting down to fresh, bare metal is the objective. Complete the soldering immediately, before the base metals have a chance to re-oxidize.

Brazing and Soldering Fluxes

What is the purpose of flux in soldering and brazing?
* Further cleaning the base metal surface after the initial chemical or mechanical cleaning.
* Preventing the base metal from oxidizing while heating.
* Promoting the wetting of the joint material by the braze filler material or solder by lowering surface tension and to aid capillary attraction.

How do fluxes promote wetting of the base metals?
Flux covers and *wets* the base metal preventing oxidization until the braze filler material or solder reaches the joint surfaces. Since the flux has a *lower* attraction to the base metal's atoms than the filler or solder, when the filler metal or solder melts, it slides *under* the flux and adheres to the clean, unoxidized base metal surface ready to receive it. Fluxes will not remove oil, dirt, paint or heavy oxides, so the joint surface must already be clean for them to work.

What are the main categories of brazing fluxes?
Brazing fluxes usually contain fluorides, chlorides, borax, borates, fluoroborates, alkalis, wetting agents, and water. A traditional and still common flux is 75% borax and 25% boric acid (borax plus water) mixed into a paste.

The *AWS Brazing Manual* provides specifications for brazing and brazing fluxes. This specification has 15 classifications of fluxes. Many manufacturers supply proprietary flux mixtures meeting these specifications. See Table 3–1 for abbreviated AWS flux categories and applications.

Base Metals Being Brazed	Filler Metal Type	AWS Flux Class.	Typical Flux Ingredients	Activity Temperature Range		Application
				°F	°C	
All brazeable aluminum alloys.	BAlSi	FB1-A (Powder)	Flourides Chlorides	1080-1140	560-615	For torch or furnace brazing.
All brazeable ferrous and non-ferrous metals except those with Al or Mg as a constituent. Also for brazing carbides.	BAg and CuP	FB3-A (Paste)	Borates Fluorides	1050-1600	565-870	General purpose flux for most ferrous and non-ferrous alloys. Notable exception is aluminum bronze.
Same as above.	BAg, BCuP, and BCuZn	FB3-K (Liquid)	Borates	1400-2200	760-1205	Exclusively used in torch brazing by passing fuel gas through a container of flux. Flux is applied by the flame.
Brazeable base metals containing up to 9% Al (Al-brass, Al-bronze, Monel® K500).	BAg and BCuP	FB4-A (Paste)	Chlorides Fluorides Borates	1100-1600	595-870	General purpose flux for many alloys that form refractory oxides.

Note: The letter *B* in filler metal indicates brazing usage, not the chemical element boron.

Table 3–1   Representative brazing flux categories

In working with brazing flux, what precautions must be observed?

Many fluxes contain powerful poisons with long-term and short-term actions. See the Safety section for details.

If water must be added to turn a powdered flux into a paste or to thin an existing paste, what precautions must be followed?

Distilled water must be used, since tap water may contain chemicals that would damage the joint.

What are the main types of soldering fluxes?

- Organic fluxes—consisting of organic acids and bases, after soldering they can be removed with water and are widely used in electronics.
- Inorganic fluxes—containing no carbon compounds, so they do not char or burn easily and are used in torch, oven, resistance, and induction soldering. They are not used for soldering electrical joints.
- Rosin-based fluxes—easily cleaned from parts after soldering. They are usually non-corrosive; available as powders, pastes, liquids, and as a core within soldering wire. They are used in electrical and electronics applications.

How are brazing fluxes applied?

- Flux can be applied by spraying, brushing, or dipping.
- Flux is adhered to the end of a brazing rod by heating of the rod's end with the torch and dipping it into the flux. The flux is a dry powder.
- Some brazing (and braze welding) rod comes from the factory with flux already applied to the outside.
- Sheets, rings, and washers of flux can be inserted in the joint before assembly.
- Special guns can inject flux (or mixtures of flux and filler metal) directly into the joint.
- Flux can be dissolved in alcohol and supplied within the fuel gas stream directly to the brazing joint eliminating the manual operation of adding flux. This process automatically controls the amount applied. See Figure 3–14.

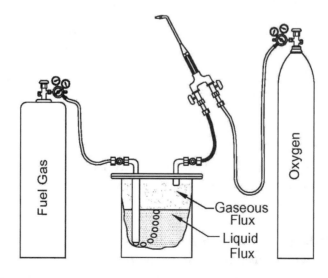

Figure 3–14   Gas fluxing unit for oxyfuel brazing

How are soldering fluxes applied?
Soldering fluxes are brushed, rolled, or sprayed. Many solders have flux cores, so no separate fluxing step is needed.

Brazing Filler Materials and Soldering Alloys

What properties must brazing filler materials have?
They must have the ability to make joints with mechanical and chemical properties for the application.

- Melting point below that of the base metals being joined and with the right flow properties to wet the base metals and fill joints by capillary attraction. See Figure 3–15.
- Composition that will not allow it to separate into its components (liquation) during brazing.

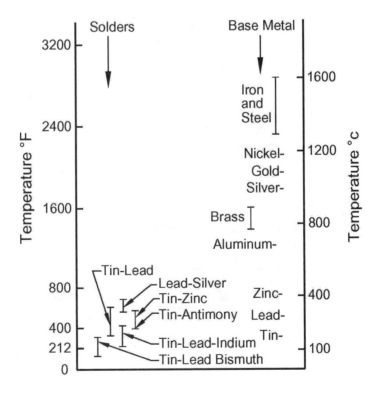

Figure 3–15   Melting points of braze filler metals and solders fall well below most base metals

What do the terms *solidus* and *liquidus* mean?
Solidus is the highest temperature at which a metal is completely solid. Liquidus is the lowest temperature at which a metal is completely liquid.

In general what can be said about the solidus and liquidus temperatures of pure metal? What can be said of alloys of two metals?
Because pure metals have an abrupt melting point, their solidus and liquidus temperatures are the same. However, in an alloy of two metals there is both a range of temperatures and a range of compositions at which both solid and liquid phases of the alloy can exist.

What is the best way to show how the melting and freezing properties of an alloy of two metals changes as its composition changes from all one base metal to all of the other base metal?
A constitutional diagram shows how the changing alloy's mix of composition affects its melt properties. See silver-copper constitutional diagram in Figure 3–16.

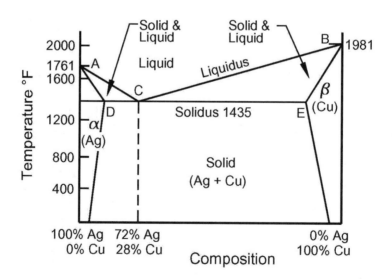

Figure 3–16   Silver-Copper constitutional diagram

What does the above constitutional diagram show?

- The *solidus* line ADEB indicates the temperature at which the alloy begins to melt for compositions of copper and silver.

- The *liquidus* line ACB indicates the temperature above which the alloy is completely melted.

- For a particular mix of the two metals, the melting point is *lower* than the melting point of either pure metal making up the alloy, or of any other mixture of the alloy. For a silver-copper alloy, the minimum melting point is 1435°F (779°C) and occurs at 72% silver - 28% copper (point C). This is called the *eutectic temperature* and *eutectic composition.* Note that copper melts at 1481°F (805°C) and silver at 1761°F (961°C), both well above the 1435°F eutectic temperature.

- For alloy mixtures other than the eutectic, there is a range of temperatures in which both solid and liquid phases of the alloy can exist together (area ADC and area CEB). In these areas of temperature and composition the alloy is mushy or slushy, while at the eutectic it has a sharp melting point and is as fluid as a pure metal.

How can the information in a constitutional diagram suggest brazing and soldering alloys for specific applications?

- By using a eutectic alloy, we can minimize the temperature at which we perform the brazing (or soldering in the case of tin-lead or tin-antimony alloys) and still have a fluid composition that can easily make its way into the brazing joints.

- By using a non-eutectic alloy, we can achieve an alloy that is slushy or less fluid than at the eutectic. The wider the difference between the solidus and liquidus lines in the constitutional diagram, the more sluggish the alloy is in this temperature range. This would be helpful where too fluid an alloy would not stay in place in an inverted joint and where we want capillary attraction to prevail over gravity. A good example of this is soldering a fitting or seam upside down.

Do other soldering and brazing two-metal alloys have the same general pattern as the silver-copper constitutional diagram?

Yes, see the tin-lead constitutional diagram in Figure 3–17.

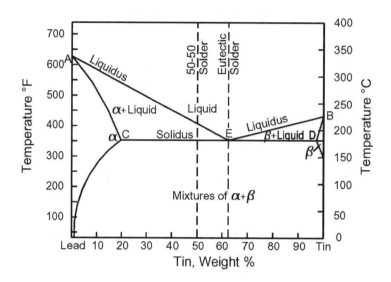

Figure 3–17   Tin-lead constitutional diagram

What are the most common brazing filler metals?

Filler materials covered by AWS specifications are grouped as:

- Aluminum-silicone
- Copper
- Copper-phosphorus
- Copper-zinc
- Heat-resisting material
- Magnesium
- Nickel-gold
- Silver

What determines the choice of braze filler metal?

The filler choice is determined by base materials or metals.

What is the purpose of shims or washers of filler metal preplaced in the work?

Preplacing filler metal permits the parts to be brazed or soldered in an oven or by other means without need of human attention to feed in the filler metal at the right time and place. See Figure 3–18.

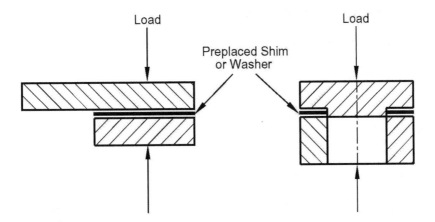

Figure 3–18   Method of preplacing brazing filler metal

What is *stop-off* used for?
Stop-off is used to outline the area not to be brazed. It prevents the flux from entering that area.

What are the most common solders?
Tin-lead alloys are the most common solders. It is customary to indicate the tin percentage first, then the lead content and the same with other two-metal alloys. A 40/60 tin-lead solder is 40% tin and 60% lead.

Tin-lead solders 35%/65%, 40%/60% and 50%/50% are popular because of their low liquidus temperatures. Tin-lead solders 60%/40% and the 63%/37% eutectic are used when low-processing temperatures are needed. Tin-silver, tin-copper-silver, and tin-antimony alloys are used where lead must be eliminated for health reasons as in stainless steel fabrication for kitchens, food processing equipment, and copper potable water systems. *Never* use lead-containing solder on potable water systems.

What are some other less common solders?
• Bismuth-containing solders provide alloys with very low melting temperatures for sprinkler heads and heat detectors for alarms.
• Indium alloys with liquidus temperatures as low as 230°F (138°C) are used for glass-to-glass and glass-to-metal seals in electronics.

Troubleshooting Brazing & Soldering Processes

Problem: No flow or no wetting.

Causes:

- Wrong braze filler
- Temperature too low
- Time at temperature too short
- Parts not properly cleaned
- Parts fit poorly
- Heat source in wrong location

Problem: Excess flow or wetting causes hole plugging or brazing wrong joints.

Causes:

- Temperature too high
- Time at temperature too long
- Too much filler material
- No stop-off used

Problem: Erosion—Braze filler material eats away parent metal.

Causes:

- Temperature too high
- Time at temperature too long
- Excessive braze filler metal
- Cold worked parts

Safety

What special chemical hazards do brazing and soldering present and what precautions must be taken?

Base metals and filler metals may contain toxic materials such as: antimony, arsenic, barium, beryllium, cadmium, chromium, cobalt, mercury, nickel, selenium, silver, vanadium, or zinc. These will be vaporized during brazing or soldering and cause skin, eye, breathing, or serious nervous system problems. Some of these toxic materials are cumulative such as lead and may be absorbed through the skin. The following precautions are essential:

- Keep your head out of the brazing or soldering plume.
- Perform brazing or soldering in a well ventilated area.
- On failure of normal ventilating equipment, use respiratory equipment.

Many brazing and soldering fluxes and heating bath salts contain fluorides. Others contain acids and aluminum salts. The following precautions apply:

- Avoid direct contact with skin.
- Do not eat or keep food near these materials.
- Do not smoke around these materials.
- Insure MSDSs are affixed to containers of these materials and major equipment using them so they are visible to you and others.

What is an excellent source of information about these hazards in addition to the MSDSs?

See the AWS booklet *Z49.1, Safety in Welding, Cutting and Allied Processes.*

What eye protection is needed for brazing and soldering?

- For soldering, wear safety glasses or face shields to protect the eyes from external injuries caused by sparks, flying metal, or solder splashes.
- For brazing, using a number 5 tinted lenses will protect against internal (retinal) eye damage caused by viewing the radiation coming off hot metal. Some brazing requires darker lens shades of up to number 8.

What other safety precautions must be taken while soldering or welding?

- Skin protection from sparks and hot metal prevented by gloves and non-flammable clothing.
- Fires from the welding process can be prevented by moving flammables away from the weld zone and having water or fire extinguishers close at hand.
- Use adequate ventilation when using cleaning solvents to prepare the joints; chlorinated hydrocarbons are toxic and may create phosgene gas when heated.
- Always wear chemical-type eye goggles or face shields, rubber gloves, and long sleeves while using cleaning solutions, pickling solutions, or acids. Note that chemical-type goggles do not have ventilation holes above the eyes where splashes could enter.

Chapter 4

Common Welding Elements

Experience is the name everyone
gives to their mistakes.
Oscar Wilde

Introduction

There is a lot of material common to the electrically-based welding processes —SMAW, GMAW, FCAW, and GTAW, so it makes sense to cover them all at once. The joints and positions are also common to OAW. These items are:

- Joint types
- Joint edge preparation
- Parts of a weld
- Selection of joint preparation
- Welding positions
- Other types of welds
- Welding terminology
- Welding cables
- Safety equipment
- Safety practices

Joint Types

What are the five basic joint types?

- Lap joint
- Butt joint
- Corner joint
- T-joint
- Edge joint

See Figure 4–1.

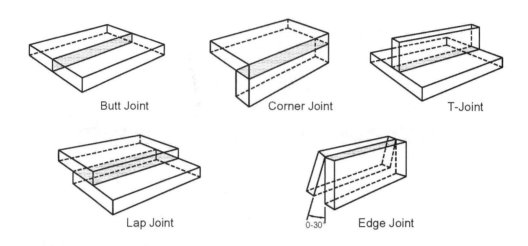

Figure 4–1   Joint types

Joint Edge Preparation

Why is edge preparation done?
Joint preparation provides access to the joint interior. Without it the entire internal portion of the joint would not be fused or melted together making the joint weak. Remember that a properly made, full-penetration joint can carry as much load as the base metal itself, but full penetration will only occur with the correct joint preparation.

How are edge shapes for weld joint preparation made?
Usually, they are made by flame cutting, plasma arc cutting, machining or grinding however, castings, forgings, shearing, stamping and filing are also common methods used to prepare material for welding.

What edge shapes used in preparation for welding?
See Figure 4–2.

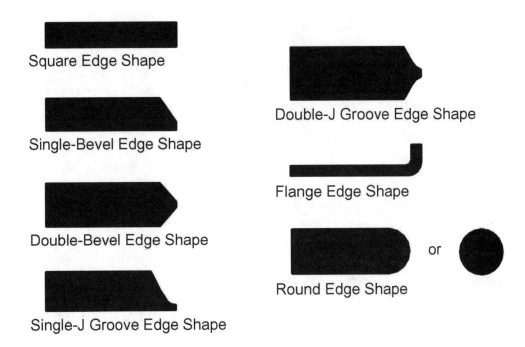

Figure 4–2   Edge shapes for weld preparation

Edge Preparation Terminology

What are the principal parts of V-and U-groove joint preparations?
- Depth of bevel
- Size of root face
- Root opening
- Groove angle
- Bevel angle

See Figure 4–3.

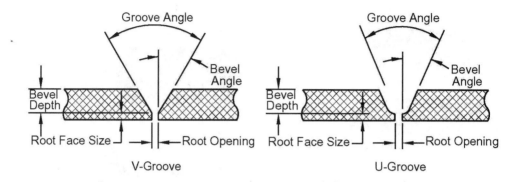

Figure 4–3   Parts of V- and U-groove joint preparations

Edge Shape Combinations

What are the most common combinations of joint preparations for butt joints?
See Figure 4–4.

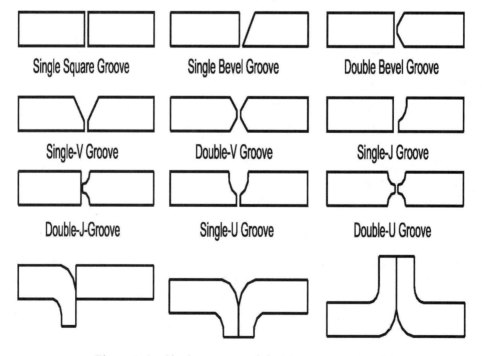

Figure 4–4   Single-groove and double-groove weld joint

What common preparations are used on corner joints?
See Figure 4–5.

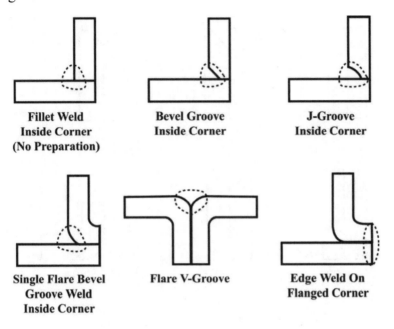

Figure 4–5   Weld preparation for corner joints

What weld preparations are used on T-joints?
See Figure 4–6.

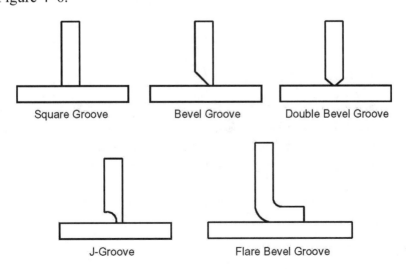

Figure 4–6   Weld preparations for T-joints

What are the weld preparations for edge joints?
See Figure 4–7.

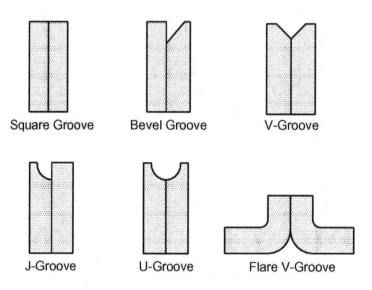

Figure 4–7   Weld preparations for edge joints

What are the weld preparations for lap joints?
See Figure 4–8.

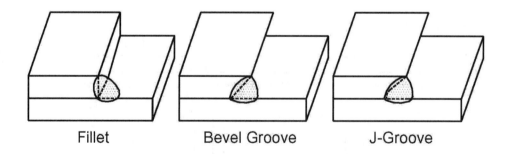

Figure 4–8   Weld preparations for lap joints

Selection of Joint Preparation

What determines which joint preparation geometry to use?
The factors are:
- Kind of joint loads (tension, compression, shear, or torsion).
- Level of joint loading.
- Static or dynamic loading.
- Thickness and type of the metals joined.
- Welding position.
- One side/both sides access to the weld.
- Skills of welders.
- Trade-offs between joint preparation costs/filler metal costs/welding labor costs; each joint and geometry carries its own combination of total cost.

What is an excellent source of weld joint design and preparation information for steel?
The *Structural Welding Code—Steel, ANSI/AWS D1.1* is a document with over 50 detailed drawings for pre-qualified joint designs using SMAW, GMAW, and FCAW for steel plate. Joints for submerged arc welding (SAW) are also included. There is additional information for joining tubing, weld testing, and inspection. The document is available from the AWS.

Joint Preparation

What are the most common weld preparations?
See Figure 4–9.

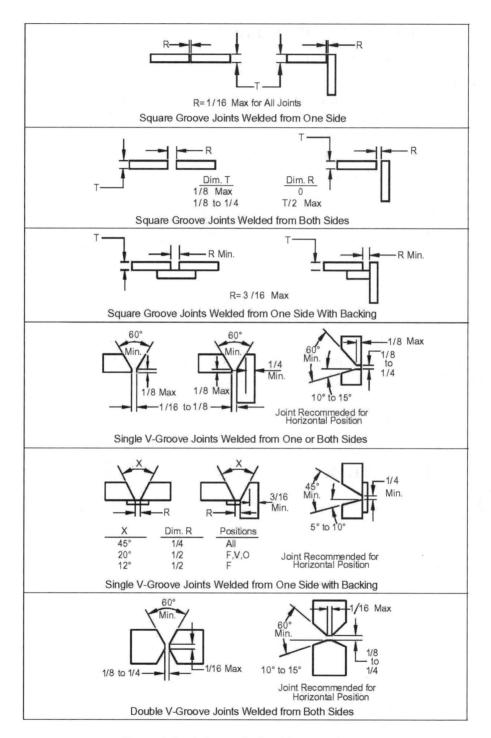

Figure 4–9   A few typical weld preparations

Parts of the Weld

What are the terms used to describe the parts of a groove weld?
- Effective throat or size of weld
- Face
- Toe
- Face Reinforcement
- Root Reinforcement

See Figure 4–10.

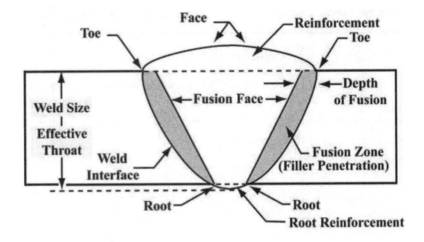

Figure 4–10   Parts of a groove weld

What are the terms used to describe the parts of a fillet weld?
- Leg or size of weld
- Face
- Toe
- Convexity
- Root penetration
- Fusion zone
- Leg of a fillet weld
- Root of the weld
- Actual throat and theoretical throat

See Figure 4–11.

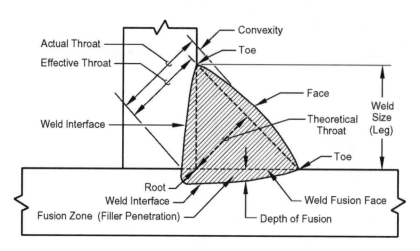

CONVEX FILLET WELD

Figure 4–11A   Parts of a convex fillet weld

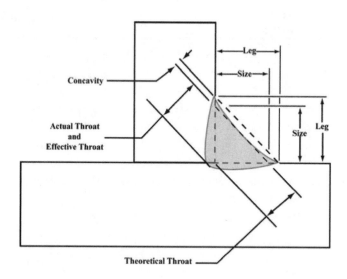

CONCAVE FILLET WELD

Figure 4–11B   Parts of a concave fillet weld

Welding Positions

How are welding positions plate designated?

They are divided by the position of the axis of the weld with respect to the horizontal and whether they are made on plate or pipe. They are used to designate positions for testing of welders and the application of a specific process. See Figure 4–12.

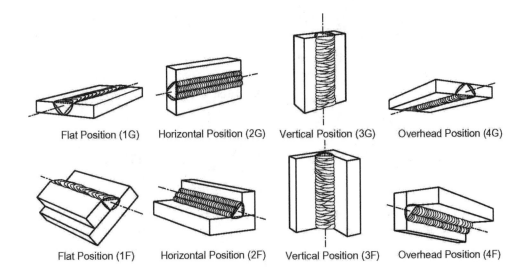

Figure 4–12   Groove weld (upper) and fillet weld positions (lower)

What are the welding positions for pipe?

See Figure 4–13. Note the difference between welding positions A , and C: In position A, (1G) the pipe may be rotated about its longitudinal axis to provide access to any part of the weld joint allowing the welder the opportunity to weld the entire pipe in the flat (1G) position; in position C, (5G) the pipe is fixed and *cannot* rotate forcing the welder to weld upward or downward vertically, flat on the top and overhead on the bottom; position B is pipe in a vertical position and welded on the horizontal plane; pipe in D is on a 45° angle and all positions (flat, horizontal, vertical and overhead) are welded when pipe is in this position; the final position is pipe at a 45° with a restrictor in place (the restrictor allows the welder to weld only from one side of the restrictor) making this the most difficult of all welding positions.

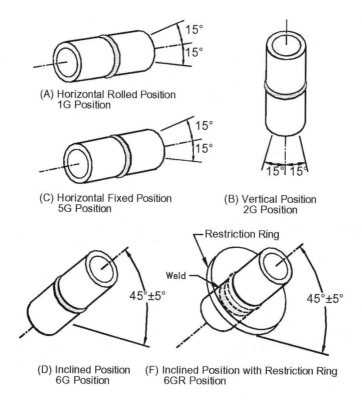

Figure 4–13   Pipe weld positions

Plug and Slot Welds

What are *plug* and *slot welds* and why are they used?
They join two (or more) parts together by welding them at a point *other* than
their edges. Plug and slot welds are used to secure multilevel parking garage
and ship decks from shearing forces. They are particularly useful in sheet metal
and auto-body work where welds can be completely concealed by grinding and
painting. A hole or slot is made in the work-piece facing the welder and weld
is made inside the hole. Filler metal completely fills the hole or slot and penetrates into the lower work-piece(s) securing them together. Plug welds are
round and slot welds are elongated and rounded at the ends.
See Figure 4–14.

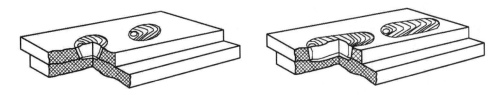

Figure 4–14   Examples of plug and slot welds

Intermittent Welds

What is the difference between a *chain intermittent* and a *staggered inter-mittent* weld?
See Figure 4–15.

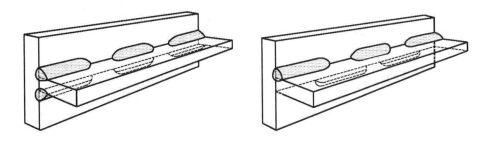

Figure 4–15   Chain intermittent fillet weld (left) and staggered intermittent fillet
weld (right)

Welding Terminology

What terms describe the position of the electrode with respect to the weld?
They are:
- Axis of the weld—an imaginary line drawn parallel to the weld bead through the center of the weld.
- Travel angle—is the smallest angle formed between the electrode and the axis of the weld.
- Work angle—for a T-joint or corner joint, the smallest angle formed by

a plane, defined by the electrode (wire) and the axis of the weld, and the work piece.

- Push angle during forehand welding—this is the travel angle during push welding when the electrode (wire) is pointing *toward* the direction of weld progression.
- Drag angle during backhand welding—this is the travel angle during drag welding when the when the electrode (wire) is pointing *away* from the direction of weld progression. See Figure 4–16.

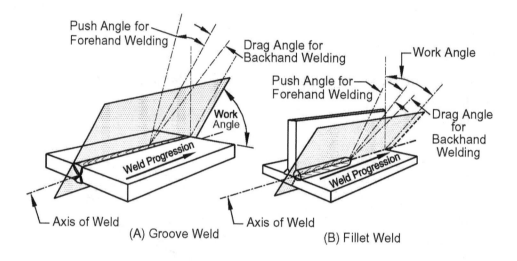

Figure 4–16   Orientation of the electrode

What is *travel speed*?
It is the velocity or speed of the electrode (wire) along the travel axis, usually in inches/minute or cm/minute.

What is a *tack weld*?
Welders place small, initial welds along joints to hold the work pieces in place so the parts remain in alignment when they are welded. Tack welds work hold work firmly in position, but can be broken with a cold chisel in the Event further adjustment is needed. Beginning welders tend to make them too small. One inch is the standard length of a tack weld. A tack should be as strong as the weld itself as it becomes and integral part of the finished weld.

What is a *joggle joint*?

See Figure 4–17. Joggle joints are used where a strong joint and flat surface is needed to join two pieces of sheet metal or light plate. There are hand tools available to put the joggle into sheet metal. They are useful whenever a finished surface concealing the weld is needed and where a butt joint would not work with thin sheet metal.

Figure 4–17   Joggle weld joint preparation

What is the difference between a *stringer bead* and a *weave bead*?

In a stringer bead the path of the electrode is straight, with no appreciable side to side movement, and parallel to the axis of the weld, while a weave bead has a side-to-side motion which makes the weld bead wider (and the heat-affected zone larger) than that made with a stringer bead.

What is *padding* or *overlaying* and what is it used for?

Padding is when weld filler metal is applied to a surface to build up a plate or shaft, to make a plate thicker, or to increase the diameter of a shaft. It is used either to restore a dimension to a worn part or to apply an extra hard wear surface. See Figure 4–18A shaft, bar or pipe and 4–18B is resurfacing a plate.

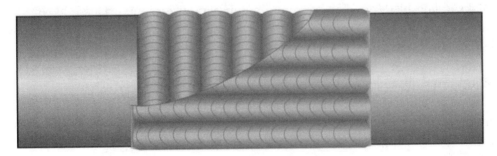

Figure 4–18A   Resurfacing on shaft, bar or pipe axial and circumferential welds

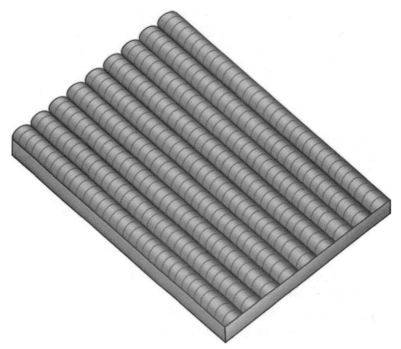

Figure 4–18B   Resurfacing on a plate

What is the purpose of *surfacing welds*?
Surfacing, also called *hard surfacing*, is the application of extra hard weld metal (padding) to surfaces subject to severe wear and abrasion. The teeth, buckets, and blades of earth moving equipment are often surfaced, as are the interior chutes of rock crushers. SMAW, GTAW, FCAW, and GMAW processes can all perform surfacing given the proper electrode metal composition.

What does the term *boxing* mean?
Boxing is when a fillet weld is continued *around* a corner. Normally a fillet weld is made from one abrupt end of the joint to the other abrupt end of the joint. See Figure 4–19.

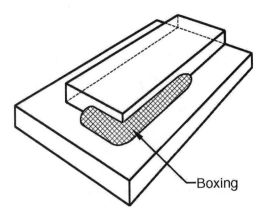

Figure 4–19   Boxing weld

Welding on Thick Plates

What is a *root pass* weld and where is it used?
A root pass uses weld filler metal to close the root space between the weld faces. It is especially helpful in welding pipe and thick plates where only one side of the weld is accessible and no backing material is used.

What is a *back weld*?
A back weld is applied after a groove weld is completed. The back weld is made to insure full penetration through the material being joined. Before we apply the back weld we must grind or gouge into the bottom of the groove weld until we reach sound weld metal then we may apply the back weld to the bottom of the groove weld. See Figure 4-20.

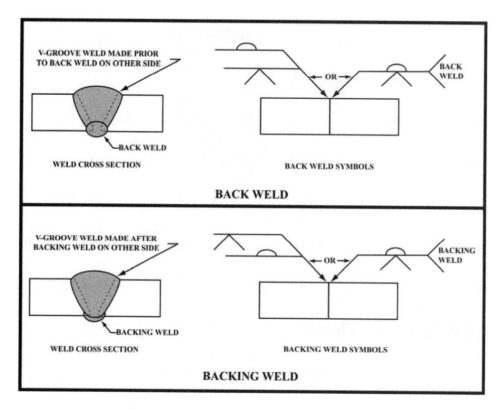

Figure 4–20-21   Back weld and backing weld

What is a backing weld?

A backing weld is applied to the bottom or root of a groove weld before the groove weld is applied. Because the root or bottom of the weld is made first it becomes a backing for the groove. The difference between a back and backing weld is the sequence of welding. Before the groove weld is completed the backing weld must be ground or gouged to sound weld. See Figure 4–20-21.

Why are weld *backing plates* used and what materials are used for them?

A backing plate is used to contain the large weld pool when joining two thick sections that are accessible from only one side. It takes the place of a root pass. The backing plate also shields the weld pool from atmospheric contamination coming in from the back of the weldment. Backing plates are usually tack welded to the two sections of the weld, but there are proprietary ceramic tapes and metal-glass tapes that perform the same function and do not need to be tacked into place. Copper and other materials are also used as backing plates. See Figure 4–22.

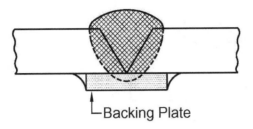

Figure 4–22   Weld backing plate

What is a *runoff plate* or *tab* and why are they used?
It is a plate of the same material of the work being joined which is tack welded to the joint at the start and/or end of the groove joint. The runoff plate contains a groove like the pieces being joined. It prevents the discontinuities caused by beginning and ending the welding process. See Figure 4–23.

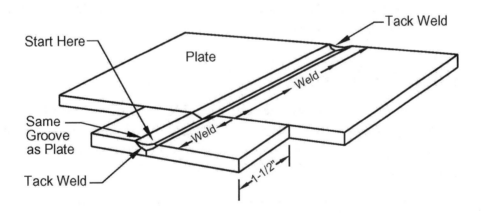

Figure 4–23   Runoff plate or tab

How can a large weld be made when the electrode deposition is much smaller than the weld width?
By using multiple passes of parallel weld beads. See Figure 4–24.

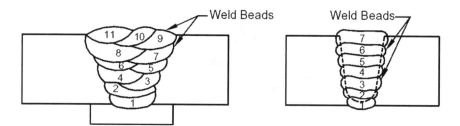

Figure 4–24   Multiple passes to join thick material

Slag Removal

Why is slag removal between weld bead applications important?
Slag must be fully removed between weld passes or the remaining slag will form inclusions *within* the weld metal and weaken it. Slag is usually removed with a slag hammer and wire brush angle grinders or pneumatic peening tools may also be used. Sometimes a wire wheel is used. Pipe welding, grinders and power wheels are used between each welding pass to assure a slag-free surface on which to begin the next pass.

Welding Cable Sizing

How do you determine what size welding cable is adequate for a task?
For copper cables, look up the cable size required in welding lead sizing chart based on the power supply-to-work distance and the current setting in Table 4–1. Tables are also available for aluminum conductors.

Copper Welding Lead Sizes in AWG							
Amperes➡	100	150	200	250	300	400	500
Feet⬇ Meters⬇							
50 15	2	2	2	2	1	1/0	2/0
100 30	2	1	1/0	2/0		3/0	
150 46	1/0	2/0	3/0	4/0		4/0	
200 61	1/0	3/0	4/0				
250 76	2/0	4/0					
300 91	3/0						
400 122	4/0						

Table 4–1   Copper welding lead sizing

Work-Lead Connections

What are the common designs of work-lead connections?
- Spring-loaded
- Screw-clamp
- Magnetic attachment
- Tack welded connectors

See Figure 4–25.

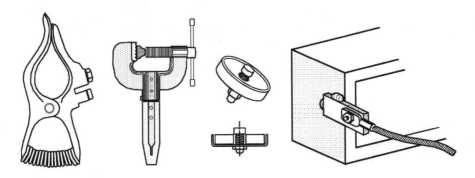

Figure 4–25   Work-lead connection methods include: spring-loaded, screw-type,
magnet type, and tack-welded connections

Why is it important to have a good work-lead connection?
A poor work-lead connection will generate heat between the connection and the work. It is best to make a solid connection on freshly cleaned base metal. Use a grinder or wire wheel to get through rust, paint, and mill scale. If the welding electrode holder overheats, this is an indication of a poor work-lead connection.

Welding Hand Tools

What hand tools is the welder likely to need?
- Chipping hammer to remove welding slag
- Wire brush for cleaning welds
- Hammer and cold chisel to break tack welds
- Pliers for moving hot metal safely
- Wire cutters to trim electrode wire (GMAW and FCAW only)

See Figure 4–26.

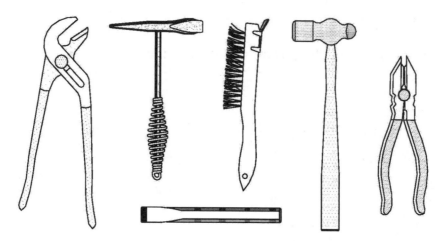

Figure 4–26   Various hand tools

Safety Equipment

- Welding helmet with the proper lens shade for the process and amperage.
- Leather capes and sleeves or jacket called *skins* or *leathers*, to protect the welder's clothing from sparks, especially while welding overhead.
- Welder's cap to protect from sparks getting behind the welding helmet and into the welder's hair. See Figure 4–27.

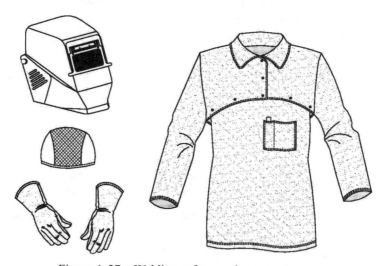

Figure 4–27   Welding safety equipment

- Breathing apparatus to provide the welder with fresh air in confined spaces with inadequate ventilation. Safety glasses *under* the welding helmet.

What is and what are the advantages of the electronic faceplates?
- An electronic faceplate or lens is one of the most recent and important safety devices developed in the welding industry. These devices are designed to be clear, or nearly clear, then darken the instant arc is established; when purchasing be aware of the time the lens takes to darken 1/25,000 of a second or faster is recommended.
- The welder does not need to raise and lower his helmet when performing a series of welds: he can always see where he is with the helmet down.
- The beginner does not have to master holding his electrode steady when he drops his helmet. This permits beginners to perform better welds earlier in their training.
- Electronic faceplates offer continuous eye protection from infrared radiation coming off red-hot metal even when they are *not* in the darkened mode. It is just easier on the eyes and the welder is less likely to incur eye injury from inadvertent arc strikes.

Safety

What is *the* authoritative source of welding safety information?
Consult the AWS (American Welding Society) booklet *Safety in Welding Cutting and Allied Processes, Z49.1.*

What safety issues must we remember with SMAW, GMAW, FCAW, and GTAW?
- Protection of face and eyes from sparks and radiation with a helmet and lens of appropriate shade number (darkness).
- Protection of *all* of the welders skin from arc and weld material radiation by covering it with cotton, wool, specially treated canvas jackets or leather garments; ultra violet radiation is carcinogenic.
- Personnel in the welding area must be protected from the welding arc and sparks by protective screens. Never view the welding being performed through the protective screens alone; the only way to safely view welding is through the proper shade lens and welding helmet or goggles.
- Beware of hazards from gases and insure adequate ventilation; inert shielding gases may cause suffocation in confined areas.
- Provide adequate ventilation from welding process smoke and the metal

vapors, particularly heavy metals like zinc and cadmium that are toxic; keep your head out of the welding plume.

- Leathers or specially treated canvas jackets must be worn when welding vertically or overhead to protect the welder from the falling hot metal, sparks and slag.
- A welder's hat will prevent both radiation burns to the head and hot sparks, falling slag and hot metal burns.
- High-top boots can prevent hot sparks and slag from burning your feet.
- Never weld with pant cuffs sparks falling into cuffs will burn pants.
- Make sure your welding gloves are dry and have no holes.
- Keep hands and body insulated from both the work and the metal electrode holder.
- Do not change the polarity switch position while the machine is under welding current load.
- Welding machines must be turned off when not attended.
- Must not stand on a wet surface when welding to prevent electric shock.
- Welding cables and electrode holders must be inspected for broken insulation regularly to prevent electric shock.
- Welding power supplies on AC lines must be properly grounded and emergency shut-off switch location known and accessible.
- Welding area must be dry and free of flammable materials.
- Protect your ears from welding and grinding noise with ear plugs or ear protectors.
- Any compressed gas cylinders must be properly secured and out of the spark stream.
- Must avoid wrapping welding cable around their arms or bodies in case a vehicle snags the cables.
- Never cut or weld on containers without taking precautions, see Chapter 1, Safety.
- Shielded metal arc welding the welder must plan for disposal of electrode stubs: they are hot enough to cause burns and to start fires and must not be dropped from heights because of the hazard to others.

Chapter 5

Shielded Metal Arc Welding

I hear and I forget.
I see and I remember.
I do and I understand.
Confucius

Introduction

Most accounts trace the history of electric welding to 1801 when Sir Humphrey Davy discovered an arc could be created with high-voltage electric current by bringing two terminals near each other. In 1881 Augustine de Mertines began experimenting with arc welding; these experiments were used to weld various parts of a lead battery plate using a carbon arc as the heat source. Carbon arc welding patents were issued to N. de Benardos, a Russian, May 17, 1887. Between 1889 and 1908 the N. G. Slavinoff system of arc welding, where the carbon electrode was replaced with a bare metal electrode was developed. American patents on this metal arc process were issued to Charles Coffin in 1889. Oscar Kjellberg, founder of ESAB of Sweden, applied a coating to the bare electrode between, 1907 and 1910, which began the development of coated metal arc welding electrodes.

By 1912 Lincoln Electric, and three other well-established manufacturers, offered the first arc welding machines. It remains an important process for structural steel and pipe line welding. Decision makers in these critical industries use it because it is so reliable. Even with thousands of welds made in adverse conditions, on startup the trans-Alaska pipeline did not have a single weld failure. Arc welding is popular for industrial, automotive, and farm repair because its equipment is relatively inexpensive and can be made portable. More welders have learned this process than any other. Although it will be around for many years, and its annual filler metal poundage continues to grow, it is declining in importance as wire feed welding processes continue to gain popularity and market share. We will cover theory, equipment, electrode rod classification and selection, and safety.

Process Names

What is the AWS designation for welding with a flux coated electrode?
Shielded metal arc welding (*SMAW*) is the AWS designation.

What are other common names for SMAW?
SMAW is also commonly called *arc welding*, *stick welding*, and *stick electrode welding*.

What equipment makes up a SMAW setup?
- Constant-current (CC) welding power supply
- Electrode holder, lead, and its terminals
- Work lead and its terminals
- Welding electrodes. See Figure 5–1.

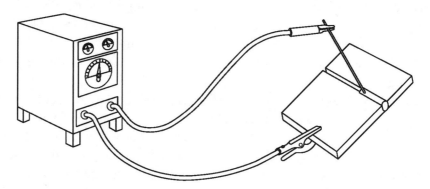

Figure 5–1   Shielded metal arc welding machine

What additional equipment may be found in an SMAW process welding station?
- Booth walls or light shields
- Jigs, fixtures, and positioning devices
- Stool
- Ventilation system
- Workbench

Process

How does the SMAW process work?

Shielded metal arc welding is a manual process. An electric circuit is established between the welding power supply, the electrode, the welding arc, the work, the work connection, and back to the welding supply. Electrons flowing through the gap between the electrode and the work produce an arc that furnishes the heat to melt both the electrode metal and the base metal. Temperatures within the arc exceed 6000°F (3300°C). The arc heats both the electrode and the work beneath it. Tiny globules of metal form at the tip of the electrode and transfer to the molten weld pool on the work. As the electrode moves away from the molten pool, the molten mixture of electrode and base metals solidifies and the weld is complete.

In the flat or horizontal position electrode metal transfer is affected by gravity, gas expansion from the electrode shielding materials, electromagnetic forces, and surface tension. In other positions gravity opposes these forces. The electrode is coated with a flux. Heat from the electric current causes the flux's combustion and decomposition. This creates a gaseous shield to protect the electrode tip, the work, and the molten pool from atmospheric contamination. The flux contains materials that coat the molten steel droplets as they transfer to the weld and become slag after cooling. This slag also floats on the weld puddle's surface, solidifies over the weld bead when cool where it protects the molten metal and slows the cooling rate. The flux coating on some electrodes contains metal powder to provide additional heat and filler to increase the deposition rate. The electrode flux and metal filler electrode determine the chemical, electrical, mechanical, and metallurgical properties of the weld as well as the electrode handling characteristics. Only 50% of the heat power furnished by the power supply heats the weld; the rest is lost to radiation, the surrounding base metal, and the weld plume. See Figure 5–2.

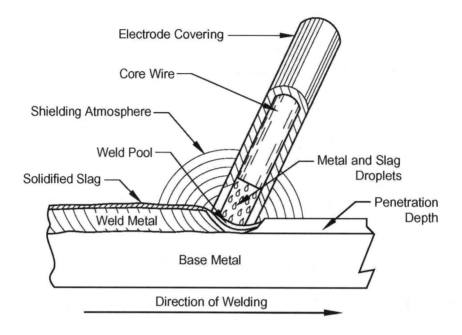

Figure 5–2   Shielded metal arc process

Applications

What metals can SMAW weld?
- Aluminum
- Bronze
- Carbon steel
- Cast iron
- Hard-facing
- High-strength steels
- Low-alloy steels
- Malleable iron
- Nickel
- Stainless steels

Little aluminum welding is done by SMAW as other, better processes exist. Note that some of these metals require preheat, post-heat, or both to prevent cracking.

For what thicknesses of metal is SMAW best suited?
Welds from 1/8 inch (3 mm) to unlimited thicknesses are possible. Thicknesses less than 1/8 inch (3 mm) can be joined but require much greater skill, while those over 3/4 inch (19 mm) are more economically done by other methods like FCAW.

Advantages

What are the advantages of the SMAW process?
- Low cost equipment
- Can weld many different metals including the most commonly used metals and alloys
- Relatively portable and useful in confined spaces
- Same equipment welds thicknesses from 1/16 inch (16 gauge or 1.5 mm) to several feet in thickness with different current settings
- Welds can be performed in any position
- The process is less affected by wind and drafts than gas-shielded processes
- There is no upper limit on thickness of metal to be welded
- It can be performed under most weather conditions

Disadvantages

What are the disadvantages of the SMAW process?
- Not suitable for metal sheets under 1/16 inch (1.5 mm) thickness
- Operator duty cycle and overall deposition rate are usually lower with SMAW than with wire-fed processes, since the SMAW process must be stopped when the electrode is consumed and needs to be changed
- Not all of the electrode can be used; the remaining stub in the electrode handle must be discarded wasting one to two inches of electrode
- Frequent stops and starts during electrode changes provide the opportunity for weld defects

In what weather conditions must SMAW *not* be performed?
SMAW is not permitted in rain, snow, and blowing sand, and when the base metal is below 0°F (-18°C). Temporary shelters known as dog houses around the weld site permit welding during these conditions, increase the comfort of the welder, and help raise the weld steel temperature. Preheating the weld base metal to 70°F (21°C) with multi-flame (or rosebud) tips allows welding to proceed when the outside temperature is below 0°F (-18°C).

Welding Power Supplies

What type welding machine is best for the SMAW process: Constant-current or constant-voltage? Why?
Constant-current (CC) welding power supplies are best for SMAW. The slope of the voltage versus current curve reduces current with increasing arc length. This characteristic both gives the welder control over weld pool size and still limits maximum arc current. Some variation in arc length is inevitable as the welder moves the electrode along the weld while the CC welding supply insures a stable arc as these variations happen. See Chapter 13 for SMAW power supply details.

What is the range of current used for SMAW?
Usually between 25 and 600 amperes are the current variables.

What open-circuit voltages are used on SMAW?
The potential voltage across the welding machine output when no welding is being done runs from 50 to 100 volts.

What range of operating voltages appears across the SMAW arc?
Between 17 and 40 volts is the normal variable depending on arc length.

What effect does open-circuit voltage have on the SMAW process?
The higher the open-circuit voltage, the easier it is to strike an arc, but the greater the risk of electric shock.

What types of currents are used for SMAW welding?
- Alternating current (AC)
- Direct current electrode negative (DCEN = DCSP)
- Direct current electrode positive (DCEP = DCRP)

In general, which is better to weld with AC or DC power?
DC always provides the most stable arc and more even metal transfer than AC. Once struck, the DC arc remains continuous. When welding with AC, the arc extinguishes and re-strikes 120 times a second as the current and voltage reverse direction. The DC arc has good wetting action of the molten weld metal and uniform weld bead size at low welding currents. For this reason it is excellent for welding thin sections. DC is preferred to AC on overhead and vertical welding jobs because of its shorter arc. Sometimes arc blow is a serious prob-

lem and the only solution may be to switch to AC. Most combination electrodes designed for AC *or* DC operation work better on DC.

What electrode polarity and characteristics does *DCEN* have?

DCEN stands for direct current electrode negative. This is also called *DCSP*, direct current straight polarity. DCEN produces less penetration than DCEP, direct current electrode positive, but has a higher electrode burn-off rate. See Figure 5–3.

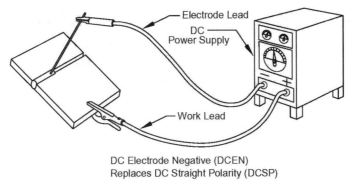

DC Electrode Negative (DCEN)
Replaces DC Straight Polarity (DCSP)

Figure 5–3   DCEN

What electrode polarity and characteristics does *DCEP* have?

DCEP stands for DC-electrode positive. This is also called *DCRP*, direct current reverse polarity. DCEP is especially useful welding aluminum, beryllium-copper, and magnesium because of its surface cleaning action which permits welding these metals without flux. DCEP produces better penetration than DCEN, but has a lower electrode burn-off rate. See Figure 5–4.

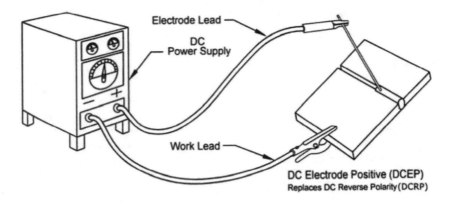

DC Electrode Positive (DCEP)
Replaces DC Reverse Polarity (DCRP)

Figure 5–4   DCEP

How do the three SMAW welding polarities rank in heat into the weld at the same current rating?

Going from highest to lowest heat into the weld:

- DCEN
- AC
- DCEP

See Figure 5–5.

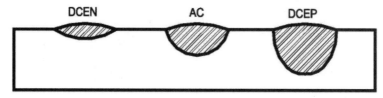

Figure 5–5   Effect of weld bead penetration at same welding current

Are there any advantages in SMAW to using an AC arc over a DC arc?

There are no particular advantages to using AC over DC in the SMAW process, except the lower cost of an AC welding power supply and a possible way to get around arc blow. Equipment cost aside, AC has no great advantages.

What is *arc blow*?

Whenever current flows, it creates a magnetic field around the conductor. Any shape of wire other than a straight line will produce an asymmetrical field along the wire—points where the field is either stronger or weaker than average. At currents above 600 amperes, the force created by the unevenness of the field may even cause the wire itself to move. More often, welders see the arc drawn in a particular direction like smoke in the wind. This is caused by residual magnetism in the part or the uneven magnetic field caused by current flowing through the part to the work connection. Some welders call this arc blow. See Figure 5–6.

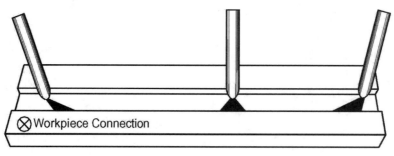

Figure 5–6   Arc blow on left and right arc

How can the effects of arc blow be reduced?

Ways to reduce arc blow are:
- Use AC, instead of DC welding current.
- Move the welding work lead to a position where are you welding away from the work lead connection.
- Use a shorter arc length.
- Clamp a steel block over the far end (unfinished end) of the weld.
- Weld away from the base metal edge, or toward a heavier tack or weld.
- Change welding direction.

How can you determine the current setting of the welding machine pictured in Figure 5–7?

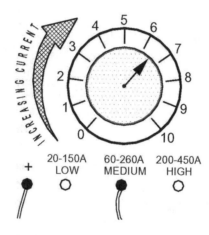

Figure 5–7   Can you describe the amperage setting of this welding machine?

Use the following steps:
1. Note the **MEDIUM** tap or coarse setting is in use with an output from 60 to 260 amperes.
2. This means that this tap will provide a minimum current of 60 amperes with the rotary knob set to zero. In the rest of the 60 to 260 ampere range an additional 200 amperes are gained by the ten steps on the knob.
3. Each of the ten knob points adds 20 amperes to the setting (200 amperes/ 10 knob steps = 20 amperes/each of the knob steps).
4. The knob is set to 6.5 steps, or 6.5 3 20 amperes plus the base of 60 amperes, or 190 amperes.

Remember that these settings are likely to be approximate and output current will vary with power line voltage variations, cable length, cable size, and other factors. The best output current setting may be slightly above or below the dial setting.

What factors determine how much amperage to set on the welding machine?

Once an electrode class and diameter is selected for a task, use Table 5–1 to determine the manufacturer's recommended amperage range. Note that there will be some differences between recommended ranges for the same electrode class and diameter so it is important to consult the manufacturer's specification sheet for the electrode you are using. Also the material thickness will affect the current setting.

If the required current for an electrode lies within the range of two current settings, should you use the upper or lower range setting?

Usually it is best to use the *lower* current range setting. This is because the top of a current range is likely to have less current ripple (AC riding on the top of the DC) than the same current output setting from a higher current range, and will provide a smoother arc.

How can you tell when you have just the right arc length for your electrode?

Although the correct arc length is usually 1/8" one should never exceed the diameter of the electrode, the sound of the arc is an excellent indicator of arc length. The proper arc length sounds like the crackling of bacon and eggs frying. A short an arc, makes a sputtering sound; too long an arc, a humming sound.

What maintenance should be performed on the welding power supply monthly?

The power supply should be blown out with compressed air to remove slag, grinding dust, and other dust from its interior.

Electrode Holder and Work-Lead Connection

Besides holding the electrode, what are the functions of the electrode holder?

- Electrically insulate the welder from the welding power supply voltage.

- Thermally insulate the welder from the conducted heat of the electrode.
- Make a secure electrical connection between the welding cable and the electrode with a minimum of voltage drop.

Why is it important to keep electrode holders in good condition?
Poor gripping of the electrode by the electrode holder's jaws will cause poor conduction of electricity, and power intended for the weld will be lost in the holder. The holder will also heat up and prevent the operator from holding it. Eventually the heat will deteriorate its electrical insulation properties exposing the welder to possible electric shock. Most electrode holders have notches where the electrode conductor end fits and will not move.

What size electrode holder is best?
The smallest size that can be used without overheating is best as it will be the lightest weight and least tiring.

SMAW Electrode Functions

What is the construction of SMAW electrodes?
SMAW electrodes contain a solid or cast metal core wire covered by a thick flux coating. Every change of flux composition and thickness alters the operating characteristics. These changes plus current type and polarity determine how the rod will handle and what type bead it will deposit. They are made in lengths from 9 to 18 inches (230 to 460 mm). Some electrodes are made with a metallic tube containing a mixture of metal powders.

What are the specific functions the rod coating performs during welding?
- The principal function of the coating is to provide a gas stream to shield the molten weld pool from atmospheric oxygen, hydrogen, and nitrogen contamination until it solidifies
- Supplies scavengers, deoxidizers, and fluxing agents to clean the weld and prevent excessive grain growth in the weld material
- Provides chemicals to the arc that control the electrical characteristics of the electrode: Current type, polarity, and current level
- Covers the finished weld with slag, a protective covering to control weld cooling rate, protect the cooling materials of the weld from the atmosphere, and control bead shape
- Adds alloying materials to the weld pool to enhance weld properties

What are some typical examples of the above functions?

- Iron powder added to the coating increases the weld temperature and deposition rate
- Potassium in the coating makes the gases of the plasma readily ionized which improves AC operation. This is when the arc extinguishes and reignites 120 times a second. Without this additive, AC operation would not be possible
- The shielding gas and the deoxidizers prevent the pickup of oxygen and nitrogen by the filler metal and molten weld pool

SMAW Electrode Classification

What are the two main systems for classifying welding electrodes?

- The ANSI/AWS carbon and low alloy steel electrode classification system based on ANSI/AWS A5.1. Table 5–2 explains the coding and Table 5–3 describes the electrodes in each classification
- Section IX of the *ASME Pressure Vessel Boiler Code* provides another way to classify the same group of electrodes

While the AWS classification system provides 16 classification groups, the ASME boiler code has four. Both provide convenient and effective ways of looking at the welding electrodes for carbon steel

How are electrodes marked to identify them?

Electrodes for carbon and low alloy steel have the ANSI/AWS classification number stamped directly on the flux of the electrode in one or more places.

Does the AWS electrode classification system use color coding for welding steel?

Color coding SMAW *coated* electrodes is obsolete, but uncoated electrodes used for surfacing are still color coded. Some manufacturers use color codes to identify their electrodes and are explained in their data sheets.

Lincoln Electric Product Name	AWS Class	Electrode Polarity	Sizes and Current Ranges (Amperes)					
			3/32"	1/8"	5/32"	3/16"	7/32"	1/4"
Fleetweld® 5P	E6010	DCEP	40-70	75-130	90-175	140-225	200-275	220-325
Fleetweld 35	E6011	AC	50-85	75-120	90-160	120-200	150-260	190-300
		DC±	40-75	70-110	80-145	110-180	135-235	170-270
Fleetweld 7	E6012	DC–	---	80-135	110-180	155-250	225-295	245-325
		AC	---	90-150	120-200	170-270	250-325	275-360
Fleetweld 37	E6013	AC	75-105	110-150	160-200	205-260	---	---
		DC±	70-95	100-135	145-180	190-235	---	---
Fleetweld 47	E7014	AC	80-100	110-160	150-225	200-280	260-340	280-425
		DCEN	75-95	100-145	135-200	185-235	235-305	260-380
Jetweld® LH-70	E7018	DC+	70-100	90-150	120-190	170-280	210-330	290-430
		AC	80-120	110-170	135-225	200-300	260-380	325-530
Jetweld 3	E7024	AC	---	115-175	180-240	240-315	300-380	350-450
		DC±	---	100-160	160-215	215-285	270-340	315-405
Jetweld 2	E6027	AC	---	---	190-240	250-300	300-380	350-450
		DC±	---	---	175-215	230-270	270-340	315-405
Jetweld LH-3800	E7028	AC	---	---	180-270	240-330	275-410	360-520
		DC+	---	---	170-240	210-300	260-380	---

Table 5-1 Recommended current ranges for various diameter electrodes

E 60 10

Electrode ————————
Strength in kpsi ————————
Position ————————
Type of Coating and Current ————————

E 8018-B1H4R

Electrode ————
80,000 PSI Min. ————
All Position ————
For AC or DCEP ————
Chemical Composition of ————
Weld Metal Deposit

Diffusible Hydrogen Designator Indicates the Maximum Diffusible Hydrogen Level Obtained with the Product.

Moisture Resistant Designator Indicates the Electrode's Ability to Meet Specific Low **Moisture Pickup Limits** under Controlled Humidification Tests.

Position

1. Flat, Horizontal, Vertical, Overhead
2. Flat and Horizontal Only
3. Number 3 Position Is Not Designated
4. Flat, Horizontal, Vertical Down, Overhead

Types of Coating and Current

Digit	Type of Coating	Welding Current
0	Cellulose Sodium	DCEP
1	Cellulose Potassium	AC or DCEP or DCEN
2	Titania Sodium	AC or DCEN
3	Titania Potassium	AC or DCEP
4	Iron Powder Titania	AC or DCEN or DCEP
5	Low Hydrogen Sodium	DCEP
6	Low Hydrogen Potassium	AC or DCEP
7	Iron Powder Iron Oxide	AC or DCEP or DCEN
8	Iron Powder Low Hydrogen	AC or DCEP

DCEP - Direct Current Electrode Positive
DCEN - Direct Current Electrode Negative

Chemical Composition of Weld Deposit

Suffix	%Mn	%Ni	%Cr	%Mo	%V
A1				1/2	
B1			1/2	1/2	
B2			1-1/4	1/2	
B3			2-1/4	1	
C1		2-1/2			
C2		3-1/4			
C3		1	.15	.35	
D1& D2	1.25-2.00			.25-.45	
G[1]		.50	.30 min	.20 min	.10 min

(1) Only one of the listed elements required.

Table 5–2   The AWS electrode classification system for carbon and low alloy steel electrodes

AWS Class	ASME System	Current and Polarity	Welding Positions	Type of Covering	Type of Arc	Degree of Penetration	Surface Appearance	Type of Slag	Character of Slag
EXX10	F-3	DC, positive electrode	All	High-cellulose, sodium	Digging	Deep	Flat, wavy	Organic	Thin
EXX11	F-3	AC or DC, electrode positive	All	High-cellulose, potassium	Digging	Deep	Flat, wavy	Organic	Thin
EXX12	F-2	Ac or DC, electrode negative	All	High-titania, sodium	Medium	Medium	Convex, rippled	Rutile	Thick
EXX13	F-2	AC or DC, either Polarity	All	High-titania, potassium	Soft	Shallow	Flat or concave, smooth ripple	Rutile	Medium
EXX14	F-2	DC, either polarity or AC	All	Iron-powder, titania	Soft	Medium	Flat, slightly convex, smooth ripple	Rutile	Easily removed
EXX15	F-4	DC, electrode positive	All	Low-hydrogen, sodium	Medium	Medium	Flat, wavy	Low-hydrogen	Medium
EXX16	F-4	AC or DC, electrode positive	All	Low-hydrogen, potassium	Medium	Medium	Flat, wavy	Low-hydrogen	Medium
EXX18	F-4	DC, electrode positive or AC	All	Low-hydrogen, potassium, iron powder	Medium	Shallow	Flat, smooth, fine ripple	Low-hydrogen	Medium
EXX20	F-1	DCEN or AC for horizontal fillets; DC, either polarity, or AC for flat work	Horizontal fillets & flat	High-iron oxide	Digging	Medium	Flat or concave, smooth	Mineral	Thick
EXX22	F-1	DC, either polarity or AC	Horizontal, flat	High-iron oxide	Soft, smooth	Medium	Flat or slightly convex	Mineral	Medium
EXX24	F-1	AC or DC, either polarity	Horizontal fillets and flat	Iron powder, titania	Soft	Shallow	Slightly convex, very smooth, fine ripple	Rutile	Thick
EXX27	F-1	AC or DC, electrode negative	Horizontal fillets and flat	High-iron oxide, iron powder	Soft	Medium	Flat to slightly concave, smooth fine ripple	Mineral	Thick
EXX28	F-1	AC or DC, electrode positive	Horizontal fillets and flat	Low-hydrogen, potassium, iron powder	Medium	Shallow	Flat, smooth, fine ripple	Low-hydrogen	Medium
EXX48	F-4	Ac or DC, electrode positive	Flat, horizontal, vertical-down, overhead	Low-hydrogen, potassium, iron powder	Soft	Shallow	Concave. Smooth	Low-hydrogen	Thin

Table 5–3   The AWS electrode classification system for carbon and
low alloy steel electrodes

How does the ASME system for classifying coated electrodes for proper electrode selection (of carbon and mild steel only) work?
By classifying electrodes into four main groups:
- F-1, High Deposition Group (also called *Fast–Fill*)
- F-2, Mild Penetration Group (also called *Fill–Freeze*)
- F-3, Deep Penetration Group (also called *Fast–Freeze*)
- F-4, Low Hydrogen Group

The *F* numbers come from the classification system used in Section IX (Class SFA) of the *ASME Pressure Vessel Boiler Code* and in the AWS Code, Section A5.1 or A5.5.

What are the properties of the High Deposition Group or F-1?
- The electrode coating contains 50% iron powder by weight, so these electrodes produce a higher weld deposition per electrode than any members of the other groups
- Dense slag and slow cooling make this group useful only for flat and horizontal fillet
- Produces smooth ripple-free bead with little spatter
- Heavy slag produced is easily removed

What are the properties of the Mild Penetration Group or F-2?
- Electrodes have a titania, rutile, or lime-based coating
- They are excellent for welding sheet steel under 3/16 inch (5 mm) thick where high speed travel with minimum skips, slag entrapment, and undercut are required. These are often used with DCEN

What are the properties of the Deep Penetration Group or F-3?
- Electrodes have a high cellulose coating that produces deep penetration and a forceful arc. They may also contain iron powder, rutile, and potassium
- The weld solidifies rapidly for use in all positions
- Excellent for welding mild steel in fabrication and maintenance work
- Best choice on dirty, painted, or greasy metal
- Light slag
- Especially good for vertical-up, vertical-down, and overhead and open root welding

What are the properties of the Low Hydrogen Group or F-4?
- Resistance to hydrogen inclusions and underbead cracking in medium to high carbon steels, hot cracking in phosphorus-bearing steels, and porosity in sulfur-bearing steels

- Excellent for x-ray quality welds and mechanical properties
- Less preheat than other electrodes
- Reduced likelihood of underbead and micro-cracking of high carbon and low alloy steels and on thick weldments
- Can produce excellent multiple pass, vertical, and overhead welds in carbon steel plate
- Best choice for galvanized metal

What are the common dimensions of SMAW electrodes?
The most common length is 18 inches (460 mm). The common diameters (of the core wire) are in inches (mm):

$1/16$ (1.6 mm)	$3/32$ (2.3 mm)	$1/8$ (3.2 mm)	$5/32$ (3.9 mm)
$3/16$ (4.7 mm)	$7/32$ (5.6 mm)	$1/4$ (6.3 mm)	$5/16$ (7.9 mm)

Welding electrodes are usually supplied in 50 lb. containers are cardboard or sealed metal cans to keep out moisture.

What problem does moisture entering the electrode coating cause?
Water absorbed out of the atmosphere by the flux on the electrode will introduce hydrogen into the weld causing cracking and brittleness. Dry electrodes can take from 30 minutes to four hours to pick up enough water (and the water's hydrogen) to affect weld quality. For this reason, low-hydrogen electrodes must be kept in drying ovens after removing from sealed electrode dispensers until immediately before use. Be sure to determine which electrodes are to be heated and which are not. The minimum drying time after exposure to the atmosphere and the maximum time out of the original container before use is specified in the manufacturer's instructions for the specific electrode. Low hydrogen electrodes end in the numbers 5, 6, and 8. See Table 5–4.

Tensile Strength EXX Series	Hours Usable after Oven Drying Cycle
E70XX	4
E80XX	2
E90XX	1
E100XX	1/2

Table 5–4   Electrode usability after oven drying cycle

What can be done with electrodes exposed to the atmosphere beyond the time allowable?

All electrodes having low-hydrogen coverings conforming to AWS A5.1 shall be rebaked for at least two hours between 500°F (260°C) and 800°F (430°C). Those low-hydrogen electrodes conforming to AWS A5.5 shall be rebaked for at least one hour at temperatures between 700°F (370°C) and 800°F (430°C).

In addition to the carbon and low alloy steel electrodes discussed above, what other classes of SMAW electrodes exist and what ANSI/AWS specifications cover them?

There are seven other specifications beside that for carbon steel. See Table 5–5.

Electrode Application	ANSI/AWS Specification
Carbon steel	A5.1
Low alloy steel	A5.5
Corrosion resistant steel	A5.4
Cast iron	A5.15
Aluminum and its alloys	A5.3
Copper and its alloys	A5.6
Nickel and its alloys	A5.11
Surfacing	A5.13 & A5.21

Table 5–5   ANSI/AWS specifications for covered electrodes

Electrode Selection

What factors must be considered in selecting an electrode?

This is not a simple decision since there are trade-offs between speed, total welding cost, and weld strength. Electrode selection is a matter of matching the operating characteristics of the electrode to the job requirements as well as to the possible need for low hydrogen electrodes. Here are the major factors in the selection:

- Welder's skill
- Properties of the base metal
- Position of the weld joint
- Type of joint
- Type of power supply
- Tightness of the joint fit-up
- Total amount of welding needed

In general, how should the steel of the electrode wire and the steel of the base metal compare?

The electrode rod steel should have the same or higher tensile strength as the base metal and similar chemical properties.

What is the value of including iron powder in the electrode coating?

- Welding heat is used to melt the core and coating, not excess areas of the base metal
- Iron in the flux coating adds to the weld deposit and increases deposition rate
- Drag technique of welding is used

How do you go about selecting an electrode for carbon steel?

1. Determine the joint, position, and metal thickness illustration in Figures 5–8 through 5–12, which best illustrates the task you have.
2. Read the electrode recommendation above the joint.

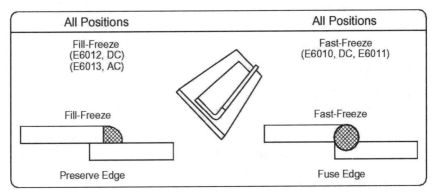

Figure 5–8   Sheet metal joints

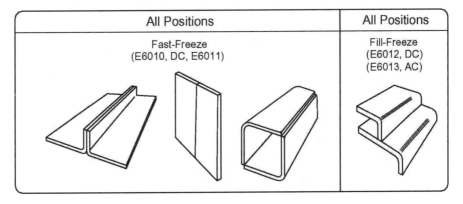

Figure 5–9   Sheet metal joints

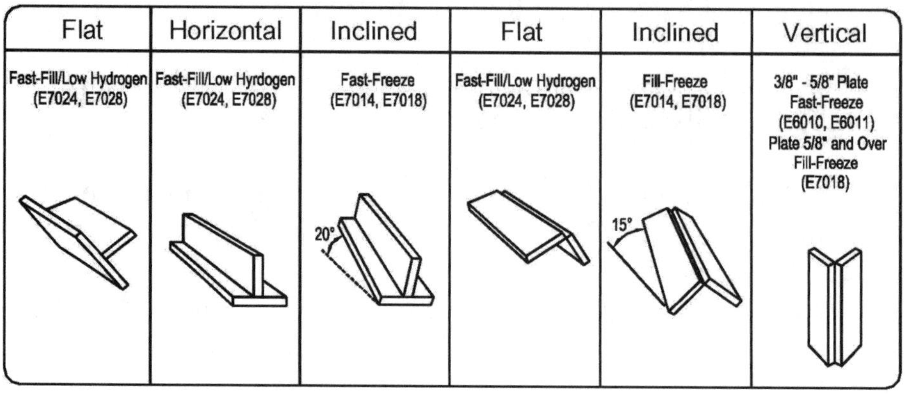

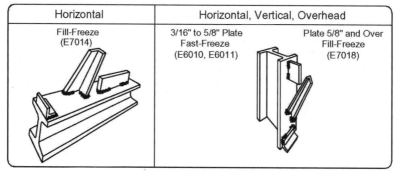

Flat	Horizontal	Inclined	Flat	Inclined	Vertical
Fast-Fill/Low Hydrogen (E7024, E7028)	Fast-Fill/Low Hyrdogen (E7024, E7028)	Fast-Freeze (E7014, E7018)	Fast-Fill/Low Hydrogen (E7024, E7028)	Fill-Freeze (E7014, E7018)	3/8" - 5/8" Plate Fast-Freeze (E6010, E6011) Plate 5/8" and Over Fill-Freeze (E7018)

Figure 5–10   Light plate joints

Horizontal	Horizontal, Vertical, Overhead	
Fill-Freeze (E7014)	3/16" to 5/8" Plate Fast-Freeze (E6010, E6011)	Plate 5/8" and Over Fill-Freeze (E7018)

Figure 5–11   Short fillets under 6 inches (150 mm) in length and having a change of direction on 3/16 inch (4.7 mm) or thicker plate

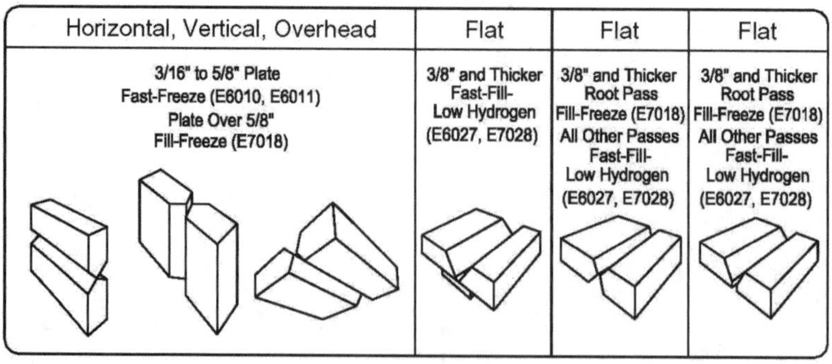

Horizontal, Vertical, Overhead	Flat	Flat	Flat
3/16" to 5/8" Plate Fast-Freeze (E6010, E6011) Plate Over 5/8" Fill-Freeze (E7018)	3/8" and Thicker Fast-Fill- Low Hydrogen (E6027, E7028)	3/8" and Thicker Root Pass Fill-Freeze (E7018) All Other Passes Fast-Fill- Low Hydrogen (E6027, E7028)	3/8" and Thicker Root Pass Fill-Freeze (E7018) All Other Passes Fast-Fill- Low Hydrogen (E6027, E7028)

Figure 5–12   Heavy plate welds

What four factors determine the size electrode to use?
- Joint thickness
- Welding position
- Type of joint
- Welder's skill

The best choice of electrode will be the one that will produce the weld required in the least time. Usually larger diameter electrodes are used for thicker materials and flat welds where their high deposition rates can be an advantage. In non-horizontal positions, smaller electrodes are used to reduce the weld pool size as gravitational forces are a factor working against the welder's skill.

SMAW Set up

What are the steps for setting up an SMAW outfit?
- Locate the welding power supply near the work and note the location of the AC power shut-off switch in case of emergency. Make sure the welding power supply is grounded. If an engine-generator is the welding power source, locate the engine shut-off switch
- Make sure the areas around the power supply and the work are dry.
- Remove flammable materials near the spark stream area and insure fire fighting materials are close at hand
- Based on the job requirements or drawings, select the electrode material and diameter
- Set welding polarity and welding current on welding machine
- Stretch out the welding cables and attach the work lead to the work securely; clean the grounding area if necessary
- Set out the electrodes, welding safety equipment (helmet, cap, gloves, leathers); you should have had safety glasses on since the first step and they should remain on under your welding helmet
- Turn on the welding machine, insert the electrode in the holder, drop your welding helmet, strike the arc, and begin welding

What are two ways to strike an SMAW arc?
- Scratch-start, usually used for AC, is performed by tilting the electrode about 15° in the direction of travel and while drawing the electrode above the work as if striking a match allowing it to momentarily touch the work; this strikes the arc. The electrode is withdrawn above the work to a height equal to its own diameter

- Tapping-start, usually used for DC, is made by lowering the electrode quickly until an arc forms and then raising the electrode its own height above the work

How should an arc be re-started on an existing weld?
Strike an arc about an inch further along the weld path than the interrupted weld bead. Then carry the arc back to the previous stopping point and begin welding again. The new weld bead will cover and eliminate the marks of the strike.

What determines the work and travel angle and welding technique when SMAW?
The joint type and welding position determine these variables. See Table 5–6.

Type of Joint	Welding Position	Work Angle (degrees)	Travel Angle (degrees)	Technique of Welding
Groove	Flat	90	5-10*	Backhand
Groove	Horizontal	80-100	5-10	Backhand
Groove	Vertical-Up	90	5-10	Forehand
Groove	Overhead	90	5-10	Backhand
Fillet	Horizontal	45	5-10*	Backhand
Fillet	Vertical-Up	35-45	5-10	Forehand
Fillet	Overhead	30-45	5-10	Backhand

*Travel angle may be 10-30° for electrodes with heavy iron coatings.

Table 5–6   SMAW orientation and welding technique for carbon steel electrodes

Filter Plate Shade Number for SMAW

What filter lens should be used in the SMAW process?

The shade of the filter lens depends on the arc current and the electrode diameter; use Table 5–7.

Welding Electrode Diameter (inches)	Welding Electrode Diameter (mm)	Filter Plate Shade Number
Up to 5/32	4	10
3/16-1/4	4.8-6.4	12
Over ¼	6.4	14

Table 5–7   Filter plate shade number based on electrode diameter

Chapter 6

Wire Feed Welding Processes— Gas Metal & Flux Cored Arc Welding

Whether you think you can or you can't — you are right.
Henry Ford

Introduction

The wire feed processes of gas metal arc and flux core welding consume over seventy percent of *total* filler materials used today and this percentage continues to grow. While this welding equipment may cost more than arc welding equipment of the same capabilities, it offers higher productivity. Not having to stop a bead, change electrodes, and restart again increases metal deposition rates and reduces weld discontinuities. Also, these processes are readily adapted to robotic/computer-controlled operations. Wire feed processes are relatively easy to learn, especially to those already trained in shielded metal arc welding, once the power source differences and voltage- amperage variables are understood. We will cover theory, equipment, materials, setup, adjustment, troubleshooting, and safety.

Gas Metal Arc Welding (GMAW)

Processes we will discuss and their names

What are the AWS designations for these processes?

Gas metal arc welding *GMAW* and flux-cored arc welding *FCAW* are the initials of the process names and the AWS designations.

What are other common names for GMAW?

GMAW is also commonly called *MIG* from Metal Inert Gas, *wire feed welding or squirt welding*. In Europe it is sometimes called *MAG* from Metal Active Gas.

Equipment

What equipment makes up a GMAW setup?

- Constant-voltage (CV) welding power supply.
- Wire feeder containing wire feed motor, spool support, wire feed drive rolls, and associated electronics (may be an integral part of welding power supply or separate unit).
- Welding gun and its cable.
- Work lead clamp, work lead (cable), and its terminals.
- Welding wire.
- Flow regulator or flow meter for shielding gas(es).
- Compressed gas cylinder.

See Figure 6–1.

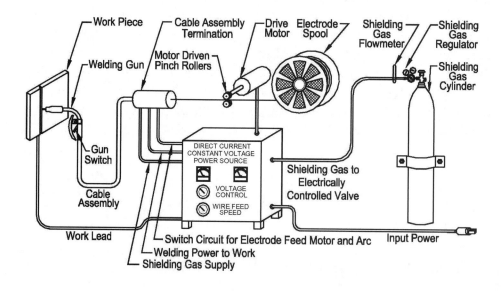

Figure 6–1   GMAW welding outfit

What additional equipment may be attached to the GMAW equipment?

- Water cooler.
- Smoke evacuator.

What is the water cooler used for?

GMAW and FCAW welding guns operated continuously at high amperage levels get so hot they literally melt the gun from the heat of the weld and the heat from the contact tip. This situation arises during spray metal transfer, which is discussed later. The solution is to feed cooling water into the nozzle to keep its temperature down. The hot water from the tip is then recirculated back to the water cooler and used again. The water cooler consists of a pump, fan, and internal spray, much like an evaporative cooler. Alternatively, city water may be used to cool the welding gun and then discharged into a drain; this approach eliminates the need for a water cooler.

What does the smoke evacuator do?

Some GMAW and FCAW guns have built-in systems for capturing and removing welding smoke at the nozzle of the welding gun. The smoke reduction adds to welder safety, comfort, and visibility, but the evacuator also adds weight to the gun and stiffness to the welding cable.

Process

How does the GMAW process work?

The welder positions the electrode wire coming from the center of the contact tip to where the weld is to begin, actually touching the welding wire to the base metal. When the welder drops his hood and squeezes the trigger on the welding gun three events happen simultaneously:

- The trigger turns on the CV welding power supply; welding cable conductors feeding the gun apply this voltage to the copper contact tip within the gun, and then to the electrode wire, striking the arc.
- The wire feed mechanism begins feeding the welding wire from the spool through the welding gun cable and out the contact tip into the weld pool. Note that the welding gun cable is really a bundle of cables: power cables, control cables, electrode wire liner, shielding gas line, and possibly cooling water and smoke extractor lines.
- An electrically operated valve (solenoid), opens and feeds shielding gas from the regulator/flowmeter to surround the electrode and weld pool,

shielding them from the atmosphere, particularly oxygen and nitrogen. See Figure 6–2.

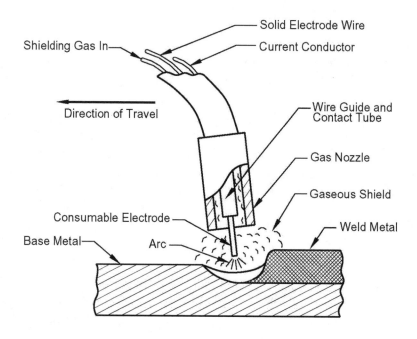

Figure 6–2   GMAW gun

Welding begins as the section of electrode wire between the tip and the base metal is heated and deposited into the weld. As the wire is consumed, the feed mechanism supplies more electrode wire at the pre-adjusted rate to maintain a steady arc. The welder manipulates the gun and lays down the weld in the desired pattern. To stop the process when the weld is completed the welder releases the trigger shutting off the welding current, wire feed, and shielding gas.

Applications

What metals can GMAW weld?
- Aluminum
- Carbon steel
- Copper
- Low alloy steels
- Magnesium

- Nickel
- Stainless steels
- Titanium

Advantages

What are some advantages of GMAW?
- GMAW is the only consumable process that can weld most commercial alloys.
- GMAW, with its continuous wire electrode, overcomes the start-and-stop cycle of SMAW and leads to fewer discontinuities and higher deposition rates.
- Since there is no need to stop welding to change consumed electrodes, long continuous welds can be done manually.
- GMAW can easily be adapted to fully automatic welding processes.
- All welding positions can be used using the short-circuit transfer mode.
- Significantly higher utilization of filler metal than SMAW: There is no end loss as with the upper, unconsumed end of an SMAW electrode.
- With spray transfer, deeper penetration with higher deposition than SMAW and may permit smaller fillets of the same strength.
- Metals as thin as 24 gauge (0.023 inch or 0.5842 mm) may be welded.
- Easier to learn than most other welding processes.
- There are no practical thickness limitations.
- There is very little spatter and no slag with properly adjusted equipment and many manufacturers paint or plate over GMAW welds with little or no additional surface preparation.

Disadvantages

What are the disadvantages of GMAW?
- Welding equipment can be more expensive, more complicated and slightly less portable than SMAW.
- GMAW is more difficult to use in tight quarters as the gun is large and the gun cable is somewhat stiff and inflexible.
- The large size of the GMAW gun combined with the $1/2$ to 1 inch (12.7 to 25 mm) stickout of the electrode wire makes it harder to see the arc and achieve quality welds.
- GMAW outdoor use is limited to very calm days or where shielding screens can be used to prevent the shielding gas from being blown away. GMAW

cannot be performed outdoors in greater than a 5 mile/hour (8 km/hour) breeze.

Welding Variables

What are the major welding variables in GMAW affecting penetration, bead shape and weld quality?

- Always DCRP polarity
- Arc voltage
- Electrode diameter
- Electrode extension
- Electrode orientation

- Shielding gas
- Travel speed
- Weld joint position
- Welding current
- Wire feed rate

How is the welding wire feed rate determined?

On many wire feed machines, wire feed rate is marked on the speed control or by a digital read-out, but the best way to check the speed is to run out electrode (being careful not to permit an arc) for one minute, then measure the wire output. Usually, each number on a wire feed speed adjustment scale represents an increment of 70 inches/minute (1.8 m/minute). In general, both the wire feed rate and the arc voltage will need adjustment to get the optimum weld.

What effect does changing welding current have in GMAW welding?

Welding current is determined by the wire feed rate: the higher the electrode wire feed rate, the more current the CV welding power supply must provide. The welding current (amperage) on a CV power supply cannot be directly adjusted. With all other welding variables held constant, an increase in wire feed speed (and thus welding current) will:

- Increase depth and width of weld penetration
- Increase deposition rate
- Increase weld bead size

Refer to Figure 6–3 and Table 6–1.

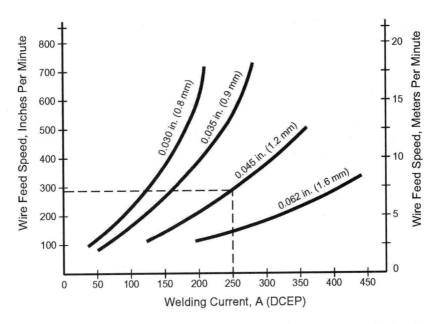

Figure 6–3   Welding currents versus wire feed speeds for carbon steel electrodes

How do we determine amperage (current) with wire speed settings?

Welding steel in the short-circuit-transfer mode (GMAW-S) using .035 (0.9 mm) or smaller welding wire variables for the correct machine adjustment for the wire feed are:

AMPERAGE TO WIRE SPEED GMAW SHORT-CIRCUIT TRANSFER

INCHES PER MINUTE (IPM)	APPROXIMATE AMPERAGE
70	30–70
140	70–100
210	100–140
280	140–165
350	165–180
420	180–200

Table 6–1   Relationship of IPM to amperage GMAW-S

In the short-circuit-transfer mode wire speeds should never exceed 420 IPM and the voltage will be less than 24 volts. Remember on the common industrial wire feeder each number represents 70 inches per minute (IPM) so 420 IPM would be at number six on the dial.

What are the amperage to wire speed settings for GMAW spray transfer?
See Table 6–2.

AMPERAGE TO WIRE SPEED GMAW SPRAY TRANSFER

INCHES PER MINUTE (IPM)	APPROXIMATE AMPERAGE
420	330
490	350
560	400
630	430
700	450

Table 6–2   Relationship of IMP to amperage in the GMAW spray transfer mode

In the spray transfer mode the optimum wire diameter is .045 (1.2 mm) and larger; spray is primarily a flat position welding process on materials over 1/8" thick. It is not a good choice for open root or poorly fit-up joints. Spray is a high production process using an argon rich blend of shielding gas. Voltage range for spray transfer will be well over 24 volts.

What type and polarity welding current is used for GMAW?
DCEP (also called DCRP) is *always* used. This is because the three metal transfer modes that actually move metal ions through the arc (short-circuit, globular, and spray transfer modes), move positively charged metal ions. These positive ions must travel through a positive to negative voltage field—just what DCEP provides.

What effect does arc voltage have on the weld?
We really want to control arc length, but it is hard to measure and closely related to arc voltage, so we choose to control arc voltage. Most welding specifications indicate arc voltage setting, not arc length. This is because with all other variables held constant, arc length is proportional to arc voltage. Arc voltage depends on materials welded, shielding gas, and metal transfer mode.
- Too short an arc may produce stubbing—cold wire pushes into the workpiece upsetting the smooth flow of shielding gas and pumping air into the arc stream causing porosity and cracking from atmospheric nitrogen.
- Too long an arc can cause the arc to wander, degrading both penetration and weld bead quality. In general, higher voltages tend to flatten the weld bead and increase the width of the fusion zone. Very high arc voltages cause porosity, undercut, and spatter.

The best way to set arc voltage is to begin at the recommended voltage based on welding tables from AWS or the electrode manufacturer's data sheets. The base metal, type of metal transfer, and shielding gas will yield a starting point and exact refinement will come from several trial runs.

For short-circuit –transfer you should set the common industrial (CV) constant voltage (constant potential) power supply at 18 volts and your wire feed speed at 210 IPM as a beginning setting then observe what is happening to the weld area if:

- welding wire coming from the contact tip melts but the metal you are welding is not melting increase the voltage.
- welding wire is coming out to fast and pushing into the material (stubbing) increase your voltage setting, Increase a notch at a time do not increase a full number, until the short-circuiting begins and the sound of the welding is similar to the sound of bacon frying.
- welding wire forms droplets and splashes into the molten pool decrease your voltage.

What effect does travel speed have on the weld?
Travel speed measures the arc's rate of progress along the weld line in inches/minute (mm/minute). If all other variables are held constant, the greatest weld penetration will be at an intermediate speed. Too low a travel speed, and the electrode melts on the top of the weld pool and penetration is poor. At slightly higher travel speeds, the welding wire melts on touching the base metal, increasing penetration. Still higher speeds reduce the heat input to too low a level for good penetration.

What is *electrode extension* or *stickout* and what effect does it have on the weld?
Stickout is the distance between the end of the contact tip and the end of the electrode wire optimum stick-out for GMAW-S is 3/16" (5 mm). See Figure 6–4.

- If stickout is too long, the wire overheats with the current flowing through it on the way to the weld.
- If stickout is too short, the wire will not get hot enough to make a properly fused bead.

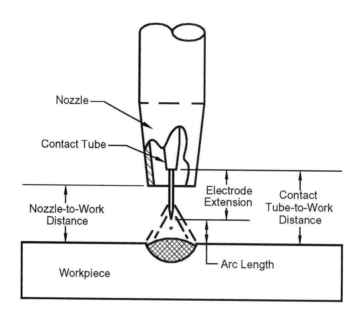

Figure 6–4   GMAW-S electrode and gun dimensions

What effect does electrode orientation have on the weld?
How the gun is held affects the weld bead:

- In the *forehand technique* with a lead angle, the gun is tipped toward the direction of welding. Going from perpendicular to a lead angle gives less penetration and a wider, thinner weld. On some materials, particularly aluminum, a lead technique will produce a cleaning action just ahead of the weld pool. This promotes wetting—the visible inter-melting and fusion of the wire feed electrode with the base metal—and reduces base metal oxidation.
- In the *torch technique*, the gun is perpendicular to the direction of welding. It has neither a lead nor a drag angle.
- In the *backhand technique* with a drag angle, the gun is tipped in the direction of welding. Maximum penetration occurs in the flat welding position with a drag angle of 25° from perpendicular. For all other positions, the backhand technique is used with a drag angle of 5 to 15°.

See Figure 6–5 for comparison of these three positions and resulting bead width and penetration.

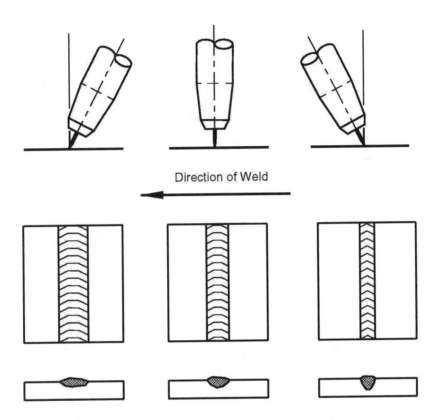

Figure 6–5   Forehand technique (left), torch perpendicular (center) and backhand technique (right) and their results

What effect does weld joint position have on the weld?

- Spray metal transfer is usually done in the flat position because it is easier to control a large weld pool when flat.
- Pulsed and short circuit metal transfer may be done in any position.
- Welding fillets done with spray transfer in the flat position are more uniform than the same welds done in the horizontal position.
- Welding electrode wire 0.045 inches (1.2 mm) or less is usually used in the overhead and vertical positions to make weld pool control easier by making the pool smaller.
- Sometimes welding downhill at about 15° can increase welding speed, decrease penetration, and reduce weld thickness. This tactic is helpful in welding thinner sheet metals.

Welding Variable ⬇	Result Wanted							
	Penetration		Deposition Rate		Bead Size		Bead Width	
	✳	❋	✳	❋	✳	❋	✳	❋
Wire Feed Rate (Current)	✳	❋	✳	❋	✳	❋	⇔	⇔
Voltage*	Little Effect	Little Effect	⇔	⇔	⇔	⇔	✳	❋
Travel Speed	Little Effect	Little Effect	⇔	⇔	❋	✳	✳	❋
Stickout	❋	✳	✳	❋	✳	❋	❋	✳
Wire Diameter	❋	✳	❋	✳	⇔	⇔	⇔	⇔
Electrode Orientation	Back Hand	Fore Hand	⇔	⇔	⇔	⇔	Back Hand	Fore Hand
Shield Gas %CO_2	✳	❋	⇔	⇔	⇔	⇔	✳	❋

✳ = Increasing   ❋ = Decreasing   ⇔ = No Effect

* Voltage settings can be important for sheet metal.

Table 6–3   Effects of changing weld parameters

What effect does electrode diameter have on the welding process?

In general, the larger the electrode diameter the higher the *minimum* welding current needed to perform a given metal transfer process beginning with short-circuit transfer, then globular and finally spray transfer as the welding current increases. See Table 6–1 that summarizes the impact of changing weld parameters.

Transfer Mechanisms

What are the four types of GMAW metal deposition processes?
- Short-circuit transfer
- Globular transfer
- Spray transfer
- Pulsed-spray transfer

What factors determine the type of transfer process?
- Magnitude and type welding current
- Electrode diameter
- Electrode composition
- Electrode extension
- Shielding gas

For a given electrode diameter, in what order do the transfer mechanisms occur going from small to large welding currents and how do they work?
- Short-circuit transfer—This transfer occurs at the lowest current ranges and electrode diameters. It produces a small, fast-freezing weld pool suitable for joining thin sections, making open root passes, and performing out-of-position welds. Metal transfers only when it is in contact with the work at between 20 to over 200 times/second. There is no transfer of metal through plasma. This process is detailed below.
- Globular transfer—This transfer mode occurs as the short-circuit mode transitions to the spray transfer mode. Its poor penetration and tendency to produce spatter limits its production applications.
- Spray transfer—There are three requirements for spray transfer: argon-rich shielding gas, DCEP polarity, and a welding current above the critical value, called the *transition current*. Below the transition current the transfer is globular, above it transfer is by spray mechanism. Spray transfer produces a discrete stream of metal droplets that are accelerated by electrical forces to overcome gravity effects. The result is no short-circuiting, so no spatter, excellent penetration, and the ability to weld out of position. High deposition rates are possible.
- Pulsed-spray transfer—Spray transfer's high deposition rate is hard to utilize on thin sections as the high currents needed to produce spray transfer often lead to burn through. The solution is to use a pulsed waveshape from the welding power supply. The transfer during the pulses provides the spray transfer characteristics of high deposition without spatter while the time

between the pulses provides for a cooling period to prevent burn through. See Power Supplies, Chapter 13. See Figure 6–6.

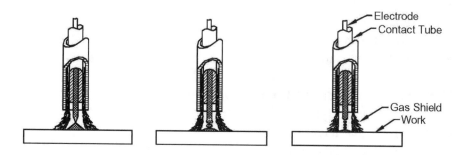

Figure 6–6   The three types of GMAW metal transfer: short-circuit (left), globular (middle) and spray (right)

	Transfer Mode			
	Short Circuit	**Globular**	**Spray**	**Pulsed Spray**
Conditions for this mode	• Low current levels • Small diameter electrodes	• Current levels just above short-circuit transfer • Occurs between short circuit and spray transfer	• Occurs at above transition current • DCEP • Argon-rich shielding gas	• Same as spray but using pulsed power source
Characteristics	• Low deposition rates	• Spatter • Poor penetration	• High deposition rates • Spatter-free • Excellent penetration	• Same as Spray, but can perform all positions.
Applications	• < 12 gauge • Root pass	• >12 gauge • Vertical down	• >1/8 inch • Flat horizontal	• >1/8 inch • All
	• Vertical up • Vertical down • Overhead • Flat horizontal			positions

Table 6–4   Conditions for GMAW metal transfer and applications

Under what conditions will short-circuit metal transfer occur and when is it used?

Short-circuit metal transfer occurs at the lowest current settings and on the smaller electrode diameters. It is characterized by a small, fast-freeze weld pool, suitable for thinner materials, bridging large root openings, and out of position welds. No metal is transferred across the arc gap. Metal only transfers when the electrode contacts the weld pool. Assume we are looking at the process after it has been established. To understand short-circuit metal transfer process follow the explanation for Figure 6–7.

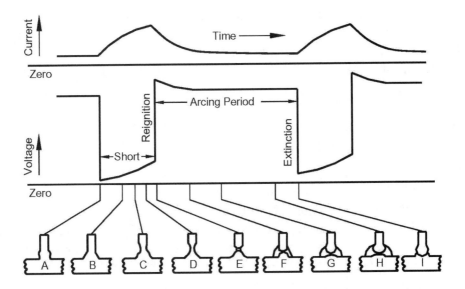

Figure 6–7   Short circuit metal transfer process

A. While the arc operates, both the base metal and the electrode wire are at a high temperature and the wire feed mechanism steadily advances wire. When the wire touches the weld puddle, suddenly three events happen simultaneously as shown in Figure 6–7:
- Heavy current flows through the electrode into the weld and the base metal causing the end of the electrode to begin melting.
- As a result of the short, the current rises rapidly.
- The voltage drops from the arc voltage to a much lower value with the beginning of the shorted electrode, and the arc goes out.

B. Current continues to rise as the constant-voltage welding power supply attempts to maintain its voltage setting and raise the voltage. The arc stays out. The melting electrode continues to deposit metal into the weld pool.

C. As the voltage and current continue to rise, the electrode begins to form an inverted cone on the weld. More electrode wire is pushed onto the inverted cone by the continuous wire feed action.

D. Voltage and current continue to rise, the cone gets bigger as additional electrode wire is fed into it, and the electrode wire on top of the inverted cone begins to neck down and prepares for the pinch off of the next step. This electromagnetic pinch-force results from the welding current flowing through the short circuit between the electrode and the deposited metal on the weld and acts radially inward toward the electrode.

E. Just after point D, the electrode separates from the cone formed over the weld, the arc reignites, the current begins to fall off and the voltage rises to arc voltage from short circuit voltage. Metal transfer ceases to the weld.

F. The arc heats both the electrode and the weld.

G. During the arcing period the arc heat smoothes out the cone, flattening it into the weld and heats up both the weld and the electrode preparatory to the next cycle.

H. The cycle is complete when the voltage drops below arc voltage and the electrode wire touches the weld and extinguishes the arc.

Note that:

- Electrode metal is transferred to the weld *only* under short circuit conditions; no metal ions travel through the arc.
- The short circuit metal transfer cycle repeats 20–250 times per second.
- The magnetic field from the welding current is the major influence in the short circuit metal transfer; gravity is not a major factor, so all welding positions may be used.
- The electromagnetic field around the welding electrode causes the pinch effect. The drop of molten metal on the tip of the electrode transitions to a molten spherical drop of metal attached to the tip which is pinched off by the electromagnetic forces to become a free drop of metal driven into the weld pool.
- Short-circuit metal transfer is used with electrodes under 0.045 inch (1.1 mm) diameter.
- Larger diameter electrodes are used on thicker stock and use the spray transfer method for deposition efficiency.
- Short-circuit metal transfer has a low heat input to the base metal making it ideal for welding thinner sheet metals without burn-through, typically metals less than 3/16 inch (4.7 mm) thickness.

Under what conditions does globular metal transfer occur and when is it used?

Globular metal transfer occurs at an average current range just above short circuit metal transfer. The electrode metal transfers in droplets 2–4 times the electrode diameter. They fall in an irregular pattern with no set frequency, and produce a lot of spatter. Most often this type transfer occurs when carbon dioxide is the shielding gas. Because the droplet transfers under the influence of gravity, globular metal transfer is limited to the flat position.

Low voltage settings produce too short an arc length, allowing the drop to get too big, short to the work and then disintegrate, producing spatter. The voltage must be high enough to lengthen the arc to ensure the transfer of the drop to the weld before the drop grows big enough to cause a short and the resulting spatter. Too high a voltage may cause lack of fusion and penetration that greatly limits its use in most production situations. Globular transfer is *not* a good choice for GMAW welding.

What is a *buried arc*?

Under some conditions while using carbon dioxide shielding gas, the current may be slightly increased to produce buried-arc transfer. This creates a deep weld pool beneath the surface causing good penetration and trapping the weld spatter. A combination of both short circuit and globular transfer occurs under these conditions and increases deposition rates above those for short-circuit metal transfer.

Under what conditions does spray metal transfer occur and when is it used?

Three conditions must be present for spray transfer to begin:
- Argon or argon-rich shielding gas must be used.
- DCEP polarity.
- Current level must be above the transition current when globular metal transfer changes to spray metal transfer. See Figure 6–8.

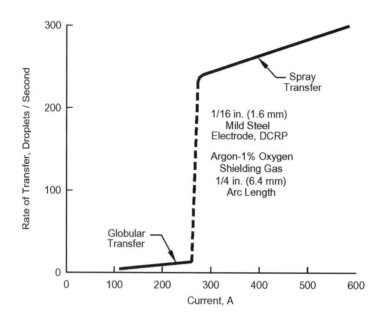

Figure 6–8   Weld metal drop size vs. welding current

Spray metal transfer occurs at high welding currents and voltages. The droplets are small and forced across the arc in an axial (in line with the axis of the electrode wire) pattern. This transfer produces little spatter and high deposition rates. In heavy industry it is the most common metal transfer used. Spray transfer is limited in welding thin sheet metal as the higher voltages needed to initiate spray transfer may cut through the sheet metal instead of welding it. However, it can produce high deposition rates in the flat position on metal generally thicker than 1/8 inch (3 mm). See Figure 6–8.

Wire Electrode Type	Wire Electrode Diameter		Shielding Gas	Minimum Spray Arc Current (A)
	Inches	mm		
Mild Steel	0.030	0.8	98% Argon-2% Oxygen	150
Mild Steel	0.035	0.9	98% Argon-2% Oxygen	165
Mild Steel	0.045	1.1	98% Argon-2% Oxygen	220
Mild Steel	0.062	1.6	98% Argon-2% Oxygen	275
Stainless Steel	0.035	0.9	98% Argon-2% Oxygen	170
Stainless Steel	0.045	1.1	98% Argon-2% Oxygen	225
Stainless Steel	0.062	1.6	98% Argon-2% Oxygen	285
Aluminum	0.030	0.8	Argon	95
Aluminum	0.045	1.1	Argon	135
Aluminum	0.062	1.6	Argon	180

Table 6–5   Minimum current for spray transfer for various size electrodes

What is *pulsed-arc metal transfer* and when is it used?

Pulsed-arc metal transfer uses a special welding power supply. This supply produces a constant, lower-level, background current output and periodically adds a much higher current pulse on top of the background one. See Figure 6–9. The idea is that the background current sustains the arc and keeps the electrode and base metal hot, but does not transfer any metal to the weld. This is because the background current is below that required for spray metal transfer, and not long enough (more than one-tenth of a second) for globule transfer to begin. During the peak portion of the higher current pulse, spray metal transfer occurs directing a stream of small metal droplets into the weld pool.

Pulsed-arc metal transfer allows the use of spray metal transfer at substantially lower *average* currents (and therefore heat input), than would be possible with constant DCEP spray metal transfer. High currents are necessary to produce spray metal transfer and such currents tend to burn-through or cut thinner metals. Hence, there are distinct advantages to pulsed-arc metal transfer: welding thinner gauge sheet metal with more control of the weld pool, the use of spray transfer mode in not only flat but vertical position welding, and freedom from spatter when joining thick weldments. For pulsed-arc transfer mode, the voltage ranges from 25 to 45 volts. See Chapter 13, Power Supplies.

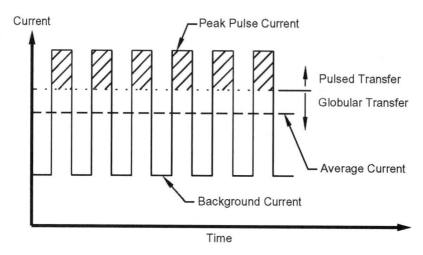

Figure 6–9   Pulsed arc power supply waveform

Shielding Gases

What effect does shielding gas have on the weld?

Shielding gas prevents atmospheric oxygen and nitrogen from getting into the weld pool and contaminating the weld. Various shielding gases may also:

- Promote high welding speeds without burn-through
- Reduce spatter and improve weld appearance
- Reduce heat affected-zone
- Improve wetting of base metal

In selecting a shielding gas, consider cost, weld appearance, and weld penetration. In some cases, mixes of different gases are best for the job.

What inert shielding gases are used for GMAW?

- Argon (Ar)—often used for out-of-position welding because of its lower heat conductivity; also it is ten times heavier than helium so it blankets the weld better than helium.
- Helium (He)—this gas is an excellent heat conductor bringing heat from the arc into the weld area. It used where high heat input is needed as in joining thick sections and when welding copper and aluminum which are excellent conductors of heat and remove heat rapidly from the welding zone.

These inert gases will not react with other chemical elements.

What reactive gases are used for GMAW?

- Carbon dioxide (CO_2)—causes better metal transfer, lower spatter, more stable arc and improved flow of metal to reduce undercutting.
- Oxygen (O_2)—blended with argon in small percentages reduce volts.
- Hydrogen (H_2)—in small percentage concentrations removes light rust and eliminates surface cleaning.
- Nitrogen (N_2)—used on copper.

Reactive gases will form compounds with other chemical elements and are used to achieve specific objectives at low concentrations in GMAW.

What effect does adding oxygen-containing gases to argon have on the weld?

Oxygen and carbon dioxide added to argon reduce chances of undercutting, smooth the arc, and increase penetration. See Figure 6–10.

What gas mixtures are most commonly used for GMAW?

- Argon + Helium
- Argon + Oxygen
- Argon + Carbon dioxide
- Helium + Argon + Carbon dioxide

Why are mixtures of inert and reactive gases used?

Mixtures produce better arc stability, lower spatter and better bead structure than any one gas used by itself. By mixing gases we can get the best performance properties each component gas can produce.

What are the characteristics produced by the various shielding gases/combinations?

See Figure 6–10.

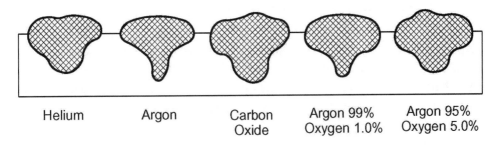

| Helium | Argon | Carbon Oxide | Argon 99% Oxygen 1.0% | Argon 95% Oxygen 5.0% |

Figure 6–10   Bead shape and penetration with various shielding gases

Electrode Wire Handling

What are the three most common methods of moving electrode wire from the spool to the welding area?

- Drive wheels in the wire feeder—one or more pairs of drive wheels in the wire feed machine grip the electrode wire and push it through the wire feed tube to the welding gun. This is by far the most common arrangement and works well except for very soft electrodes like aluminum. See Figure 6–11.

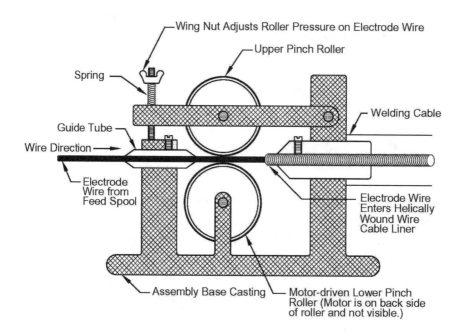

Figure 6–11   Typical wire feed mechanism using drive wheels

- Spool gun/torch—the torch assembly contains both a small spool of electrode wire and a DC motor to drive it out the contact tube. These work well for handling soft wires like aluminum, but the penalty is the increased weight the welder must manipulate. See Figure 6–12.

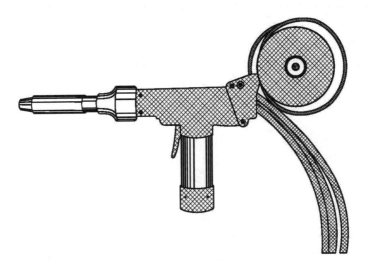

Figure 6–12   Spool gun for soft electrode wires

Pull gun—the torch assembly contains a small DC motor and drive wheels, which assist in pulling the soft wire through the cable liner and out the torch, tip. This has the advantage of reduced torch weight with the ability to handle soft electrodes. See Figure 6–13.

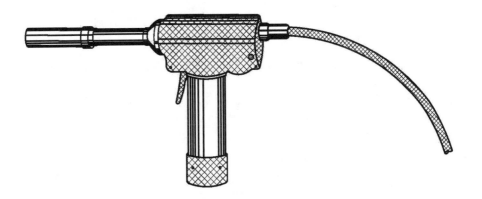

Figure 6–13   Pull gun

How does welding current get from the welding power supply to the wire electrode?

Welding current travels down the GMAW cable to the welding gun assembly via a copper cable and then transfers to a copper transfer tube, which puts it onto the electrode. See Figure 6–14.

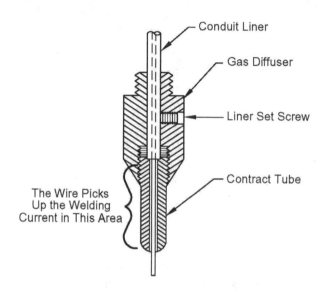

Figure 6–14   Detail of GMAW gun tip showing copper transfer tube which feeds welding current into the electrode wire

Process Setup

If you lack specific instructions on wire feed speed, how should you initially set the speed?

With the way most constant-voltage wire feed machines are set up 210 inches/ minute (5.3 meters/minute) will be at the ten o'clock position on the speed control or the number 3 position. Set the voltage to 18 volts. Try to weld on a scrap of the same type and thickness as the work. If the wire stubs, turn the voltage up until it does not; if the wire globs, turn the voltage down. This approach will work on a wide range of material thicknesses—22 gauge to 1/4 inch (0.7 to 6 mm) thickness. Remember, we want to set the voltage and feed rate so the welder sounds like frying eggs. Once you learn the crackling sound of a good GMAW weld, you will find setting parameters easy.

What is the procedure to set up GMAW equipment?

1. Make sure the welding power-supply and wire feeder power are turned OFF.
2. Mount the electrode wire spool on the wire feeder.
3. Without letting go of the wire (it will uncoil and make a snarl if you do), open the feed roller gate, insert the electrode wire into the feed roller mechanism, and into the start of the liner tube. Close the gate, securing the wire. Make the wire feeder rolls match the wire diameter.
4. Turn on the power for the wire feed motor, then use the JOG or INCH button to load the wire into the welding cable liner and up to and through the welding gun.
5. Use the JOG or INCH button to actuate the electrode wire drive motor, then adjust feed roller pressure until you can no longer stop the wire feeding into the rollers by pinching the electrode wire between your thumb and index finger. Erratic wire feeding usually results from too much roller pressure which causes the electrode wire to flatten (and loose feed) as it passes through the rollers.
6. Properly secure the compressed shielding gas cylinder, crack the valve to remove any dirt. Then put the flow regulator or flow meter (or both) on the cylinder valve and secure other end of the gas line into the welding machine.
7. Turn welding power supply ON and using the PURGE button (or if no PURGE button, then squeeze the welding gun button being sure the gun is not touching anything or anybody) for several seconds to remove air from the lines and fill them with shielding gas. While performing this purge, set the flow meter or flow regulator for the recommended purge gas flow rate; 20 ft^3/hour (9.5 l/min) for 0.035 inch (0.9 mm) wire would be a good starting point. On larger diameter electrode wire, begin with 20 ft^3/hour (9.5 l/min) on horizontal and 30 to 35 ft^3/hour (14 to 16.5 l/min) on out-of-position welds.
8. Set proper polarity (DCRP), voltage and wire feed speed. Get this info from manufacturer's data sheets for the electrode wire or from tables attached to the welding machine.
9. Attach the work lead to a clean spot on the work.
10. Use wire cutters to trim the electrode stickout to proper length—about 1/2 inch (12 mm).
11. Make sure welding area is dry and free of flammable materials as well as volatile fumes and that other personnel are protected from arc radiation and sparks.

12. Put on your welding helmet (your safety glasses should already on and remain on *under* the helmet), position the electrode against the work, squeeze the trigger and begin welding.

GMAW Electrodes

What is the difference between *wire feed rate* and *weld deposition rate*?
The wire feed rate, measured in inches/sec, or pounds/hour, is the speed at which the electrode wire emerges from the welding gun. The deposition rate is the weight of welding wire going into the weld. It is nearly always less then the wire feed rate. The difference between the two is slag, spatter, or fumes from the electrode wire. The deposition efficiency is the ratio of the weld deposition rate to the wire feed rate. GMAW with an argon shielding gas can reach 98% deposition efficiency.

Why does a constant voltage (CV) power supply maintain a constant stick-out in GMAW when the welder manipulates the gun and varies the gun-to-work distance?
We want to maintain the arc length initially set on the machine to maintain optimum weld penetration, geometry, and appearance. Moving the gun along the welding line increases the weld-to-gun distance and thereby increases the arc length. Stick-out length changes in response to the arc length.

With the increase in the stick-out length, the electrical resistance along it also increases. Because a longer stickout wire has longer dwell time, the electrical current flowing through it heats it more rapidly. Current flowing through both the stick-out and the arc increase the voltage across the power supply. The power supply senses this increase in voltage and reduces current as its characteristic curve demands. The electrode wire begins to burn back more rapidly and the stick-out gradually returns to the original, pre-disturbance length. The characteristics of the welds can be disturbed by the welder changing the gun-to-work length.

What is the construction of GMAW electrodes?
They are usually solid, bare wire similar in composition to the base metal being welded. Some electrode wire for welding ferrous metals has a thin copper plating to facilitate the drawing of the wire when made at the factory. Some wire makers would have you believe this coating is added to prevent rust and to make the wire run better; it is not.

Silicon is also added to maintain metal integrity at high arc temperatures. As an aggressive scavenger, it combines with unwanted elements and forms a glaze on the weld surface. It pops off when cool and is another reason to wear safety glasses at all times.

What are the common dimensions of GMAW electrode wires?
The common diameters in inches are:

0.020	0.062 (or 1/16)	0.313 (or 5/16)
0.025	0.094 (or 3/32)	0.375 (or 3/8)
0.030	0.125 (or 1/8)	0.500 (or 1/2)
0.035	0.188 (or 3/16)	
0.045	0.250 (or 1/4)	

How is GMAW welding electrode wire supplied?
- On one pound spools
- On ten pound spools
- On twenty pound spools
- On fifty pound spools
- In large cardboard drums (about the size of 55-gallon steel drums)
- In coils on 48" x 48" pallets

In general, how should the steel of the electrode wire and the steel of the base metal compare?
They should be metallurgically similar. Often additional elements—deoxidizers and alloying elements—are added to adjust the weld metallurgy for the damage or changes done by exposure of the weld pool to atmospheric nitrogen and oxygen.

What specifications apply to GMAW electrodes?
See Table 6–6.

Base Material Type	AWS/ASTM Specifications
Carbon Steel	A5.18
Low-alloy Steel	A5.28
Aluminum Alloys	A5.10
Copper Alloys	A5.7
Magnesium	A5.19
Nickel Alloys	A5.14
Stainless Steel (300 & 400 Series)	A5.9
Titanium	A5.16

Table 6–6   AWS/ASTM specifications applicable to GMAW electrode wire

How does the American Welding Society identify the GMAW welding wire?
See Table 6-7A.

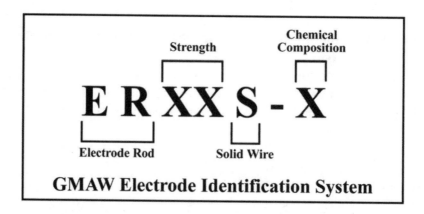

Table 6–7A   The AWS designation for GMAW welding wire

"ER" designates the wire as being both an electrode and a rod, meaning it may conduct electricity (electrode), or simply be applied as a filler metal (rod) when used with the GTAW process.

What are the characteristics of the most common GMAW electrodes for carbon steel meeting AWS/ASTM Specifications A5.18?
See Table 6–7B.

AWS Designation	Characteristics	Shielding Gas(es)
ER70S-2	Contains deoxidizers that permit welding on thin rust coatings. Can weld in any position. Excellent for out-of-position short-circuit welding. Makes excellent welds in all mild steels.	Ar-O_2 Ar-CO_2, CO_2
ER70S-3	Single or multipass beads. Base metal must be clean. Preferred for galvanized metals. Used on autos, farm equipment, and home appliances. Has better wetting action and flatter beads than E70S-2.	Ar-CO_2, CO_2
ER70S-4	Used for structural steels like A7 and A36. Used for both short-circuit and spray. Used in ship building, pipe welding, and pressure vessels. Flatter and wider beads than E70S-3.	Ar-O_2 Ar-CO_2, CO_2
ER70S-5	Used on rusty steel. Not recommended for short-circuit transfer mode. Flat position only.	Ar-O_2 Ar-CO_2, CO_2
ER70S-6	General purpose wire used on sheet metal. Will weld through thin rust. Welds all positions. Used with high welding currents.	Ar-O_2 Ar-CO_2, CO_2
ER70S-7	Used on heavy equipment and farm implements. Welds in all positions.	Ar-CO_2, CO_2

Table 6–7B   GMAW electrode wires for mild steel

What polarity does nearly all GMAW electrode wire use?
Direct current electrode positive (DCEP) or reverse polarity (DC+).

Slope

What is the *slope* adjustment on a CV welding power supply?
The *slope* adjustment optimizes the short circuit metal transfer in the GMAW process to minimize spatter. Slope controls the amount of the short circuit current—the amperage flowing when the electrode is shorted to the work.

Increasing slope on the power supply adds inductance in series with the arc, retarding rapid arc current changes and limiting maximum short-circuit current, reducing spatter. Slope adjustment is not needed for the other metal transfer modes and little slope is used on them. Many newer welding machines do not have slope adjustments.

In Figure 6–15, there are three output curves for a typical constant-voltage (CV) welding power supply for short-circuit transfer mode. Curve A has the least slope and curve C has the most. While all three curves are shown running the same current level of 130 amperes, curve A has short-circuit current of 300 amperes, curve B has short-circuit current of 250 amperes and curve C has the lowest short-circuit current with a current of 200 amperes. Excessive short-circuit current causes spatter and can be limited by adding slope. You cannot measure short-circuit current directly using the power supply ammeter since short-circuit current occurs for only a small part of the cycle and the meter *averages* the current it sees.

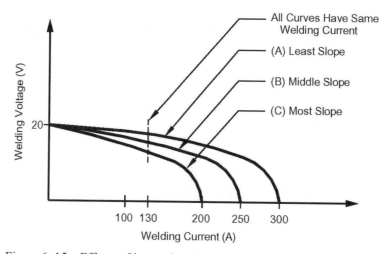

Figure 6–15   Effects of increasing slope on maximum current during
short-circuit conditions

Here are some facts to remember when adjusting slope:
- If you do not know where to set slope, try starting with the adjustment in the middle position.
- When the slope is set too flat (low), the arc will be harsh and explosive.
- When the slope is set too steep (high), the arc will lack power and stub.
- When you reduce slope, the current is increased.
- When electrode size decreases, the slope may need to be increased.

Shielding Gases

What are the functions of shielding gas?

The main function of shielding gas is to prevent the atmospheric oxygen and nitrogen from reaching the molten metal of the weld. Most metals in the molten state combine with atmospheric oxygen and nitrogen to form oxides and nitrides. These two classes of compounds can leave the weld with porosity, embrittlement or trapped slag and greatly reduce its mechanical properties.

- Shielding gases may also:
- Stabilize the arc.
- Control the mode of metal transfer.
- Establish the penetration level and weld bead shape.
- Enhance welding speed.
- Minimize undercutting.
- Clean oil or mill scale just prior to welding.
- Controls weld metal mechanical properties.

What are the two different types of shielding gas metering systems?

- Pressure regulator with high- and low-pressure gauges.
- Pressure regulator with ball in tube flow meter.

See Figure 6–16.

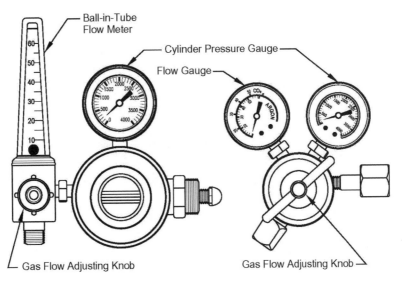

Figure 6–16   The two types of shielding-gas metering systems: regulator with ball-in-tube flow meter (left) and regulator with gauges (right) both control flow in cubic feet per hour these are not pressure gauges

How should the shielding gas be adjusted?

Adjust gas flow according to either AWS specification or manufacturer's data sheet. If neither is available begin at about 25 ft^3/hour (11.8 l/min); increase gas flow until visible signs of weld porosity cease.

What happens when the shielding gas flow rate is too high?

Getting an effective inert gas shield from the atmosphere depends on both having enough flow and maintaining a laminar (turbulence free) flow. Turbulence from too high a gas flow rate will bring air into the area we want to shield ruining the weld. It also wastes shielding gas.

What gases are commonly used for GMAW shielding?

The most common mixtures are argon and carbon dioxide in 75–25%, 80–20% and 85–15% mixtures. Some of the other more common gases and gas mixes are:

- Ar
- Ar + He
- Ar + O_2
- CO_2
- He
- He + Ar + CO_2

Many other gas mixes are in use, some combine as many as four gases.

What is a good general purpose shielding gas mixture?

Carbon dioxide is commonly used on mild steel for minimum cost.

75% argon + 25% carbon dioxide is used for general welding of carbon steel, and low alloy steel with excellent results.

Joint Preparation

What types of joint preparations are needed with GMAW?

- Metals from 0.005 to 3/16 inch (0.130 mm to 4.8 mm) can be welded without edge preparation.
- Metals from 0.062 inches to 3/8 inch (1.6 to 10 mm) can be welded in a single pass *with* joint edge preparation.
- Multipass welding is required above 3/8 inch (10mm) and preparation is needed; there is no limit to the thickness of metal that can be welded.

What changes in V-groove design can be made in going from SMAW to GMAW process?

Because the GMAW electrode has a smaller diameter than the SMAW electrode, it can get to the bottom of a narrower V-groove. Therefore the preparation for V-grooves may be made to a smaller angle. This saves as much as 50% on welding filler metal and welding time. There is no problem using an SMAW V-groove design for GMAW; it will just take longer than necessary. See Figure 6–17.

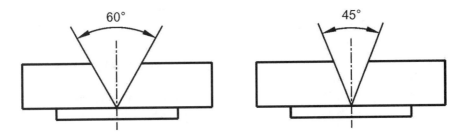

Figure 6–17   V-groove joints for SMAW (left) and GMAW (right)

GMAW Troubleshooting

What are typical GMAW problems and solutions?

Table 6–8 provides some good starting points for troubleshooting, but is by no means comprehensive.

Problem	Possible Cause	Remedy
Undercutting	1. Travel speed too high	1. Use slower travel speed
	2. Welding voltage too high	2. Reduce voltage
	3. Excessive current	3. Reduce wire feed speed
	4. Insufficient dwell	4. Increase dwell at puddle edge
	5. Gun angle	5. Change gun angle so arc force helps in placing metal
Porosity	1. Inadequate shielding gas coverage	1. Gas flow too high or to low Shield work from drafts Decrease torch to work distance Hold gun at end of weld until metal solidifies
	2. Gas contamination	2. Use welding grade shielding gas
	3. Electrode contamination	3. Use clean and dry electrodes
	4. Workpiece contamination	4. Remove all dirt, rust, paint, moisture from works surface
	5. Arc voltage too high	5. Reduce voltage
	6. Excess contact tube-to-work distance	6. Reduce stick-out

Incomplete Fusion	1. Weld zone surfaces not clean	1. Clean weld surfaces carefully
	2. Insufficient heat input	2. Increase wire feed speed and voltage Reduce electrode extension
	3. Too large a weld puddle	3. Reduce weaving, increase travel speed
	4. Improper weld technique	4. When weaving, dwell momentarily at edge of groove
	5. Improper joint design	5. Use joint design with wide enough angle to all access to groove bottom
	6. Excessive travel speed	6. Reduce travel speed
Incomplete Joint Penetration	1. Improper joint design	1. Joint design must provide access to bottom of groove
	2. Improper weld technique	2. Reduce root face Keep electrode normal to work Keep arc on leading edge of the puddle
	3. Inadequate welding current	3. Increase wire feed speed
Excessive Melt-through	1. Excessive heat input	1. Reduce wire feed speed
	2. Improper joint penetration	2. Reduce root opening Reduce root face dimension
Weld Metal Cracks	1. Improper joint design	1. Modify joint design to provide adequate weld metal to overcome constraint conditions
	2. Too high a weld depth-to-width ration	2. Decrease arc voltage, decrease wire feed speed or both
	3. Too small a weld bead (particularly in fillet and root beads)	3. Decrease travel speed
	4. Hot shortness	4. Use electrodes with higher manganese content Increase groove angle to increase filler metal in weld Change filler metal
	5. High restraint of joint members	5. Use preheat Adjust welding sequence

Heat Affected Zone (HAZ) Cracks	1. Hardening in the HAZ
	2. Residual stresses too high
	3. Hydrogen embrittlement
		1. Preheat to retard cooling rate
		2. Use stress relief heat treatment
		
		3. Use clean electrode
		   Use dry shielding gas
		   Hold weld at elevated temperature for several hours to permit diffusion of hydrogen out of weld

Table 6–8   GMAW troubleshooting

GMAW Safety

In addition to the safety measures outlined in Chapter 4, what safety issues must we remember with GMAW?

- Protection of face and eyes from sparks and radiation with a helmet and lens of appropriate density number. Note that the heat of GMAW can be great enough to crack a filter lens and for this reason, all welding masks should have an appropriate glass or plastic filter cover *both* in front of and behind the welding glass filter.

- The total radiated energy both visible and invisible of GMAW is much higher than SMAW processes, so extra precautions must be taken to protect the eyes and skin. Also, because the GMAW process produces less smoke than the SMAW process, more of the radiation produced is available to harm the welder. To protect yourself, first, use the Table 6–7 to determine the filter glass shade to use on GMAW: try the darker shade for the current you are using, and drop to the next lighter shade until you can see the welding action clearly. *Never* drop to a shade lighter than the lowest recommended one. Second, protect your skin from ultraviolet, invisible radiation with dark leather or wool clothing. Pay particular attention to both the direct and reflected light from the arc on the arms (cover with a shirt or leathers), the neck area (add a commercially available, leather flap to the bottom of your welding helmet) and the top of your head (wear a welder's cap). See Table 6–9.

Welding Current (A)	Minimum Protective Shade	Suggested Shade for Comfort
> 60	7	—
60-160	10	11
160-250	10	12
250-500	10	14

Table 6–9   Lens shade selection chart for GMAW and FCAW

- GMAW can be a noisy process and hearing protection may be necessary for both comfort and safety. Continuous exposure to relatively low noise levels (especially high frequencies) can cause permanent hearing loss. Ear plugs and ear muffs may be needed.
- Wear safety glasses to protect against silicon popping off the weld surface and other hazards.

Flux Cored Arc Welding (FCAW)

Process Name

What is the AWS designation for *flux cored arc welding*?
The AWS designation is *FCAW* but when no external shielding gas is used it is called self-shielded and has the AWS designation *FCAW-S*. If a shielding gas is used, it is called *gas-shielded* and has AWS designation *FCAW-G*.

What are other common names for FCAW?
FCAW-S is also commonly called *flux core* or *Inner-Shield*. FCAW-G is commonly called Outer-Shield.

Equipment

What equipment comprises an FCAW welding outfit?
- FCAW requires a constant-voltage (CV) power supply, the same as GMAW.

- Depending on the welding electrode wire, external gas shielding from a cylinder/ flow regulator/flow meter may or may not be needed. Using an external shielding gas when the electrode is designed to work without one can produce a defective weld, so be sure to consult the manufacturer's data sheets.

- Wire feeders for larger diameter electrodes, usually have four feed rollers instead of two. This reduces the feed roller pressure on the electrode to avoid damage; FCAW electrode wire is more fragile than the solid GMAW electrode wire.

- Because very high welding currents are used—they can exceed 250 amperes—many FCAW guns have a sheet metal heat shield to protect the operator's hands from the intense heat of the weld. Although most are air-cooled, some FCAW guns are water-cooled to operate on a 100% duty cycle. See Figure 6–18.

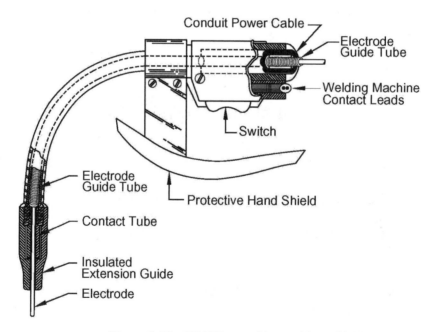

Figure 6–18   FCAW gun with metal heat shield

Process

How does the FCAW process work?

The structure and chemical composition of FCAW electrode wire makes the difference between GMAW and FCAW. Unlike GMAW, the FCAW electrodes consist of a thin-walled metal tube filled with flux, not a solid wire. The powdered flux provides alloying elements, arc stabilizers, denitriders, deoxidizers, slag formers, and shielding gas generating chemicals. For many electrodes the volume and forcefulness of the shielding gas eliminates the need for external shielding gas. The slag produced also shields the weld pool from atmospheric oxygen and nitrogen and retards its cooling rate to reduce martensite formation. Working in tandem, the shielding gas and slag allow FCAW to be used successfully in field conditions. Because of the FCAW-S heavy smoke no physical shielding is needed to protect the welding from wind. Depending on the electrode, both DCEP and DCEN are used; AC is not.

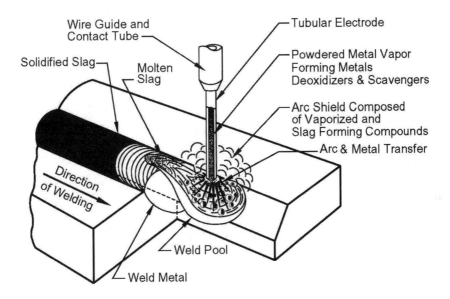

Figure 6–19   Self-shielded FCAW process

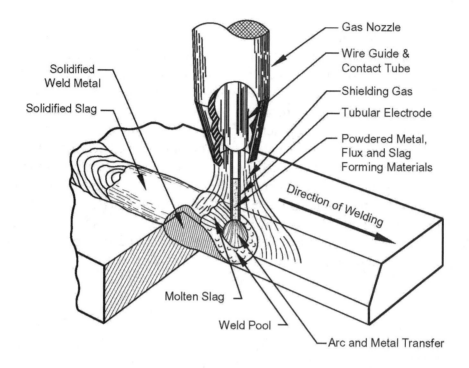

Figure 6–20   Gas shielded FCAW process

Applications

On what metals can FCAW be used?

Only ferrous metals and nickel-based alloys: all low- and medium-carbon steels, some low-alloy steels and stainless steels can be welded with FCAW.

How automated can FCAW be made?

Most FCAW is *semi-automatic*, meaning the welder manually positions the feeder gun and the wire feed rollers continuously feeds the electrode wire into the weld. However, the process has been made *fully automatic* with a computer-driven robot manipulating the torch along a preset path. Very complex parts can be made in this way.

Advantages

What are the advantages of FCAW?

- Because of its excellent penetration, FCAW can use groove angles as narrow as 30°, saving as much as 50% of the filler metal needed for SMAW which requires larger groove angles and in some cases can avoid joint beveling in metal up to 1/2 inch (13 mm) thick.
- Different thickness material can be welded with the same electrode by power supply adjustment.
- Easier to use than SMAW.
- Excellent weld pool control.
- FCAW has better welder visibility of the weld pool than GMAW because the gas diffuser nozzle is not needed and can be removed.
- FCAW is often permitted by codes for critical welds on boilers, pressure vessels and structural steel.
- FCAW oxidizers and fluxing agents permit excellent quality welds to be made on metals with some surface oxides and mill scale. Often metal from flame cutting can be welded without further preparation, a major cost saving.
- High deposition rates of more than 25 lb/hour (11.3 kg/hour) are possible compared with 10 lb/hour (4.5 kg/hour) for SMAW with 1/4 inch (6 mm) diameter electrodes.
- Self-shielded FCAW electrodes work better in windy field conditions than gas-shielded GMAW electrodes.
- There is no stub loss as in SMAW where it can average 11% of the electrode; nearly all the electrode is used.
- Unlimited thicknesses can be joined with multiple passes.
- Welds in all positions can be made.
- Wire feeder and welder must be close to the point of welding.

Disadvantages

What are the disadvantages to FCAW?

- Slag removal requires an additional production step, similar to SMAW.
- FCAW generates large volumes of fumes and smoke, requiring additional ventilation indoors and reducing welder visibility during the process.

Setup

What are the steps to set up an FCAW outfit?

- Same as GMAW.
- Be sure to consult the electrode manufacturer's recommendations for polarity (most are DCEN), wire feed speed and voltage setting.
- If a shielding gas is needed it is likely to be carbon dioxide; the most common mixture of gases 75% argon and 25% carbon dioxide.

FCAW Electrode Wire Classification

How is FCAW wire for welding mild and low alloy steel classified by the AWS?

See Figure 6–21.

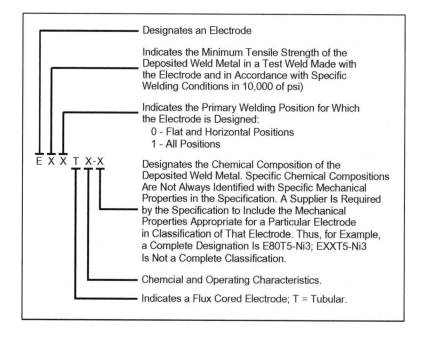

Figure 6–21   FCAW wire classification for mild and low alloy steel

Which FCAW electrodes are self shielding and which are gas shielded?
Table 6–10 will show which electrode requires gas shielding.

AWS Classification	External Shielding Medium	Current and Polarity
EXXT-1 (Multiple-Pass)	CO2	DCEP
EXXT-2 (Single-Pass)	CO2	DCEP
EXXT-3 (Single-Pass)	None	DCEP
EXXT-4 (Multiple-Pass)	None	DCEP
EXXT-5 (Multiple-Pass)	CO_2	DCEP
EXXT-6 (Multiple-Pass)	None	DCEP
EXXT-7 (Multiple-Pass)	None	DCEN
EXXT-8 (Multiple-Pass)	None	DCEN
EXXT-9 (Multiple-Pass)	None	DCEN
EXXT-10 (Single-Pass)	None	DCEN
EXXT-11 (Multiple-Pass)	None	DCEN
EXXT-G (Multiple-Pass)	*	*
EXXT-GS (Single-Pass)	*	*

* As agreed upon between supplier and user.

Table 6–10   Shielding and polarity requirements for mild steel FCAW electrodes

Troubleshooting

What are some typical problems and their solutions in FCAW?
See Table 6–11.

Problem	Possible Cause	Corrective Action
Porosity	1. Low gas flow	1. Increase gas flow setting Clean spatter-clogged nozzle
	2. High gas flow	2. Decrease to eliminate turbulence
	3. Excessive wind drafts	3. Shield weld zone from wind or draft
	4. Contaminated gas	4. Check gas source Check for leak in hoses and fittings
	5. Contaminated base metal	5. Clean weld joint faces
	6. Contaminated filler wire	6. Remove drawing compound on wire Clean oil from rollers Avoid shop dire Rebake filler wire
	7. Insufficient flux in core	7. Change electrode
	8. Excessive voltage	8. Reset voltage
	9. Excess electrode stick-out	9. Reset stick-out & current
	10. Insufficient electrode stick-out	10. Reset stick-out & current
	11. Excessive travel speed	11. Adjust speed
	Excessive voltage	Reset voltage
	Insufficient electrode stickout (self-shielded electrodes only)	Reset stickout and balance current
	Excessive travel speed	Reset stickout and balance current

Incomplete fusion or penetration	1. Improper manipulation
	2. Improper parameters
	
	3. Improper joint design
		1. Direct electrode to joint root
		2. Increase current
		   Reduce travel speed
		   Decrease stickout
		   Reduce wire size
		   Increase travel speed (self-shielded electrodes)
		3. Increase root opening
		   Reduce root face

Cracking	1. Excessive joint restraint
	
	2. Improper electrode
	   Insufficient deoxidizers or inconsistent flux fill in core
		1. Reduce restraint
		   Preheat
		   Use more ductile metal
		   Employ peening
		2. Check formulation of flux

Electrode feeding	1. Excessive contact tip wear
	2. Melted or stuck contact tip
	3. Dirty wire conduit in cable
		1. Reduce drive roll pressure
		
		2. Reduce voltage
		   Replace worn liner
		3. Change conduit liner
		   Clean out with compressed air

Table 6–11   FCAW trouble shooting

FCAW Safety

What specific safety precautions apply to FCAW?
- Same as GMAW except needs better ventilation indoors since FCAW generates more smoke and fumes—about the same amount as SMAW.
- Lens shade same as GMAW. See Table 6–9.

Chapter 7

Non-Consumable Electrode Welding Processes— Gas Tungsten & Plasma Arc Welding

Mistakes are a fact of life.
It is the response to the error that counts.
Nikki Giovanni

Introduction

Gas tungsten arc welding was developed by Russell Meredith at Northrop Aviation in the early 1940s and came into widespread use during WWII with the need to replace riveting for aluminum and magnesium in aircraft. Although this process requires more skill than most other processes and does not have high metal deposition rates, improvements in shielding gas mixtures, torch design, and power supply electronics have made it an indispensable tool in many industries where high quality welds are essential on aluminum, magnesium, or titanium. It can weld most metals, even dissimilar ones. Plasma arc welding is a close relative of gas tungsten. Although scientists had been working on plasma for welding applications since the early 1900s, it was not until the 1960s that commercial equipment was introduced. This equipment can weld nearly all metals in all positions and offers better directional control of the arc than gas tungsten arc with a smaller heat affected zone. We will cover the theory, equipment, setup, applications, troubleshooting, and safety for these two processes.

Gas Tungsten Arc Welding (GTAW)

Process Name

What is the AWS designation for *gas tungsten arc welding*?
GTAW from the initials of the process name is the AWS designation.

What are other common names for GTAW?
GTAW is commonly called *TIG* welding from tungsten inert gas welding. It was also called Heliarc™ early in its development. In Europe it is called *WIG* for wolfram inert gas—Wolfram is the German name for tungsten.

Equipment

What equipment makes up a GTAW setup?
- Constant-current (CC) welding power supply, either DC or AC/DC
- Tungsten electrodes
- GTAW torch and associated cable
- Shielding gas, cylinder, regulator, or flow meter and hoses
- Work clamp and lead
- Filler metal (optional)
- Foot pedal or finger-tip power control (optional)
- Cooling water source and drain (optional)
- Cooling water recirculating system (optional)

See Figure 7–1.

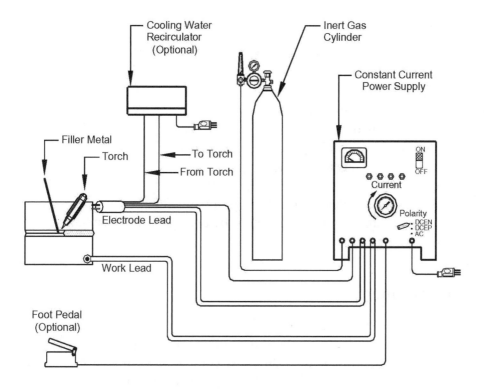

Figure 7–1   GTAW equipment

Process

How does the GTAW process work?

A low-voltage, high-current, continuous arc is formed between a tungsten electrode on the welding torch and the work through an inert gas, argon or helium. The intense heat of this arc, approximately 10,000°F (5500°C), melts the surface of the base metal forming the weld pool. On thinner metals, edge joints and flange joints, no metal is added; this is called *autogenous welding*. On thicker materials, filler metal is added in the form of a wire or rod fed into the arc. The inert gas supplied through the welding torch not only provides a suitable arc, it displaces air, shielding the weld pool and the electrode from atmospheric contaminants. No metal is transferred across the arc, so there is no spatter and little or no smoke. Since the welder may continuously control welding current with the fingertip or foot control, there is excellent control of the weld pool resulting in high-quality welds.

Advantages

What are some advantages of GTAW?

- Welds are of high quality
- Welds nearly all metals and alloys
- All weld positions are possible
- No slag developed
- Excellent welder visibility of arc and weld pool
- Little post-weld cleaning needed
- There is no spatter
- Provides excellent welder control of root pass weld penetration
- Allows heat source and filler metal to be controlled independently
- Joins dissimilar metals

Disadvantages

What are the disadvantages of GTAW?

- Higher welder skills are required than other welding processes
- Lower deposition rate and productivity compared with other processes
- Equipment is more complex and expensive than other more productive processes
- Low tolerance for contaminants in filler or base metals
- Possible problems welding in drafty environments

Torches and Cables

What are the three principal torch designs?

- Small, gas-cooled manual welding torches rated up to 200 amperes, see Figure 7–2. The relatively cool inert shielding gas on its way to the arc cools the torch.

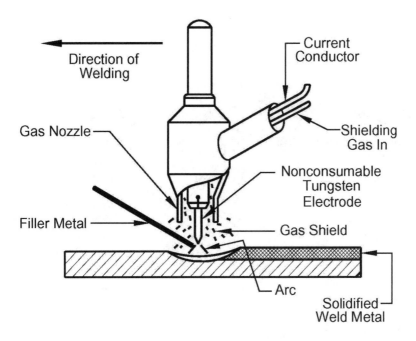

Figure 7–2   Gas-cooled GTAW torch

- Large water-cooled manual welding torches rated over 200 amperes, see Figure 7–3. Either water from a tap or a recirculating cooler prevents heat from destroying the torch.
- Automatic welding torches are similar in the cooling mechanics but the configuration is different. See Figure 7–4.

Generally, manual torches have the electrode mounted at an angle to the cables to relieve the welder of having to support the twisting force of cables coming out of the back of the torch as in Figures 7–2 and 7–3. However, some jobs are more easily done with pencil-style or straight-line torches where the cables come directly out of the back of the torch like those designed for automatic applications. These smaller, pencil-style manual designs usually do not have water cooling.

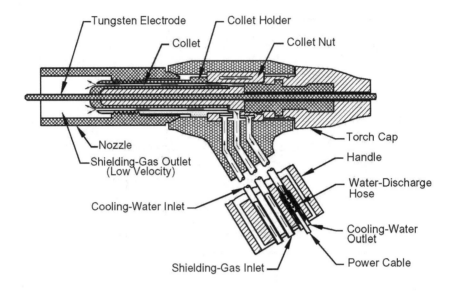

Figure 7–3   Water-cooled GTAW torch

Automatic torches also have a straight-through design and are water cooled for 100% duty cycle use. Cooling water may come from a tap, be used once and discharged to a drain. Another alternative is to use an evaporative cooler with a circulating pump to provide cooling water. This permits welding away from water lines and drains and is also more economical. The heaviest torches are rated at 600 amperes.

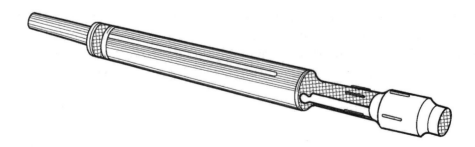

Figure 7–4   GTAW torch for automatic welding.

What functions does the *collet* in the torch perform?

All torches have precision copper collets to grip and center the electrode in the torch. The collets transfer welding current to the tungsten electrode from the welding cable. They also remove heat from the electrode to keep it from melting. There are holes or ports surrounding the electrode holder to distribute inert shielding gas evenly around the electrode and over the weld pool.

From what materials are the insulating nozzles around the electrode made and what is their function?

They are made from ceramic, metal, alumina, or fused quartz which provides see-through visibility. They can withstand both shock and intense heat. Their function is to flow gas around the electrode and into a stream at the weld pool. Their inside diameter is measured in sixteenths of an inch. If a nozzle is described as a number 5, its inside diameter is 5/16 inch, a number 10 is 10/16 or 5/8 inch inside diameter.

What is the general relationship between the inside nozzle diameter and the electrode diameter?

It should be at least three times the electrode diameter.

How long are GTAW cables?

Cables are usually 12.5 or 25 ft (4 to 8 m) in length.

Electricity and the Arc

What ranges of voltage and current are used in GTAW?

Voltage ranges from 10 to over 40 volts; current ranges from 1 to over 1000 amperes.

Since the arc temperature is well above the melting point of the tungsten GTAW electrode, why doesn't the electrode melt?

While a very small amount of the electrode does melt and ends up in the weld, the electrode resists melting because:

1. Tungsten has the highest melting point of all metals 6170°F (3420°C).
2. Tungsten has high thermal conductivity so heat can readily flow from the electrode's hot tip to the cooler collet.
3. Torch collets are designed to remove heat from the electrode and the torches themselves are gas or water-cooled.

What is the shape of the GTAW arc?
The GTAW arc is shaped much like an inverted funnel: nearly a point at the electrode and flaring out to a circle on the flat side of the work metal.

What is the relation between arc length and arc voltage?
Arc length is roughly proportional to arc voltage.

How is arc length related to electrode diameter?
Generally arc length runs from one to four times electrode diameter.

What polarities are used and what are their characteristics?
- DCEN—consider a torch connected to DCEN. When the arc begins, the electrode metal heats and as a result has enough energy to release electrons from its surface. This process is called *thermionic emission*. The voltage between the electrode and the work exerts a strong pull drawing the cloud of electrons around the surface of the electrode toward the positive work surface and accelerating the electrons to high speed. As the electrons move through the electrically excited, ionized gas, they generate heat from their friction with the inert gas atoms. When these high-speed electrons strike the work surface, their kinetic energy becomes heat. Over 99% of the current is by electron flow, the remainder by positive ions. That is, base metal ions stripped of one or more electrons become positively charged as a result and produce a current flow as they move from the work to the electrode, the direction opposite to electron flow. DCEN produces the most heat and deep weld penetration within a narrow area. However, it does *not* provide cleaning action, called *cathodic etching* that is necessary for welding aluminum and magnesium.
- DCEP—in this polarity, electrons leave the work and accelerate as they make their way through the ionized gas to the electrode under the influence of the voltage across the arc. Their impact on the electrode generates intense heat that will melt the electrode if not removed. Because this electron bombardment produces more heat at the electrode with DCEP than DCEN, a larger electrode must be used running the same current level to absorb the additional heat. In fact, DCEP can handle only 10% of the current the same sized electrode can run with DCEN. While DCEP produces less penetration than DCEN, it provides a wider area of heating than DCEN. It also provides a cleaning effect on the base metal within the arc area. Positive ion bombardment onto the work produces cathodic etching. This cleaning effect is critical when welding metals like aluminum and magnesium that instantly form surface oxides when in contact with the

atmosphere. The cleaning effect removes these oxides, and GTAW's inert gas blanket prevents them from reforming.

- AC—provides equal heat at the electrode and the work because each receives electron bombardment half the time. It provides penetration midway between DCEN and DCEP. It is used most often to weld aluminum and magnesium because its cleaning effect is essential. See Figure 7–5.

Current Type	DCEN	DCEP	AC (Balanced)
Electrode Polarity	Negative	Positive	
Electron and Ion Flow Penetration Characteristics	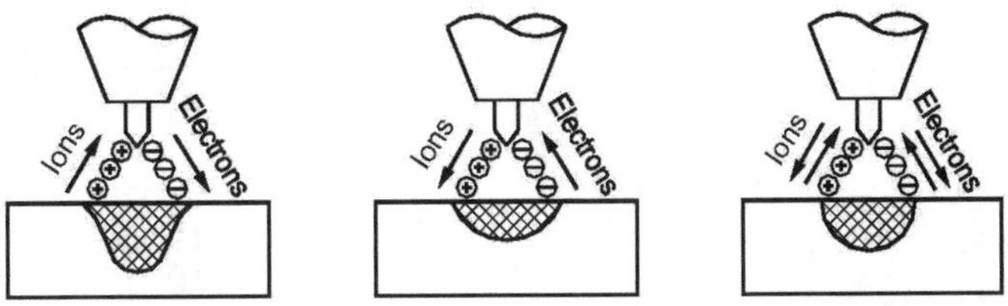		
Oxide Ceaning Action	No	Yes	Yes-Once Every Half Cycle
Heat Balance in the Arc (Approx.)	70% at Work End 30% at Eectrode end	30% at Work End 70% at Electrode End	50% at Work End 50% at Electrode End
Penetration	Deep, Narrow	Shallow, Wide	Medium
Electrode Capacity	Excellent (For Example 1/8" (3.2 mm) 400A	Poor (For Example 1/4" (6.4 mm) 120A	Good (For Example 1/8" (3.2 mm) 225A

Figure 7–5   Characteristics of current types for GTAW

What other wave shapes are commonly used in GTAW?

Unbalanced AC—provides enough cleaning effect to perform welding on aluminum and magnesium, but uses more of the cycle to put heat in the work. Unbalanced AC attempts to get the best of DCEN and DCEP. See Figure 7–6.

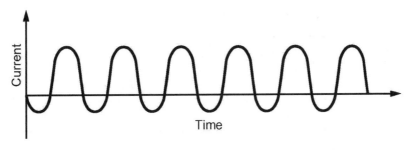

Figure 7–6   Unbalanced AC GTAW waveform to provide surface cleaning effect

Unbalanced pulses—Similar to unbalanced AC in providing a cleaning effect on aluminum and magnesium, but has faster rise and fall times for a smoother arc. See Figure 7–7.

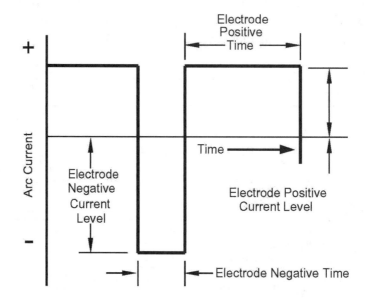

Figure 7–7   Unbalance pulses used for cleaning effect

Pulse waveforms with ramps up and ramps down—these provide a gradual starting current (the *upslope*), a stable welding current, and a gradual tapering off of the current when stopping (the *downslope*). The downslope allows the final weld pool to be completely filled and not leave a crater. Note the term *slope* as used here has nothing to do with the slope of a constant-current power supply; it refers to the shape of the current waveform when starting and stopping.

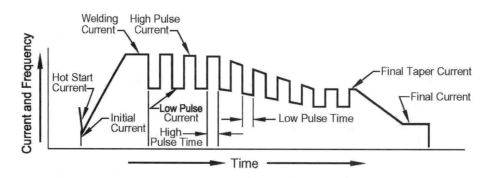

Figure 7–8   GTAW pulsed current waveform with ramps up and down

In addition to the AC or DC supplied to the arc to generate heat, what other energy is often used?
High-frequency (several Megahertz), high-voltage, low-current energy is used in addition to the low-frequency, low-voltage energy supplied to the arc. On DC polarity high-frequency is used only to start the arc and the power supply automatically shuts off the high-frequency after the arc is established. With AC polarity, high-frequency energy is supplied continuously to maintain the arc as the polarity reverses or the arc would extinguish. See Chapter 13, Power Supplies.

Is this high frequency energy dangerous?
No, but it might be uncomfortable to touch. It is 60 Hertz, high-current energy from the power supply which *does* present a shock hazard. This is why it is important to work in a dry area wearing dry gloves.

What is the rule of thumb for how many amperes of welding current is needed for each one-thousandth of an inch of aluminum thickness?
The rule is about one ampere.

Why does the US Federal Communication Commission (FCC) have regulations concerning GTAW power supplies?
By law, the FCC regulates all radio spectrum usage in the US; this includes all radio energy emissions whether intentional (like radio and TV) or unintentional (like that from welding machines). Because many GTAW power supplies use high-frequency, high-voltage pulses to start or maintain the welding arc, these pulses can cause interference with radio reception over a wide area if the systems are not properly installed, grounded, and maintained. Not only can these pulses disrupt radio and TV reception, they can interfere with critical police, fire and aircraft communications. For this reason, the FCC has established rules limiting the level of radiated power from welding machines.

Variations of GTAW Process

Are there variations on the basic GTAW process?
Yes. They are:
- *Cold wire feeder* GTAW—electrode wire is fed to the torch from a spool by a constant-speed motor eliminating the need for the operator to apply the filler metal manually.
- *Hot wire feeder* GTAW—this process is very similar to cold wire GTAW except that the filler metal wire is heated by passing a current through it

before it enters the arc. Additional shielding gas is used to protect the heated filler metal. This produces a dramatic rise in deposition rates, which are comparable to GMAW. See Figures 7–9 and 7–10.

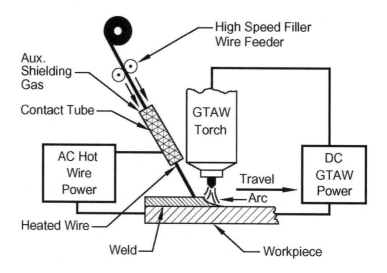

Figure 7–9   GTAW hot wire equipment

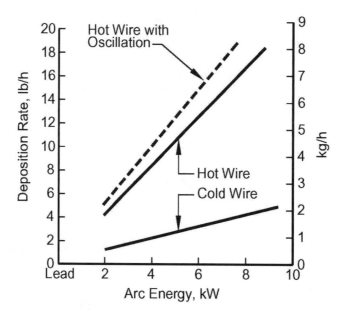

Figure 7–10   Comparison of hot and cold wire deposition rates

Welding Variables

What factors determine weld penetration?

- Arc polarity
- Arc current
- Shielding gas
- Preheat temperature, if any
- Size of part being welded
- Thermal conductivity of part
- Melting point of part's metal
- Travel speed

Tungsten Electrodes

What are the common GTAW electrode sizes?

In inches they are:

| 0.010 | 0.020 | 0.040 | 0.060 | 1/16 |
| 3/32 | 1/8 | 5/32 | 3/16 | 1/4 |

What are the seven classes of tungsten electrodes in AWS specifications?

See Table 7–1.

AWS Class.	Composition	Color Code
EWP	Pure tungsten	Green
EWCe-2	97.3% tungsten, 2% cerium oxide	Orange
EWLa-1	98.3% tungsten, 1% lanthanum oxide	Black
EWTh-1	98.3% tungsten, 1% thorium oxide	Yellow
EWTh-2	97.3% tungsten, 2% thorium oxide	Red
EWZr-1	99.1% tungsten, 0.25% zirconium oxide	Brown
EWG	94.5% tungsten, remainder not specified	Gray

Table 7–1   AWS tungsten GTAW electrode classification

What is the pure tungsten electrode (EWP) used for?

Pure tungsten electrodes contain a minimum of 99.5 wt-% tungsten, with no intentional alloying elements. The current-carrying capacity is lower than the alloyed tungsten electrodes. Pure tungsten electrodes are used primarily with alternating current (AC) for welding aluminum and magnesium alloys. Using AC the tip of the EWP forms a clean balled end, which provides good arc stability. EWP may also be used with direct current (DC) but they do not provide the arc initiation and arc stability of other alloyed electrodes.

What are the 2% cerium oxide (EWCe-2) electrodes used for?

The cerium oxide tungsten electrodes (EWCe-2) are tungsten alloyed electrodes which contain a nominal 2 wt-% cerium oxide. These electrodes were developed as a possible replacement for the thoriated tungsten electrodes because ceria, unlike thoria, is not radioactive. Ceriated tungsten electrodes, when compared to pure tungsten provide similar current levels but with improved arc starting and arc stability characteristics like thoriated tungsten electrodes. They also tend to have last longer than thoriated tungsten electrodes. They will successfully operate either AC or DC.

Why use lanthanated tungsten electrodes (EWLa-1)?

Lanthanated (EWLa-1) tungsten electrodes were developed for the same reason the ceriated tungsten electrodes were developed; lanthana is not radioactive. These electrodes contain a nominal 1 wt-% lanthanum oxide. The current levels and operating characteristics are very similar to the ceriated tungsten electrodes.

What are the characteristics of 1% (EWTh-1) or 2% thorium electrodes (EWTh-2)?

The EWTh-1 or 1% thorium oxide tungsten electrode contain a nominal 1wt-% thoria and the EWTh-2 contain a nominal 2wt-% thoria evenly dispersed throughout their length. Thoriated tungsten electrodes are used because the electrodes thoria provides higher electron emission, allowing increased current-carrying capacity about 20% higher than pure tungsten electrodes. Thoria also reduces electrode tip temperatures and provides greater resistance to contamination of the weld. These electrodes provide easier arc starting and the arc is more stable than with pure tungsten electrodes or zirconiated tungsten electrodes when using DC.

Why chose to use zirconiated tungsten electrodes (EWZr-1)?

The zirconiated tungsten electrodes (EWZr-1) contain a small amount (0.15 to 0.40 wt-%) of zirconium oxide. These electrodes have welding characteristics that fall between those of pure tungsten and thoriated tungsten electrodes. They are normally the electrode of choice for AC welding of aluminum and magnesium alloys because they combine the desirable arc stability characteristics and balled end typical of pure tungsten and have a higher current carrying capacity and better arc starting characteristics of thoriated tungsten electrodes. Zirconiated tungsten electrodes are more resistant to tungsten contamination of the weld pool than pure tungsten and are preferred for radiographic quality welding applications.

What are the two different surface finishes for tungsten electrodes?
- Cleaned (chemically)
- Polished (centerless ground)

What precautions should be followed when grinding the tips of tungsten electrodes?
- Grind with the axis of the electrode perpendicular to the face of grinding wheel. Do not grind on the side of the grinding wheel.
- Reserve the grinding wheel for electrodes only so other metals will not contaminate the tip.
- Use exhaust hoods when grinding thoriated tungsten to prevent breathing in the dust.

Why are different shapes put on electrode tips?
A more pointed tip produces a more directional and stiffer arc. Many welding procedures call for a specific shape tip. Electrons leave a tapered tip more easily than a blunt one.

What specifications apply to filler metal for GTAW?
Refer to the AWS Specification A5.12 for Tungsten Electrodes.

What are the three basic tip shapes?
- Blunt
- Tapered with balled end–A ball forms on a pure tungsten tip when it melts.
- Tapered–A ground point can be maintained on a thoriated tungsten tip for some time before erosion blunts it.

See Figure 7–11.

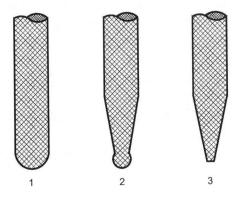

1　　　　　2　　　　　3

Figure 7–11　Electrode shapes: blunt (left), tapered with balled end (center)
and tapered (right)

Filler Metals

In what forms does GTAW filler metal come?
Rods, thin bars, and wire cut from a spool.

Shielding Gases

What inert shielding gases are most commonly used with GTAW?
- Helium—a mono-atomic inert gas is extracted from natural gas by distillation. Welding helium has a minimum purity of 99.99%.
- Argon—is also a mono-atomic inert gas extracted from the atmosphere by distillation of liquid air. Welding grade argon has a minimum purity of 99.95% which is fine for most metals except reactive and refractory ones which need 99.997% purity.

Why is argon used more frequently than helium in GTAW?
Here are the major advantages:
- Smoother and more stable arc action
- Reduced penetration
- Cleaning action on aluminum and magnesium
- Lower cost and greater availability than helium (which comes only from natural gas wells in the US)
- Lower argon flow rates than helium provide adequate shielding
- Better resistance to cross-drafts because it is heavier than air
- Easier arc starting than with helium

Argon has an atomic weight of 40 while helium has an atomic weight of just four. This makes helium much lighter than air and has the tendency to float up and away from the weld pool instead of blanketing it. Argon is closer in weight to air, so has better blanketing properties.

What are the advantages of helium over argon?
Helium provides much better penetration than argon and is often used to join metals with high thermal conductivity or to make joints on thick sections. Sometimes mixtures of both helium and argon are used to get some of the best characteristics of both gases. Helium works better welding aluminum and magnesium.

What are some ways GTAW shielding gas can be contained near the joint being welded?
See Figure 7–12.

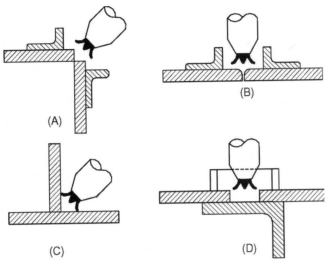

Figure 7–12   Barriers used to contain shielding gas

When is a trailing shield used?
For metals like titanium a trailing shield is necessary to keep air away from the molten metal until it has cooled enough so as not to react with it. These shields are often used when welding in a gas-purged chamber is not possible. See Figure 7–13.

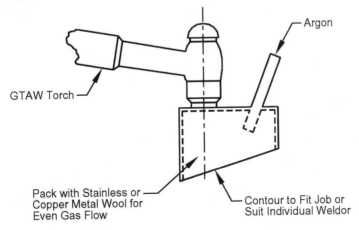

Figure 7–13   Trailing shield

Process Setup

1. Locate the welding power supply in a dry area, attach the unit to the power source and set the power switch to OFF.
2. Secure the inert gas cylinder with a safety chain so it will not tip over, remove the cylinder cap, crack the cylinder valve to remove any dirt, and using a wrench attach the regulator to the cylinder. If there is a flow meter tube, make sure it is vertical.
3. Connect the gas hoses:
   - From the flow meter or flow regulator to the welding machine
   - Between the welding machine and the torch hose
4. If water-cooled, connect the water lines:
   - From the water source to the water IN connection from the water OUT connection on the torch cable to the water drain
   - In the case of a recirculating water cooler, the IN and OUT water lines from the cooler are connected to the torch cable
5. Complete the electrical power and control connections between the torch and the welding power supply.
6. Insert the electrode in the torch collet and tighten. Install the cup if needed.
7. With the electrode and torch positioned so as not to strike an arc, turn on the power supply and apply power to the torch. Check for inert gas and water leaks and tighten fittings if needed, then adjust the gas for proper flow rate.
8. Set arc current level, polarity, and secure the work lead to the work.
9. Set any other parameters on the welding power supply; these could include:
   - Mode switch if combination GTAW/SMAW power supply
   - High frequency switch: Set to START on DC, CONTINUOUS for AC, and OFF, if pulse waveforms are used
   - Preflow and postflow gas times
   - If using a pulse waveform machine:
     - Ramp-up time (start time)
     - Ramp-down time (stop time)
     - Pulse rate
     - Background current
     - % power-on time
10. Select the weld filler metal, if used.
11. Put on your helmet over your safety glasses, gloves, and you are ready to begin welding.

GTAW Electrode Wire Classification

What specifications cover GTAW electrode rods?

See Table 7–2.

Material	AWS Specification
Aluminum	A5.10
Chromium & Chromium Nickel	A5.9
Copper & Copper Alloys	A5.7
Nickel Alloys	A5.14
Magnesium	A5.19
Surfacing Materials	A5.13
Titanium	A5.16
Zirconium & Zirconium Alloys	A5.24

Table 7–2   AWS filler metal comparison chart for GTAW

What are the correct adjustments for shielding gases?

Adjust gas flow according to either AWS specification or manufacturer's data sheet. If neither is available, begin at about 20 ft^3/hour (9.5 l/min) and increase gas flow until visible signs of weld porosity cease.

What happens when the shielding gas flow rate is too high?

Getting an effective inert gas shield from the atmosphere depends on both having enough flow and maintaining a *laminar* (turbulence free) flow. Turbulence will bring air into the area we want to shield, ruining the weld.

Joint Preparation

What joint designs are used for GTAW?

The joint design is essentially the same as SMAW.

What is the range of metal thicknesses, which need weld preparation?

- Single-pass welds can be made with no preparation on material with a thickness of 0.005 to 0.125 inches (0.13 to 3.2 mm)
- Single-pass welds can be made with preparation on material with a thickness of 0.062 to 3/16 inches (1.6 to 4.8 mm)

- Multi-pass welds can be made with preparations on material with a thickness of 0.125 to 2 inches (3.2 to 51 mm)

GTAW Troubleshooting

Problem	Cause	Solution
Excessive electrode consumption	1. Inadequate gas flow. 2. Operating on reverse polarity. 3. Improper size electrode for current needed. 4. Excessive heating in holder. 5. Contaminated electrode. 6. Electrode oxidation during cooling. 7. Using gas containing carbon dioxide or oxygen.	1. Increase gas flow. 2. Use larger electrode or change to DCEN. 3. Use larger electrode. 4. Check for proper collet contact. 5. Remove contaminated portion. 6. Keep gas flowing after arc stops for at least 10 to 15 seconds. 7. Change to proper gas.
Erratic arc	1. Base metal is dirty or greasy. 2. Joint too narrow. 3. Electrode is contaminated. 4. Arc too long.	1. Use chemical cleaners, wire brush for abrasives. 2. Open joint groove, bring electrode closer to work, decrease voltage. 3. Remove contaminated portion of electrode. 4. Bring holder closer to work.
Porosity	1. Entrapped gas impurities (hydrogen, nitrogen, air, and water vapor). 2. Defective gas hose or loose connections. 3. Oil film on base metal. 4. Too windy.	1. Blow out air by purging gas line before striking arc. 2. Check hoses and connections for leaks. 3. Clean with chemical cleaner not prone to break up arc. 4. Shield work from wind.
Tungsten contamination of workpiece	1. Contact starting with electrode. 2. Electrode melting and alloying with base metal. 3. Touching tungsten to molten pool.	1.Use high-frequency starter, use copper striker plate. 2. Use less current, or larger electrode; use thoriated or zirconium-tungsten electrode. 3. Keep tungsten out of molten pool.

Table 7–3   GTAW troubleshooting

GTAW Safety

In addition to the safety measures outlined in Chapter 4 for welding, what safety issues must we remember with GTAW?

- Because the GTAW process is smokeless and there are no visible fumes to block or absorb radiation, heat and light from its arc are intense. Proper lens shade selection is important. Also cover up all skin to prevent radiation burns.

Welding Current (A)	Minimum Protective Shade	Suggested Shade for Comfort
>20	6	6-8
20-100	8	10
100-400	10	12
400-800	11	14

Table 7–4   Lens shade selection chart for GTAW

- When welding in confined spaces such as inside tanks, provisions must be made for fresh air to replace the inert gas used in the process
- Good ventilation is vital since the arc generates dangerous ozone levels which are invisible and whose concentrations should be minimized
- Good ventilation will also prevent argon asphyxiation from slow leaks since argon is heavier than air and will tend to collect in low spots like holes and pits
- Always be sure to turn off the power supply when changing electrodes as some power supplies have open-circuit voltages as high as 85 volts, enough to be a lethal shock hazard
- Wear dry gloves without holes; never weld barehanded. Should you touch the filler wire or rod to the electrode, the full voltage of the power supply will be on the filler and you could become part of a lethal circuit. This is especially true if you have worked up a sweat and are sitting on a metal beam or working on a metal table

Orbital Welding

What is orbital welding and how does it work?

Orbital welding is an automated GTAW process for pipe and tubing, principally steel, stainless steel, and exotic alloys. The typical system welds 1/8 through 6 inch (3 through 150 mm) diameter material. The main applications are high-pressure tubing in aerospace and power station applications, high-purity tubing in semiconductor manufacturing, and sanitary tubing in food, beverage, diary, and pharmaceutical industries. While many older analog orbital welding systems are still in use, the latest designs are digital and use a microprocessor-controlled constant-current power supply and a welding head in which the welding occurs. The operator communicates with the system through a keyboard and LCD display. System software suggests a welding cycle based on the size, joint type, and work metal, but the welder can alter and fine tune the welding variables like weld speed, number of revolutions, start delay, current level, tacks, and pre/post-weld inert gas flow. A cable connects the power supply with the welding head to provide welding current, motor control lines, and inert gas. A clam shell design permits the head to open, fit over the pipe (or tubing) ends, then close to clamp them rigidly in coaxial alignment.

Once the welder clamps the ends in the welding head and turns on the welding machine, the process is fully automatic. Inert gas fills the head interior to protect the molten weld metal from the atmosphere. A DC motor drives a tungsten electrode circumferentially around the outside of the pipe (and joint) to make the weld. The microprocessor monitors the progress of the weld to assure the set parameters were accomplished. Some systems have a small printer in the welding power supply to provide a permanent record of the weld parameters. Parameters can also be stored, or transferred to other machines. Figure 7–14 pictures an orbital welding power supply and its welding head, Figure 7–15 diagrams how welding is done inside the welding head itself, and Figure 7–16 shows a menu map for a modern digital orbital welding system.

What are the advantages of orbital welding?

- The advantages of GTAW: high-quality welds without spatter, smoke, slag, or filler metal
- Rapidly produces high-quality welds with smooth interiors that prevent debris from hanging up and accumulating *inside* the pipe or tube
- This process is fully automated with welding parameters monitored and
- Verified during welding, permanently documented, all under microprocessor control

- Weld quality not dependent on welder's hand skills.
- Faster than hand GTAW process
- Process makes welds in tighter confines than hand held welding.
- Orbital welding equipment can be readily moved by one

What are the drawbacks to orbital welding?
- Requires a larger volume of welds to justify the equipment during setup
- Tube ends must be machined square, no saw cuts permitted
- High quality, low production rate process caused by setup time

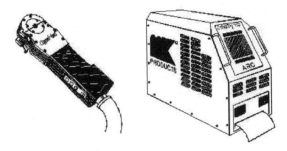

Figure 7–14   Orbital welding head (left) and power supply (right)

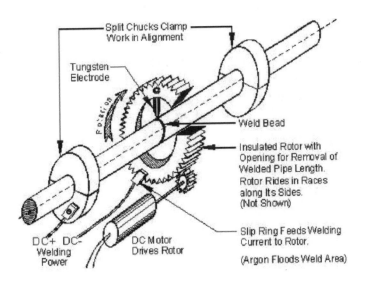

Figure 7–15   Simplified schematic of orbital welding inside the pendant

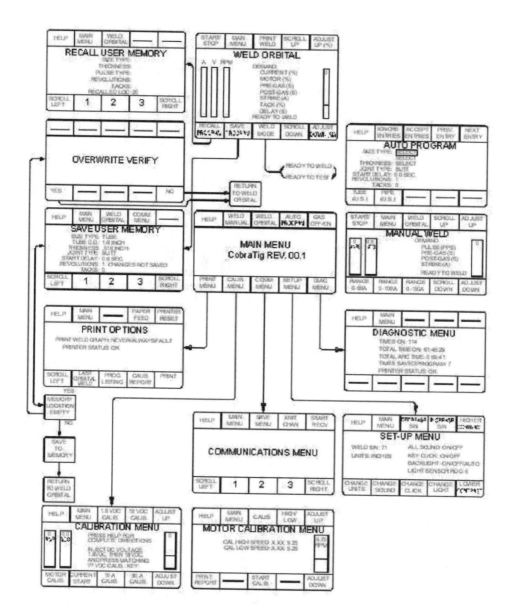

Figure 7–16   Menu map of digital orbital welding software

Plasma Arc Welding (PAW)

Process Name

What is the AWS designation for *plasma arc welding*?
It is *PAW*.

Equipment

- Constant-current (CC) DC welding power supply with drooping output characteristics
- PAW torch and associated cable
- Plasma gas cylinder, regulator, flow meter/gauge and hoses
- Shielding gas cylinder, regulator, flow meter/gauge and hoses
- Work clamp and lead
- Remote control current pedal or finger tip control
- Filler metal (optional)
- Foot pedal or finger-tip power control (optional)
- Cooling water source and drain (optional)
- Cooling water recirculating system (optional)

See Figures 7–17 through 7–18.

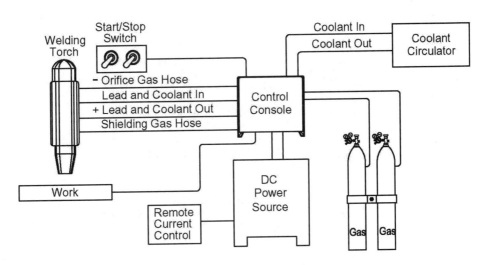

Figure 7–17   PAW equipment

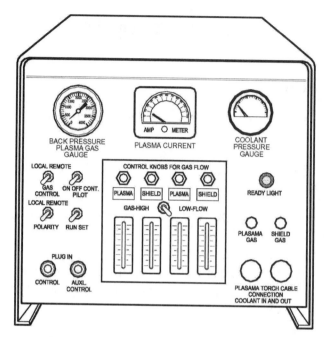

Figure 7–18   Detail of PAW control console

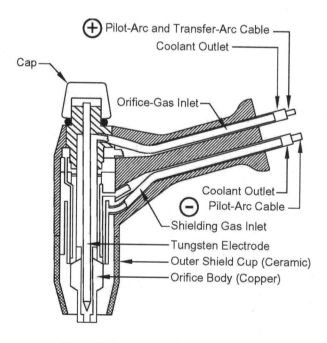

Figure 7–19   Cross section of PAW torch

Process

How does the PAW process work?

The PAW process uses two different gases, one to make the plasma and another to shield the weld pool from the atmosphere. The PAW torch uses plasma, a gas that is heated to an ionized state, as the source of welding heat. It also uses a tungsten or thoriated tungsten electrode. The PAW arc is contained and constricted by an orifice.

See Figures 7–19 to 7–21.

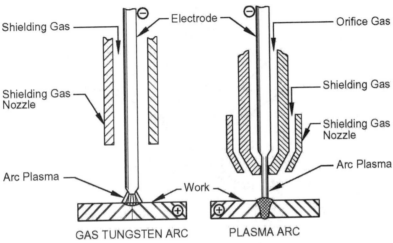

Figure 7–20   Comparison of GTAW and PAW torches

The PAW torch design constricts the arc and forces it through a smaller diameter channel while still carrying the same amount of current. This greatly increases the heat of the plasma to a temperature of about 30,000°F (16,700°C), much higher than the GTAW process. The plasma is formed using argon, argon-helium mixtures or argon-hydrogen mixtures. There are two different modes of operation:

- Non-transferred arc—operates by passing a current between the electrode and the orifice nozzle to excite the inert gases creating plasma. This design is used for plasma spraying, general heating, and the welding of electrically non-conductive materials.
- Transferred arc—operates by establishing an arc between the negative electrode and the positive work (DCEN). The transferred arc has better penetrating ability than either GTAW or the non-transferred arc. It is used for welding with currents from 0.1 to 500 amperes.

Orifice gas flow rate ranges from 0.5 to 10 ft^3/hour (0.24 to 4.8 l/hour) depending on the size of the orifice. The shielding gas is most often argon, but some tasks require helium, argon-helium or argon-hydrogen mixtures. The range of shielding gas flow rate runs from 20 to 60 ft^3/hour (10 to 30 l/hour).

Filler metal is used for most joint designs, but is not needed in flanged-type joints. PAW can be made fully automatic by feeding filler wire into the plasma and driving the torch head around the weld. Arc voltages from 21 to 38 volts are used depending on the material.

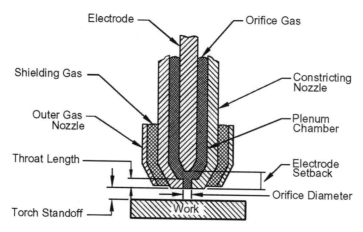

Figure 7–21 Cross sectional view of PAW torch

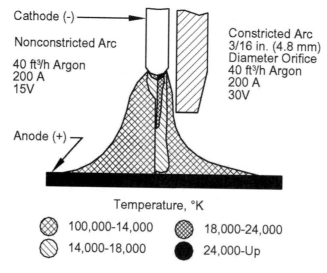

Figure 7–22 Comparison of temperature profiles of non-constricted arc and constricted arc

Applications

What metals can be welded with PAW?
Nearly all metals can be welded including aluminum, precious metals, high and low carbon steel, alloy steels, tool steels and stainless steels.

Advantages

What are the advantages of PAW?
- Welds thicker material in a single pass than GTAW— $1/4$ inch (6.4 mm) stainless steel or $1/2$ inch (12.7 mm) titanium can be single-pass butt-welded
- High heat and heat concentration permit fast travel speeds especially useful in automatic equipment
- Torch-to-work distance is less critical than with GTAW.
- The Plasma arc is more stable than GTAW and can handle greater variation in joint alignment
- PAW produces better penetration and a narrower weld than GTAW.
- Because the tungsten electrode is inside the PAW torch it is less frequently contaminated by accidental contact with the work than GTAW.
- Because of a pilot arc the welder can see the starting point with the welding hood down

Disadvantages

What are the disadvantages of PAW?
The limitations have more to do with the equipment than with the process:
- Torch is delicate
- Water cooling is required with de-ionized water recommended
- Tip and orifice alignment must be maintained
- PAW equipment is more expensive than alternative processes

Keyhole Welding Technique

What is *keyhole* welding?
Molten metal is forced back onto the top of the weld bead by the force of the plasma arc. At the same time, the arc force penetrates the bottom of the weld joint to form a keyhole. As the torch moves along the weld line, metal from in front of the keyhole moves around the plasma stream and then to the rear of the weld. See Figure 7–23.

What are the advantages of keyhole welding?

- Plasma gas flushing through the keyhole helps remove gases that could become trapped in the weld and prevents contamination from the back of the weld
- High quality butt welds can be made in $1/16$ to $3/8$ inch (1.6 to 10 mm) material without edge preparation
- Thicker material can be welded by PAW in fewer passes than by other welding methods
- Since the fusion zone has nearly vertical walls, there is a reduced tendency of transverse distortion (all standard weld preparations except butt welds have sloping sides)

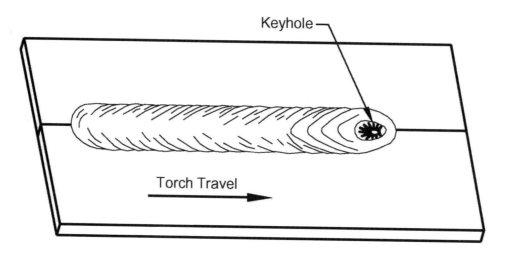

Figure 7–23   Keyhole welding

What are the drawbacks to keyhole welding?

- With more variables and narrower adjustment windows than other processes, PAW can be more difficult to manage
- Except for aluminum alloys, most PAW applications are restricted to the flat position (1G)
- PAW requires more operator skill than other processes with similar capabilities
- Plasma torches require more careful maintenance than those of other processes

Safety

What safety precautions are needed for PAW?

All of the precautions taken with GTAW should be followed with PAW but remember plasma arc welding produces intense noise and hearing protection is required.

Welding Current (A)	Minimum Protective Shade	Suggested Shade for Comfort
>20	6	6-8
20-100	8	10
100-400	10	12
400-800	11	14

Table 7–5 Lens shade selection chart for PAW

Chapter 8

Survey of Other Welding Methods

What we do for ourselves dies with us.
What we do for others and the world remains and is immortal.
Albert Pine

Introduction

Today there are more than one hundred welding processes recognized by the American Welding Society (AWS). The previous chapters covered the most important and most frequently used welding processes. These represent the majority of welding man-hours and filler metal poundage used. However, there are still many specialty welding processes which make modern life possible. This chapter looks at these less common, but still essential processes. Unlike the previous process chapters, this is a survey chapter and does not prepare the reader to use these processes without further instruction. This overview will cover principles of operation, capabilities, applications, advantages and disadvantages.

Electron Beam Welding

How does *electron beam welding* work and what are its applications?
Electron beam welding, AWS designation *EBW*, uses a focused high-energy electron beam to make welds. A heated filament, much like those found in TV picture tubes, provides an electron source. In a vacuum, a high voltage (100 kV) accelerates these electrons away from the filament toward the anode in a vacuum. Most of the electrons do not strike the anode and continue on past it where they are focused into a tight beam by a magnetic lens. The kinetic energy these electrons acquired from being accelerated by the high voltage becomes heat when the electrons strike the work metal. Power densities in the range of 0.1 through 100 MW are possible with temperatures of about 25,000°F

(13,900°C). In some welder designs, the electron beam is moved around on the weld by deflection coils and in others the work is traversed by a computer-controlled X-Y positioner. Depending on the design, welds are made in a vacuum, a partial vacuum, or at atmospheric pressure. The main advantage of EBW is excellent penetration with a very narrow weld width. Depth to width ratios of 20:1 are possible. EBW can join nearly all metals and many combinations of dissimilar metals.

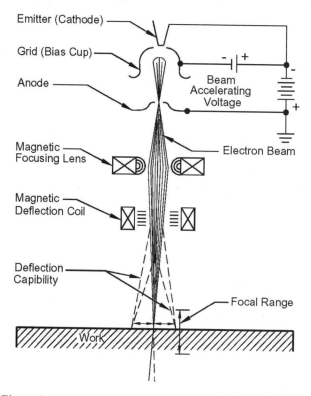

Figure 8–1   Electron beam welding machine schematic.

What are the advantages of EBW?
- Low heat input, minimizing grain growth and distortion
- Excellent penetration
- High welding speed
- High purity, fluxless process yields clean parts for vacuum service
- Shape of weld can be modified during process

What are the drawbacks to EBW?
- Joints must be positioned accurately
- Vacuum may be needed
- Vacuum chamber limits maximum part size
- Dangerous X-rays may be generated during welding
- Costly equipment

Electroslag Welding

How does *electroslag welding* work and what are its applications and characteristics?
Equipment needed to perform electroslag welding includes:
- DC power supply
- One or more electrode wires and guide tubes
- Flux
- One or more wire feeders and oscillators
- Copper retaining shoes (or molds)

Electroslag welding, AWS designation *ESW*, is not really an arc welding process since an arc is only used to start it. Here is how it works:

The process begins by striking an arc between the wire electrodes and the bottom of the joint. Powdered flux added to the joint melts from the arc heat into a liquid slag floating above the arc. Once there is adequate slag, the arc is stopped and the process continues with the heat generated by the flow of electrode current through the molten slag. Wire feeders add metal to the weld pool and an oscillator mechanism moves the electrodes from side to side of the joint to provide even filler metal distribution completing the joint. Water-cooled copper shoes on the joint faces contain the slag and the weld metal. They move up the side of the joint as the weld progresses. This process was developed to join thick sections. Metals up to 30 inches (76 cm) thick can be joined. Multiple solid or flux-cored electrodes may be used to speed the process. Welds are competed in one pass. Excellent penetration is achieved with the extreme heat from the slag.

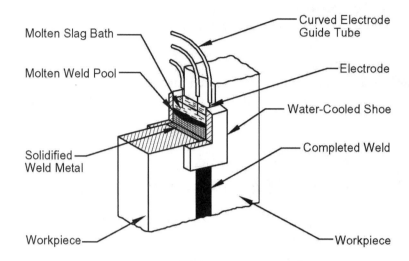

Figure 8–2　Electroslag welding schematic.

What are the advantages of ESW?
- Joins thick materials
- Has high deposition rate
- Multiple electrodes may be used
- Minimum joint preparation
- Low distortion

What are the drawbacks to ESW?
- Flat or vertical joints only
- Very complicated setup
- Cooling water needed for shoes

Friction Welding

How does *friction welding* work and what are its applications and characteristics?

In friction welding, AWS designation is *FRW*, The two parts to be joined are secured in a lathe-like machine. One of the parts remains stationary and the other part rotates. When the rotating part is brought up to speed, the surfaces of the parts to be welded are squeezed together. Friction between the moving and fixed surfaces generates heat, melting the two metals where they meet; when the heating is completed, additional force is applied along the rotating axis. When cooled, the parts are fused together. The process takes about 15 seconds and joins both ferrous and non-ferrous metals. Dissimilar metals may be welded too. FRW is particularly cost effective for securing larger parts onto the ends of shafts. See Figure 8–3.

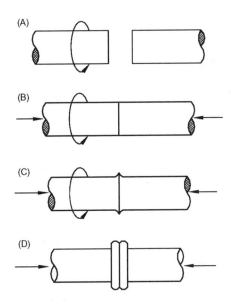

Figure 8–3   Friction welding: (A) One part is rotated, and the other part held fixed, (B) Parts are forced together producing heat, (C) Heat begins to melt the faces of the joint, and (D) Additional force is applied to complete the weld.

What are the advantages of FRW?
- No filler metal needed
- No shielding gas or flux needed
- Surface cleanliness not critical
- Joint as strong as weaker metal
- Narrow heat-affected zone
- Can join dissimilar metals
- Operator needs little manual skill
- Process easily automated

What are the drawbacks to FRW?
- One workpiece must have a symmetrical axis and be capable of rotating
- Alignment of workpieces is critical
- Equipment is costly
- Dry bearing and non-forgeable materials cannot be welded
- Free-machining alloys are difficult to weld
- Workpiece size limited by machine

Friction Stir Welding

What is the AWS designation for Friction stir welding?
The American Welding Society's acronym for Friction Stir Welding is *FSW.*

How does FSW work?
When Friction Stir Welding a cylindrical shouldered tool, with a profiled threaded/unthreaded probe (nib or pin) is rotated at a constant speed and fed at a constant traverse rate slowly plunged into a joint line between two pieces of sheet or plate material, which has been butted together so the faying edges touch. The parts must be clamped rigidly to a backing bar so the abutting joint faces are not forced apart. Frictional heat is generated between the wear resistant welding tool and the stirred material of the work-pieces. Heat causes the work-pieces to soften without reaching the melting point (the material is plasticized) and allow traversing the tool along the weld line. The plasticized material is transferred from the leading edge of the tool to the trailing edge of the tool probe and is forged by the contact of the tool shoulder and the pin profile. The process leaves a solid-phase bond between the two pieces. The process can be regarded as a solid-phase keyhole welding technique because a hole, to accommodate the probe, is generated then filled during the welding sequence. The welding of the material is facilitated by severe plastic deformation in the solid-state involving dynamic re-crystallization of the base material. See Figure 8–4.

What application for FSW?
Friction Stir Welding is primarily used in the joining of aluminum including the seams of liquid rocket booster tanks on the space shuttle, expendable launch vehicles, amphibious assault ships and wings of aircraft. Experiments have been done with welding of steel but essential problems occur with the tool material wearing rapidly, therefore depositing tool debris in the weld.

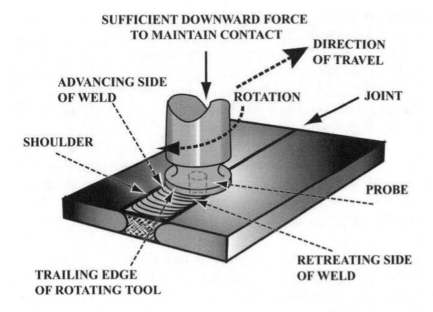

Figure 8-4 Schematic of friction stir welding

What advantages does FSW have over conventional welding?

Because the material does not melt when Friction Stir Welding, many of the problems we associate with fusion welding are eliminated; since cooling from the liquid phase is avoided, there is no problem with porosity, solidification cracking, and liquid cracking .

Other advantages over fusion welding include:

- Good mechanical properties in the as welded condition
- Improved safety due to the absence of toxic fumes
- Elimination of spatter
- No consumables- conventional tool steel can weld over 3250' (1000m) of aluminum and no filler or gas shielding is required for aluminum
- A simple milling machine can easily be automated for FSW requiring less set-up cost and less training
- Friction stir welding may be done in all welding positions because there is no weld pool
- The finished weld generally has a good weld appearance with minimal thickness over/under matching which reduces the need for machining after welding

What are the disadvantages of FSW?
Disadvantages of Friction Stir Welding include:
- An exit hole is left when the tool is withdrawn
- Heavy down forces required for this process means heavy duty clamping is necessary to hold the plates together
- The process is less flexible than manual arc processes, there is difficulty with thickness variations and non-linear welds
- Often the traverse rate is slower than other fusion welding techniques but this may be offset if fewer welding passes are required
- Less efficient than conventional arc welding when welding steels and stainless steels because the tool materials wear rapidly and debris from the tool can frequently be found inside the weld

What are important variables to consider in FSW?
There are two tool speeds to be considered in Friction Stir Welding; the speed the tool rotates and how quickly it traverses the interface. These two parameters are very important and must be chosen with care to ensure a successful and efficient welding cycle. Welding speed and heat input during the welding is complex but increasing the rotation speed or decreasing the traverse speed will result in a hotter weld. To produce a successful weld requires the material surrounding the tool be hot enough to enable the plastic flow and minimize the forces action on the tool. If the material is too cool, voids or other flaws may be present in the stir zone, in extreme cases the stir tool may break.

At the other end of spectrum too much heat input may be a detriment to the final properties of the weld; this could even result in defects due to the liquation of low-melting phases. Figure 8–5 shows a cutaway of a successful Friction Stir Welds thermo-mechanical weld zone and heat affected zone

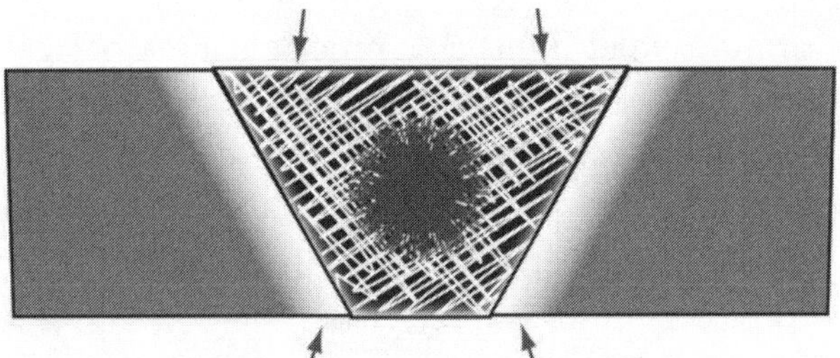

THERMOMECHANICALLY
AFFECTED ZONE

HEAT-AFFECTED ZONE

Figure 8–5   A friction stir weld cutaway showing the thermo-mechanical weld
and heat affected zones

Laser Beam Welding

How do lasers work?

Lasers draw their energy from a high-voltage power supply. The principle is simple:

1. Power supply energy excites a group of molecules of the same type and they absorb power supply energy. When these molecules release their stored energy an instant later, it is released as light with the characteristic frequency of that particular molecule. This cycle of excitation, storage, and release takes place in a tube of carbon dioxide, or a specially doped (tinted) glass-like rod.

2. The critical step in making a laser is to insert an optically resonant (tuned) circuit inside the mass of excited molecules, tuned to the frequency of the light of the laser's excited gas or rod molecules. This resonant circuit works in a similar way to an organ pipe. The organ pipe uses a stream of air to excite its resonance and generate a note. Similarly, the laser uses high voltage to exite its resonance and produce light. A pair of mirrors at each end of the cylinder of excited gas or rod forms this tuned circuit. One mirror reflects all the light striking it, while the other mirror reflects most of the light striking it, but lets a little (less than 1%) leak out. This forms the tuned optical circuit and the light leaking from one mirror, the output of the laser. This tuned circuit gets the gas (or rod) molecules to absorb and release their energy at the *same* time and at the *same* frequency. This makes laser light.

How does *laser beam welding* work and what are its applications?
The AWS designation for Laser Beam Welding is *LBW;* LBW uses laser energy (visible light or ultraviolet light) as a precise source of heat to melt materials and fuse them together. Laser welding works not because welding lasers are so powerful, but because their energy is highly concentrated in an area. See Figure 8–6.

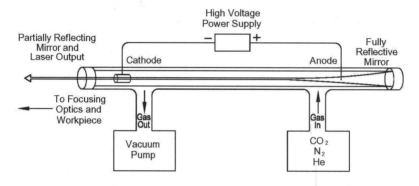

Figure 8–6   Carbon dioxide laser schematic

A combination of mirrors and lenses focuses and delivers laser energy to the weld. Most LBW is performed under computer control. Either the work is moved on an x-y (or x-y-z) computer driven stage, such as a CNC machine tool or the beam is manipulated and the work is held still.

What are the advantages of LBW?
- Minimum heat input to make the weld limits distortion and size of the heat-affected zone
- Single pass welds to 1¼ inches (32 mm) thick are possible
- No filler metal is used, no fumes or contamination
- Laser permits working around obstructions on the part
- Weld size is small, so closely-spaced parts can be welded
- Non-contact process, no tooling wear, and limited clamping distortion
- Many materials and combinations of materials can be welded
- Welds on thin material and thin wires are possible without burn-back

What are LBW drawbacks?
- Joints much greater than 0.75 inches (19 mm) depth are not considered practical for production applications
- Joints must be accurately positioned under the laser beam

- Lasers are usually less than 10% efficient in converting electric energy to laser energy
- With the rapid solidification characteristics of LBW, some porosity and brittleness occurs
- Aluminum's and copper's high conductivity and reflectivity limits their weldability

Spot & Seam Welding

How does *spot welding* work and what are its applications and characteristics?

Spot welding, AWS designation *RSW*, is a resistance welding process. The coalescense of metals is produced at the weld interface (*faying surfaces*) by the heat of an electric current passing through the electrical resistance of the work metal itself. Force applied across the weld before, during, and after the heating current is applied confines the heat by limiting the size of the weld contact area. It also holds the weld surfaces together while they solidify. See Figures 8–7 and 8–8 (left).

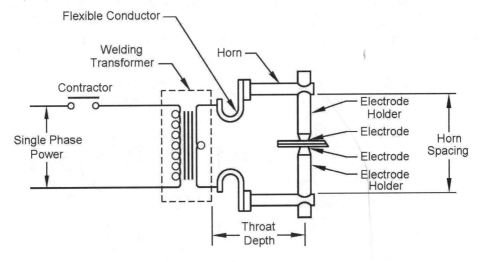

Figure 8–7   Resistance spot welding schematic

RSW can be performed one weld at a time as the welder actuates his machine as in sheet metal fabrication for heating ducts all the way up to a fully robotic welding machine which perform many welds in a predetermined sequence under computer control as in auto manufacturing.

What are RSW's advantages?
- High speed
- Adaptability to automation
- Faster than arc welding or brazing
- Low operator skill needed
- Economical in many applications

What are the drawbacks to RSW?
- Spot welds have low tensile and fatigue strength.
- Full strength of the sheet cannot be used because the loading around the spot welds is eccentric causing stress concentrations.
- Inspection is difficult.
- Workpiece surface and electrodes must be clean.
- Shop process, cannot be used in the field.

How does *resistance seam welding* work and what are its applications and characteristics?

Resistance seam welding, AWS designation *RSEW*, is similar to RSW but makes a continuous weld of overlapping weld nuggets. This is done using a pair of wheels above and below the overlapping workpieces to apply the spot welding current. The seam is gas or liquid-tight. See Figure 8–7.

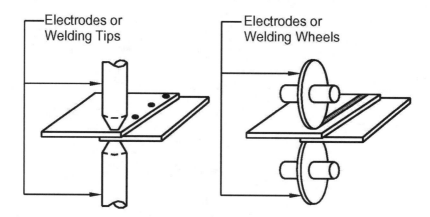

Figure 8–8   Spot welding (left) and seam welding (right)

Stud Welding

How does *stud welding* work and what are its applications?

Arc stud welding is designated by the AWS by the acronym *SW;* stud welding is used to rapidly attach studs and other fasteners to a structure's surface without piercing its metal. The power source may be DC constant-current supply or a capacitor-discharge supply. In a capacitor discharge power supply, a relatively small DC power supply slowly charges a large capacitor and the energy stored in this capacitor supplies the welding energy. The SW process begins by inserting a stud into the stud gun, placing a ceramic cup or ferrule on the end of the stud, and positioning the gun perpendicular to the surface of the structure. This break-away ceramic cup helps contain the arc heat and shields the weld metal from the atmosphere. When the gun trigger is squeezed, an arc forms between the stud and the structure metal. The control electronics determines how long this arc runs to insure molten metal on both the stud and the structure. It then releases the stud using spring pressure in the gun to force the molten surfaces together. A great variety of part shapes are available. When stud welding aluminum, a shielding gas is needed. See Figure 8–9.

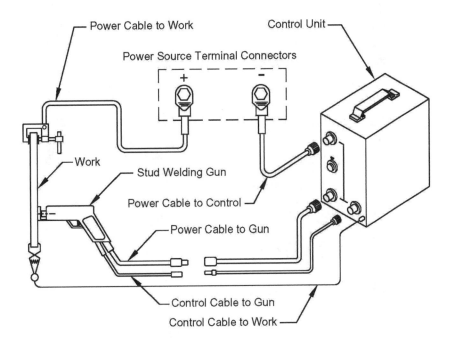

Figure 8–9   Stud welding outfit.

What type of fixtures can be stud welded?
Many types of fixtures can be stud welded including:
- Threaded bolts
- Concrete anchors
- Eye bolts
- Shear connectors

See Figure 8-9A

What are the advantages of SW?
- Simple
- Fast
- Can be automated

What are the drawbacks to SW?
- Needs clean surfaces
- Equipment sensitive to adjustment

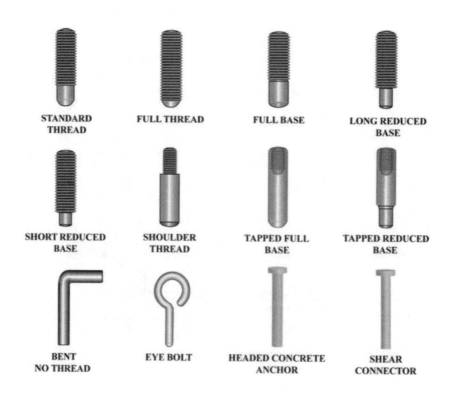

| STANDARD THREAD | FULL THREAD | FULL BASE | LONG REDUCED BASE |

| SHORT REDUCED BASE | SHOULDER THREAD | TAPPED FULL BASE | TAPPED REDUCED BASE |

| BENT NO THREAD | EYE BOLT | HEADED CONCRETE ANCHOR | SHEAR CONNECTOR |

Figure 8–9A   Shows examples of fixtures which may be stud welded

Submerged Arc Welding

How does *submerged arc welding* work and what are it applications?
Submerged arc welding, AWS designation *SAW*, uses the arc from a consumable
electrode while the arc and base metal are buried in a granular flux. The power
supply may be either constant-current or constant-voltage. There are hand-held
semiautomatic SAW machines, but the typical SAW process uses a motor-driv-
en carriage. The machine automatically strikes an arc, feeds the electrode from
large coils, distributes the flux ahead of the weld area, and finally has provisions
to gather and re-use the unmelted flux. SAW typically uses 5/64 to 7/32 inch
(2 to 5.5 mm) diameter electrodes with a current of about 1000 amperes per
electrode. Two electrodes may be used in side-by-side vertical, leading-and-
following vertical or V-shape across the line of weld making the process very
fast. Some machines use three electrodes. Single pass welds on butt joints up
to 3 inches (75 mm) thick are possible. Some flux granules melt in the arc
forming an easily removed slag which protects the weld from the atmosphere
and slows cooling. The slag also shields the operator from the arc. SAW is usually done in the flat welding position. Deposition rates approaching 200 lb/hour
(90 kg/hour) are possible with two-electrode automatic machines. This process
is ideal for ship and bridge building where long straight welds are common.
See Figure 8–10.

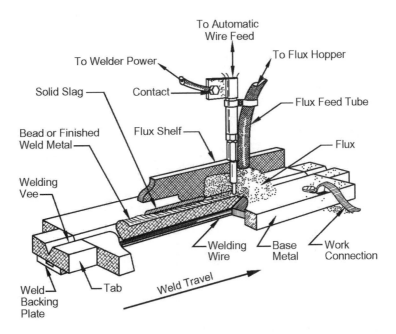

Figure 8–10   Submerged arc welding

What are the advantages of SAW?
- High deposition rate
- Deep penetration
- Mechanized process, but some equipment is hand-held
- Good for overlay of large areas
- High operator appeal as no eye protection needed

What are the drawbacks to SAW?
- Flat or horizontal fillets only
- Long setup time
- More difficult to monitor with arc out of view
- Slag removal needed

Chapter 9

Controlling Distortion

Wise men learn from other men's mistakes,
fools by their own.

H.G. Bohn

Introduction

Since welding processes expose the workpiece to high temperatures, weld-induced distortion is always present. We will investigate how distortion affects workpieces and present several methods to reduce its effects. These methods are often simple, but without them many workpieces would be ruined.

How Expansion & Contraction Cause Distortion

What happens to the dimensions of a metal cube evenly heated to a temperature below its melting point and then allowed to cool gradually back to room temperature?

Because no metal in the cube is restrained, as the temperature rises it expands at the same amount in all three dimensions. Increases in length are proportional to temperature. The metal cube returns to it's preheat dimensions when it cools to room temperature. Regardless of its shape, a metal object will return to its original dimensions on returning to room temperature if it is not restrained and gradually heated and cooled. See Figure 9–1.

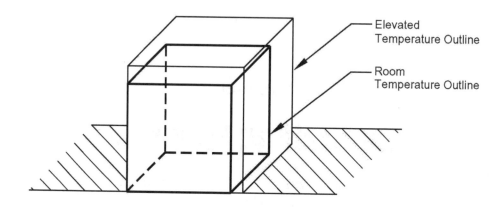

Figure 9–1   Exaggerated three-dimensional changes in unrestrained metal cube
on flat surface

What happens to the dimensions of a solid metal cube restrained in one dimension (as between the jaws of a very rigid vice) when evenly heated from room temperature to a temperature below its melting point and then allowed to cool back to room temperature?

As the temperature rises, the metal expands and the volume of the cube increases. See Figure 9–2. Since the cube cannot move horizontally (the X direction) due to the restraints of the immovable vice jaws at either end, the cube must expand *more* in the Y and Z dimensions. The amount of expansion in the Y and Z directions is greater than the elastic properties of the metal. When yield strength is exceeded, the metal has plastic and permanent flow. We say the metal of the cube has been *upset*. On returning to room temperature it is permanently fatter in the unrestrained dimension and permanently shorter in the restrained dimension.

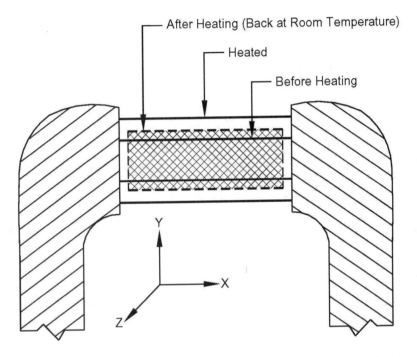

Figure 9–2   Exaggerated three-dimensional changes in a metal cylinder
restrained in one dimension

What happens to the dimensions of a metal cylinder *partially* restrained by adding a spring between the metal cylinder and the vise jaw and then temperature cycled as described in the question above?

Partial restraint produces dimensional changes between the no-restraint case and the complete-restraint case. After temperature cycling the bar is fatter and shorter, but does not show as much dimensional change as in the complete-restraint case. See Figure 9–3.

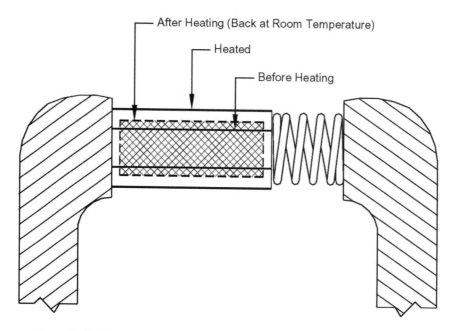

Figure 9–3   Exaggerated three-dimensional changes in metal solid with
partial restraint from spring in one dimension

What effect will raising the maximum temperature have on the size of the permanent changes in dimension seen in Figures 9–2 and 9–3 and why?
Most carbon steels rapidly loose strength above 600°F (316°C). Tensile strength falls to 30 to 40 percent of its room temperature value at 1100°F (593°C). Although the steel is far below its melting point (2750°F or 1510°C), it is greatly weakened. For this reason, the higher the peak temperatures experienced by the steel under stress by expansion forces, the greater the deformation. Also, since expansion is proportional to temperature, higher temperatures cause more expansion. For these reasons, the effects of exposure to high temperatures seen by the metal during welding can be severe and cause complications in making a satisfactory part.

What is the result of heating one edge of a piece of sheet metal with a torch to a red heat?
Before it is heated the sheet metal is square and flat. As one edge of the plate heats, it expands and softens while the cool edge does not. When the sheet cools back to room temperature, most wrinkles disappear, but the sheet is permanently shorter along the once-heated edge. Uneven heating of the plate and the restraint

offered by its cooler side cause a dimensional change called *upsetting*. Upsetting will be even more severe if a water spray cools the sheet rapidly and prevents the unheated edge from heating and expanding as the heat flows across the plate. Note that what happens during the cooling period is seldom what happens during the heating period. See Figure 9–4.

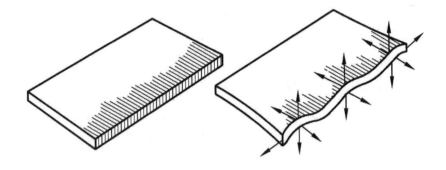

Figure 9–4   Sheet metal before heating (left) and while hot (right)

What happens to the shape of a plate after a partial cut is made in it with OFC?

Because of differential heating and the restraint offered by the uncut portion of the plate, the *hinge effect* occurs, see Figure 9–5. If two parallel cuts were made simultaneously, the metal between the cuts would show little distortion, as heating and expansion would be balanced. Dual torches are available to make such cuts.

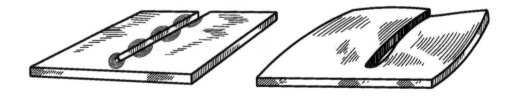

Figure 9–5   Hinge effect in partially cut plate

How can the term *distortion* be summarized?

Distortion is the permanent change in shape and dimension of metal caused by expansion and contraction. Certain welding and cutting processes are sources of distortion. It is a result of *residual stress* left by uneven heating that causes permanent shape changes. This stress not only can ruin the shape of a part, it can weaken it too. Even through the part bears no external load, the residual stress acts as an initial load on top of what is externally imposed and reduces the total load the part can withstand.

Weld Bead Distortion

Based on the distortion examples above, what will happen to a long, straight steel bar when a single weld bead is laid lengthwise along its side? See Figure 9–6. As the bead is applied, one side of the bar is heated and expands. When the bar and the filler metal on top of it cool, they contract much more than the cooler metal on the bar's opposite side. Applying the weld bead makes the bar bend toward the side of the weld bead.

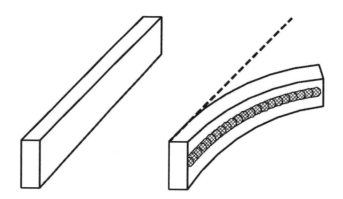

Figure 9–6   Affect of applying a weld bead on one side of a steel bar.
Straight bar, no bead applied (left). Bar bends toward bead side
with weld bead applied when cool (right)

What are the residual stresses and shape changes caused by welding in a V-groove butt weld?

There are residual stresses both along the weld axis (longitudinally) and across the weld (transversely), Figure 9–7 (left). When cooled, the weldment has permanent deformation away from the side where heat was applied along the weld line. Because this is a V-groove butt joint that has more filler metal at its top than at its root, there is more shrinkage along the top of the joint in both directions. This makes the plate dish or bend upwards, Figure 9–7 (right).

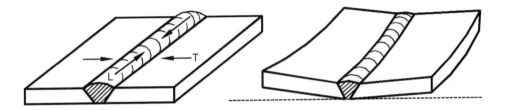

Figure 9–7   Butt weld longitudinal, L, and transverse, T, residual stresses (left)
and resulting deformation (right)

What residual stresses occur in a T-joint and what deformation do they produce?

See Figure 9–8 (left) showing both longitudinal and transverse stresses in the weld bead. There is a second weld bead on the back side of the T-joint. Because the longitudinal stresses on each side of the joint balance each other, the vertical member of the T-joint remains straight. See Figure 9–8 (right) for the distortion these residual stresses cause.

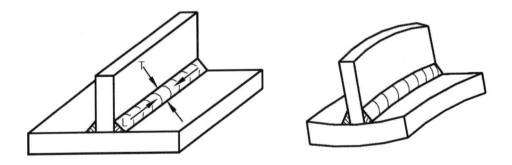

Figure 9–8   Longitudinal and transverse stresses in a T-joint (left) and
the distortion they cause (right)

Controlling Distortion

How can we eliminate the effects of distortion from welding?
There are several steps that can reduce its effect, but we can never completely
eliminate distortion.

**What is a simple way to reduce the effects of distortion that pulls a butt
joint or V-groove joint out of alignment?**
First, preset the parts. Then tack-weld the parts slightly out of position and let
residual forces bring them into proper position. See Figure 9–9 showing how a
T-joint and V-groove joint are handled.

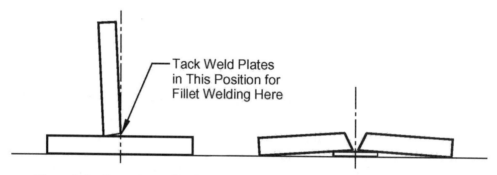

Figure 9–9   Presetting and tack welding work out of position lets weld shrinkage
bring parts back into alignment

How can distortion effects be reduced when making V-grooved welds?
By clamping, the use of restraints and wedges hold the weld joint in proper
position until the weld metal cools. This approach may not produce perfect
results, but it will help reduce distortion. See Figure 9–10.

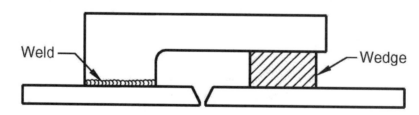

Figure 9–10   Using tack-welded restraints and wedges to hold joint in position
while welding

How can we limit heat flow from the weld joint to limit distortion?
Use chill bars. Chill bars consist of steel or copper bars clamped beside and
parallel to the weld bead. They draw heat away from the weld and reduce its
flow to the rest of the part. They also limit distortion to upsetting metal close
to the weld line and eliminate ripples completely by exerting a clamping force
which prevents ripples from forming when the work is hot, see Figure 9–11. A
groove in the lower chill bar permits the weld itself to remain hot and not have
its heat drained away by the chill bar. This groove could be flooded with shielding gas for GTAW.

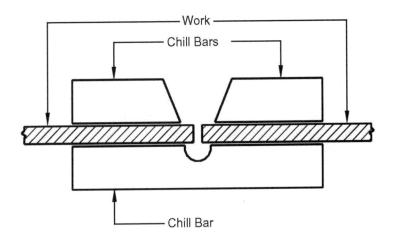

Figure 9–11   Chill bars reduce distortion by confining heat to the weld area and
by preventing work from forming ripples when hot

Excellent part alignment is needed. How can this be accomplished?
Use prestressing. Use clamps to bend the joint members in opposite direction
to the weld forces and let weld-shrinkage forces bring the parts back into position. This method works well when a jig or fixture can be used and test runs are
made to determine the amount of prestress needed. See Figure 9–12.

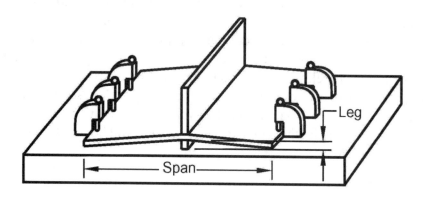

Figure 9–12   Pre-stressing weld joint to compensate for residual stress

How can distortion forces be used to reduce distortion?

Use equal distortion forces to balance each other by using two (or more) weld beads. This could be done by putting a fillet weld on both sides of a T-joint or using a double V-groove butt joint. See Figure 9–13.

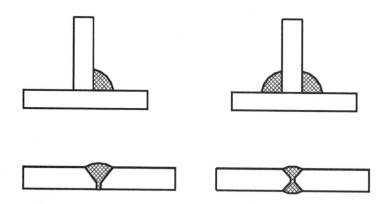

Figure 9–13   Initial joint designs (left) and balanced force designs (right)

How can we reduce total weld distortion forces as well as balance them?

Use chain intermittent or staggered intermittent weld beads. Intermittent beads not only balance one another, but also by reducing the total amount of weld bead, reduce total residual force, see Figure 9–14. Even a single intermittent weld bead will have less distortion than a single continuous weld bead and often the strength of a continuous bead is not needed.

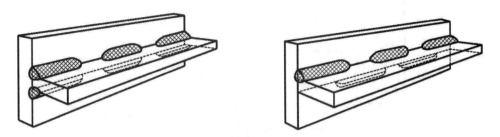

Figure 9–14   Using chain intermittent (left) or staggered intermittent welds (right)
to balance forces and reduce total weld bead metal

How can joints be redesigned to reduce distortion?

Use a V-groove and a fillet weld in place of a fillet weld alone to balance residual stress. See Figure 9–15.

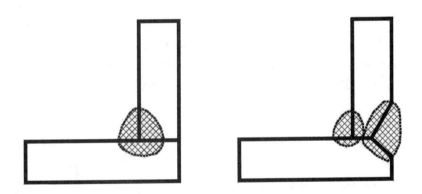

Figure 9–15   Redesigned joint can balance residual stress and reduce distortion

What method can be used to reduce the distortion in a long continuous bead?

Use back-step welding. Apply short increments of beads in the direction opposite of the end point of the weld. When applying multiple passes, start and stop the beads of each layer at different points. See Figure 9–16.

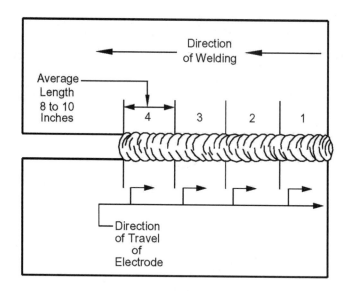

Figure 9–16   Back-step welding sequence

When making a long continuous bead, weld shrinkage forces the plates together reducing the weld root. What should be done?
The use of wedges ahead of the weld will control joint spacing. See Figure 9–17.

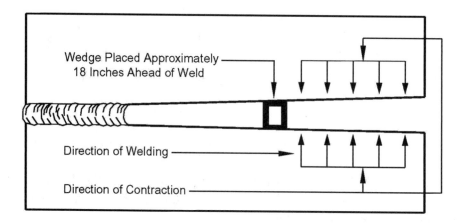

Figure 9–17   Control joint spacing with wedges

What are some other ways to reduce distortion effects?

- Preheating base metal—raising the temperature of the entire part before welding reduces temperature differences, residual stress, and distortion.
- Peening—hammering the weld metal, usually with an air hammer, slightly reshapes the metal and redistributes concentrated forces. In a multipass weld this is done between each pass. This method can be helpful, but depends on the skill and judgment of the welder; peening consistency is difficult to control.
- Stress relieving heat treatment—using an oven or electric heating coil, the entire part or the weldment area is heated high enough to remove weld-induced stress. This is commonly done in structural steel work.
- Brazing or soldering instead of welding—since brazing and soldering expose the workpiece to much lower temperatures than welding, these two processes can be used when the strength of welding is not required.

Chapter 10

Welding Symbols

Facts do not cease to exist because they are ignored.
Aldous Huxley

Introduction

Welding symbols graphically convey complete information about a weld or braze joint. This is accomplished in a standardized and abbreviated picture format which makes easily used in every country around the world. The complete symbol system is described in *ANSI/AWS A2.4, Standard Symbols for Welding, Brazing, and Nondestructive Testing*, latest edition, published by American Welding Society. Everything the welder, architect, inspector, engineer, supervisor, foreman and construction superintendent need to know about a weld is described on a welding symbol.

Symbol Information

What information is included on a welding symbol?
- Joint type
- Weld preparation shape
- Weld type
- Welding process
- Specifications or procedures
- Weld location
- Extent of welding
- Quality requirements and method of testing
- Weld sequencing
- Weld size
- Final weld configuration
- Production methods

Symbol Elements

What are the parts of a welding symbol?

A welding symbol's parts may include the following elements:

- Reference line
- Arrow
- Basic weld symbol
- Dimensions and other data
- Supplementary symbols
- Finish symbols
- Tail
- Specifications, process, or other references

Reference Line

What is a reference line?

The reference line is a *required part of a completed welding symbol* it consists of a line drawn horizontally with all of the other required information drawn on or around it. It must be placed on the drawing near the joint the attached weld symbol describes. Each of the other welding symbol elements must be placed in there proper location about the reference line, and according to the symbol standards. See the horizontal line in Figure 10–1.

Arrow

Where is the arrow placed?

The arrow is the other required part of a welding symbol and is placed at one or the other end of the reference line and comes to an arrowhead on the *arrow side* of the joint. It can point in any direction—up, down, or toward the tail. A welding symbol may even have multiple arrows.

What are the standard locations of these elements in a welding symbol?

Figure 10–1 shows a reference line and some of the other elements that may be placed on or around it. Typical weld symbols are not this complex and usually use just a few elements.

This weld symbol shows where the various pieces of information describing the weld are located:

- The *T* in the tail is where information goes that applies to this particular weld only, such as change in welding process, change in electrode, and may reference detail on the drawing. The tail may be omitted if there is no reference or specification needed.

- The *S* on the reference line can indicate, depending on the weld type:
  - Depth of preparation for a groove weld.
  - Size of a fillet weld.
  - Size of a plug or slot weld.
  - Shear strength for a spot or projection weld; this number is always to the left of the weld symbol.
- The *E* is where the *effective throat* size, or *weld size*, of a groove weld is indicated. Effective size dimensions go inside parenthesis for a groove weld, if this was a fillet weld symbol the leg sizes would appear without the parenthesis. Size and preparation is always to the left of the weld symbol on the reference line no matter where the arrow is located.

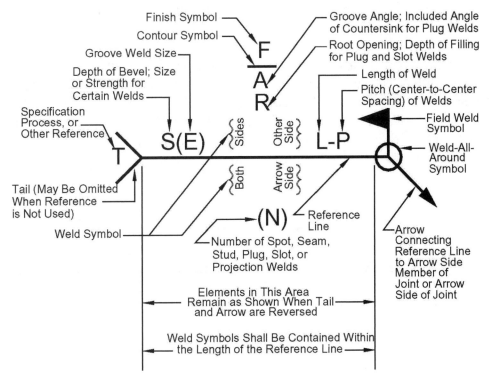

Figure 10–1   Location of welding symbol elements

- The *R* is where spacing between weld members is dimensioned. This is the root opening for a butt joint. If the weld is a plug or slot weld, this is where the depth of filling indication goes. This number should be placed within the symbol.

- The *A* indicates the groove angle (included angle) for a butt joint, or the included angle of countersink for countersunk plug welds.
- The horizontal line segment (—) between the *F* and the *A* is the location of the contour symbol for the finished weld.
- The *F* is where the method of obtaining the required contour is indicated; contours may be obtained by grinding (G), machining (M), chipping (C), hammering (H), rolling (R), or unspecified (U).
- The *L* is where the weld length is indicated; this length indication is always to the right of the weld symbol. This location is the same no matter where the arrow is located.
- The *P* is where the center-to-center (pitch) spacing of a weld is located when welding is intermittent and spacing of the welds is indicated.
- The *(N)* is the number of spot, seam, stud, plug, slot, or projection welds require

How is the arrow related to where the weld is to be placed?

The term *arrow side* of the reference line refers to the side of the weld joint that the arrow points to. Symbols placed on the *arrow side* of the reference line (above the reference line) refer to the arrow side of the joint. Symbols placed on the *other side* of the reference line (below the reference line) refer to the other side of the joint. See Figure 10–2 for examples.

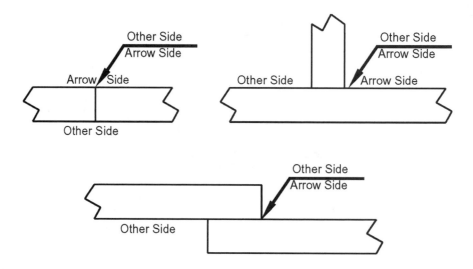

Figure 10–2   Examples of *arrow side* and *other side*

Basic Symbols

What are the basic weld symbols?

See Figure 10–3 shows basic groove and other weld symbols.

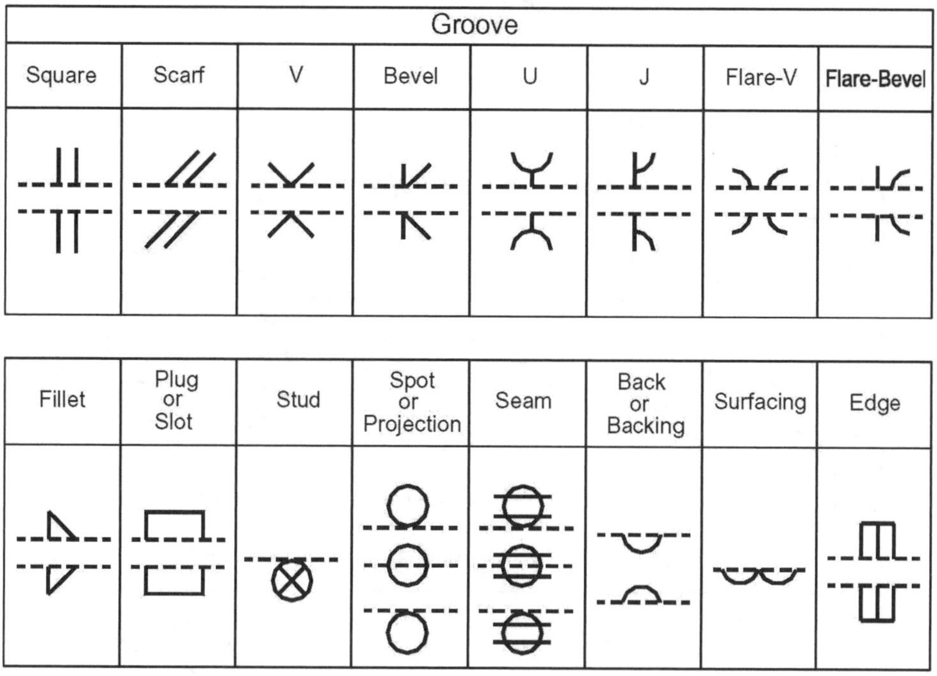

Groove							
Square	Scarf	V	Bevel	U	J	Flare-V	Flare-Bevel

Fillet	Plug or Slot	Stud	Spot or Projection	Seam	Back or Backing	Surfacing	Edge

Figure 10–3   Basic weld symbols

Now that we have defined *arrow side* and *other side*, how do we place weld symbols on the reference line?

See Figure 10–4. Note that if just one side of the joint is welded, one weld symbol for that type weld goes on one side of the reference line matching the proper side of the joint.

How would the welding symbols in Figure 10–4 be described?

- Frame one is a V-groove weld on the arrow side with the detail showing where the weld will be placed
- Frame two shows a V-groove weld on the other side with the detail showing where the weld is to be placed
- Frame three shows a double-V-groove weld and the detail shows welding both sides of the joint.

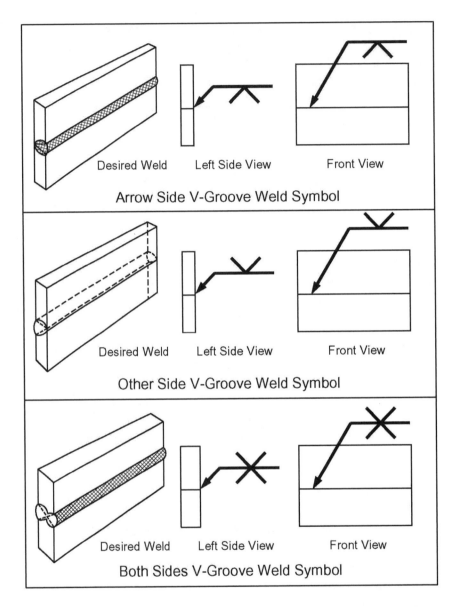

Figure 10–4   Weld symbol placement above and below the reference line
relates to arrow side, other side or both sides of the joint

Now that we know the basic welding symbols and arrow significance, what other information can be found on the welding symbol? See Figure 10–5 for examples of root openings.

- Frame one shows a welding symbol with a square-edged-groove weld symbol and the fraction $1/16$ of an in the root space which indicates the spacing between the two members (root opening) is $1/16$ of an inch the detail to the left shows the joint with the weld depicted in grey
- Frame two shows a double-bevel-groove weld with a root opening of $3/16$ of an inch and like frame one the detail shows how the weld looks
- Frame three again shows a double-bevel-groove weld with a root opening of $1/8$" but now it is in a tee joint configuration as depicted in the detail
- Frame four shows a double-V-groove weld with the number three placed in the V; the number three tells us this measurement is a metric measurement of 3 mm. Prints will be in either standard or metric numbers and should not be mixed so we must assume if this millimeter number is shown on this welding symbol all of the measurements on this print will be in metric numbers. Three millimeters is approximately $1/8$" root opening and the detail shows the joint as welded
- Frame five shows a V-groove weld on the arrow side with root space of $1/8$ of an inch and the detail of the weld is to the left of the welding symbol
- Frame six shows a U-groove weld on the other side and the number 0 appears in the "U" meaning there is no root space the members are brought tight together at the root as depicted in the detail on the left

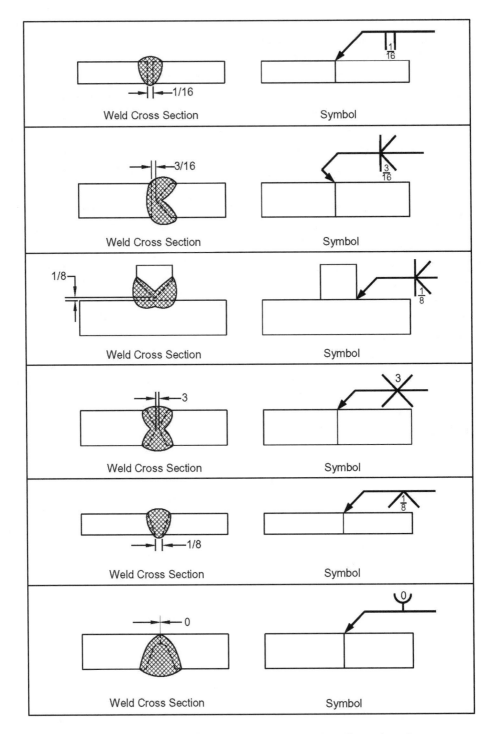

Figure 10–5   Groove weld symbols with root opening dimension placement

How is the flare-bevel-groove weld symbol used?

A flare-bevel-groove weld symbol is a weld made joining round to flat material. Note the first number "1" reading, from left to right, is a larger number than the number in parenthesis. The joint preparation number "1" is the joint preparation number measured from the center-point of the of the round stock to the outside surface of the round stock, but the weld size ⅝" is the weld size because it is not possible to make a weld nugget which will penetrate to the center-point of the round material thus the weld nugget is less than the weld preparation number. See Figure 10–6.

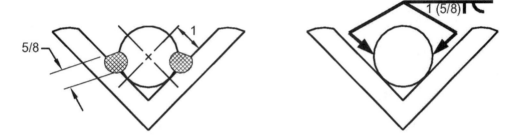

Figure 10–6   Depicts two single-flair-bevel groove welds made with round material placed inside angle iron to be welded on each side thus two arrows it also could be shown as a double-flare-bevel-groove weld if the weld symbol had been place on both sides of the reference line and only one arrow used but the two arrows leaves no question as to where the welds are to be placed as depicted in the detail

How is an edge symbol used on a flange joint?

This symbol for an edge weld is shown in Figure 10–7 this Figure shows an edge weld being made on a flange joint; the arrow is pointing to where the weld is to be placed. See Figure 10–7.

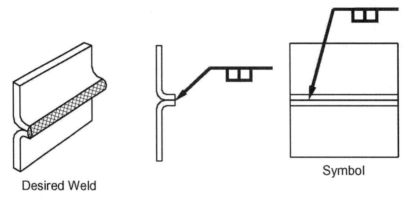

Desired Weld   Symbol

Figure 10–7   Example of arrow-side-edge-flange weld symbol

A Break in the Shaft of the Arrow

What does a break in the arrow mean?

A break in the arrow shaft indicates that only one side of the joint will be prepared for welding. The arrow points to the side of the material which must be prepared before the welding begins. See Figure 10–8.

- Frame one shows a double-bevel-groove weld with the degree of bevel angle placed outside the weld symbol, on both sides at 45° the arrow is broken and pointing to the member on the right side of the joint telling us the material on the right will be prepared with the 45° angle on both sides of the joint; when the prepared member is in place the bevel angle now is also the groove angle as depicted in the detail
- Frame two shows a V groove weld on the arrow side with a 50° groove angle meaning each bevel angle will be prepared at a 25° angle as depicted in the cross section detail
- Frame three is a double-V-groove weld with a 60° groove angle on the arrow side and a 90° groove angle on the other side
- Frame four shows a J-groove weld on the arrow side with a 20° groove angle with a broken arrow pointing to the member to be prepared for welding as depicted in the cross section detail
- Frame five shows a double-J-groove weld with a groove angle of 20° on the arrow side and 30° on the other side and the broken arrow shows which member is to be prepared as depicted in the cross section

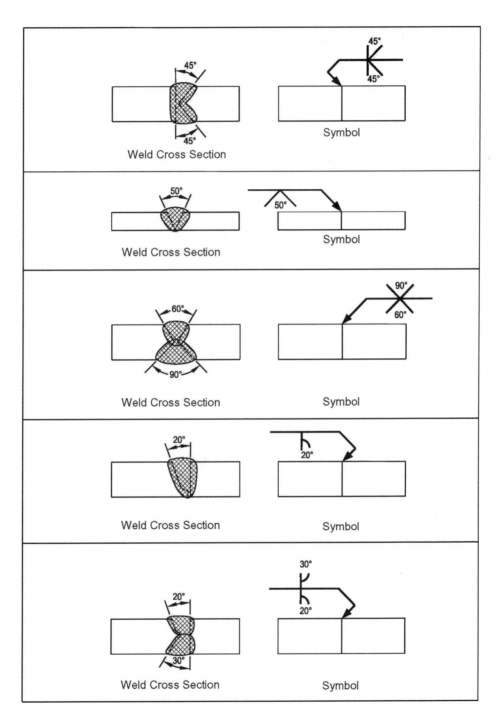

Figure 10–8   Depicted above is the break in the arrow shaft which points to
the side of joint which will be prepared prior to welding the joint

Combined Weld Symbols

What are combined weld symbols?

It is common to find weld joints requiring more than one type of weld or in the case of a double-groove welds, where welds are placed on both sides of a member, or in structural fabrication groove welds are often finished with another kind of weld such as a fillet weld. When this occurs you will find weld symbols on both sides of the reference line. When we have multiple symbols the weld are made working away from the reference line. See Figure 10–9.

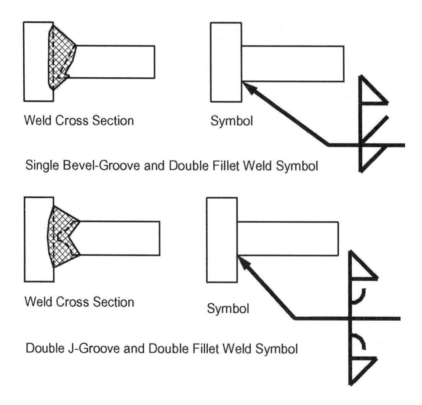

Weld Cross Section   Symbol

Single Bevel-Groove and Double Fillet Weld Symbol

Weld Cross Section   Symbol

Double J-Groove and Double Fillet Weld Symbol

Figure 10–9   Shows joints with more than one type weld; the top picture is a bevel-groove weld on the other side followed by a fillet weld on the other side and a fillet weld on the arrow side; the bottom welding symbol depicts a double-J-groove weld on both sides followed by fillet welds on both sides

Multiple Reference Lines

What are multiple reference lines and how are they used and read?

Multiple reference lines show the sequence of welding operations. The first operation is shown on the reference closest to the arrow and will be completed before continuing to the next line. Each line, moving away from the arrow, gives information about each successive operation; these operations may include other information about the welding such as inspection methods and sequence of testing. See Figure 10–10.

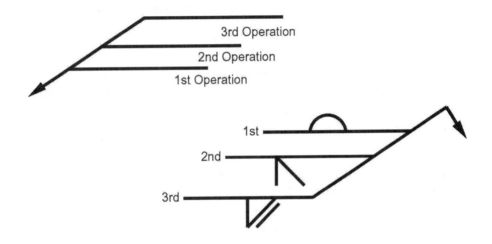

Figure 10–10   Multiple shows reference lines describing operational sequencing

Supplementary Symbols

What are supplementary symbols?

Supplementary symbols detail important information about the weld and are combined with other weld symbols. See Figure 10–11.

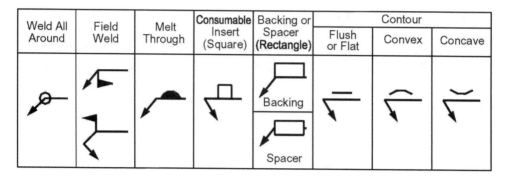

Figure 10–11   Supplementary weld symbols

What does the *all around* symbol mean?

The all around symbol is a circle around the junction of the reference line and the arrow. This symbol indicates a weld that is to be made continuously around a joint, even though there may be a change in welding direction. See Figures 10–12.

- Frame one is a fillet weld on the arrow side with the all around circle on the break between the arrow and the reference line meaning the weld will go all the way around the joint as seen in the details to the right
- Frame two is again a fillet weld on the arrow side with the all around symbol at the break between the arrow and the reference line and the joint detail shows the weld goes all the way around the joint
- Frame three shows a bevel-groove weld prepared to 1/2 the diameter of the round member welded all around followed by a fillet weld
- Frame four shows a double-bevel-groove weld as the first sequence of welding followed by a fillet weld all around the joint

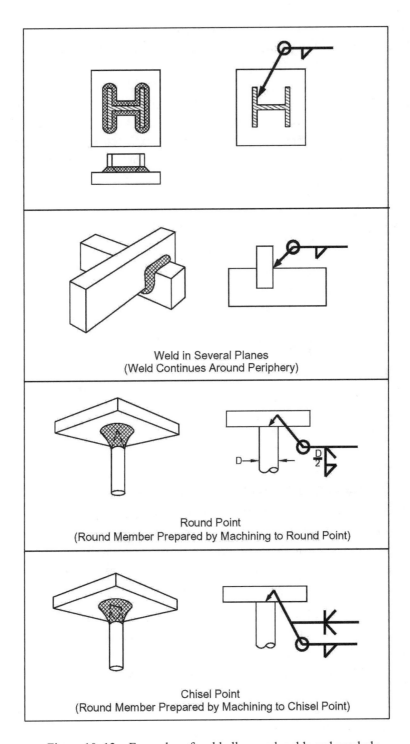

Weld in Several Planes
(Weld Continues Around Periphery)

Round Point
(Round Member Prepared by Machining to Round Point)

Chisel Point
(Round Member Prepared by Machining to Chisel Point)

Figure 10–12   Examples of weld all around welds and symbols

What does the field weld symbol indicate?

The field weld symbol is a flag at the junction of the arrow and the reference line. The flag indicates welds that require welding in the field as opposed to being pre-fabricated in a shop. Some parts are pre-fabricated in a shop prior to being taken to the field for installation and the prints may show all of the welding required so the flag shows all parties involved what must be welded in the field instead of in the fabrication shop. The flag is filled in black and may be facing any direction or be right side up or upside down. See Figure 10–13.

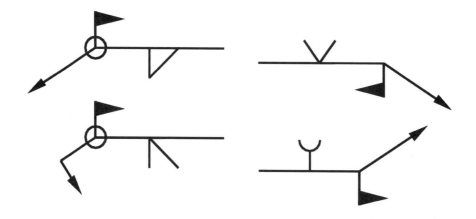

Figure 10–13   Welding symbols with the field weld flag

What does the melt through symbol represent?

The melt through symbol appears as a darkened back half moon on the other side of a groove weld symbol. The melt through symbol indicates welding from one side and melting through to the other side of the joint. This supplementary symbol may include (to the left of the symbol) a dimension indicating the amount of melt through required. See Figure 10–14.

- The first frame in Figure 10–4 shows a square-edged-groove weld symbol on the arrow side of the welding symbol with a melt-through of 1/16" on the other side detailed to the left
- The second frame welding symbol shows is a bevel-groove weld on the arrow side with a 1/16" melt-through to the other side
- The third frame welding symbol shows a V-groove weld on the arrow side with a melt through of 1/8" to the other side

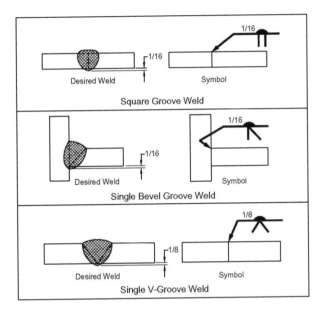

Figure 10–14   Melt through symbol examples

What does the backing symbol indicate?

The rectangular backing symbol indicates the placing of material on the back side of a groove weld. The symbol appears to be similar to a plug weld symbol, except it is placed on the opposite side of a groove weld symbol. Another difference is that the letter *R* may be placed within the symbol to indicate that material is to be removed after welding. The material specified by the backing symbol could be the same as the parent metal, copper, ceramic, glass tape, flux, gas, and other materials. The type of material required must be noted in the tail or in the structural requirement notes on the print. See Figure 10–15.

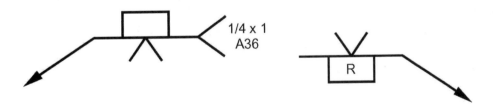

Figure 10–15   Backing material symbols: remains in place (left) and removed after welding as indicated by the "R" placed in the symbol (right)

What is the spacer symbol for?

The spacer symbol, a rectangle, placed straddling the reference line indicates materials placed within the joint and is called a *spacer* or *insert*. Like the backing material symbol, specific details about the spacer may be noted in the tail or in the structural notes on the print. See Figure 10–16 which shows information about this weld joint and should be read as; a double V groove weld prepared at a 20° groove angle (10° on each bevel angle), with a 1/2" depth of preparation on both sides, effective throat of 5/8" on each side of 1 1/4" material (making this a full penetration groove weld) with a spacer made of SAE 1010 steel dimensioned at 1/4" x 1/2" with the half inch dimension being the spacing between the weld members.

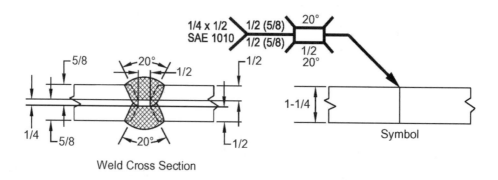

Figure 10–16   Spacer symbol indicating size and placement in the weld joint

What is the consumable insert symbol for?

The consumable insert symbol, a square placed on the reference line, indicates strips or rings of filler material placed in the weld joint. The insert or ring is completely fused into the joint. Consumable inserts are usually welded with the gas tungsten arc welding process (although other process may also call for inserts). This symbol may be placed on the opposite side of the groove weld symbol or as above straddling the reference line. The consumable insert class is placed in the tail of the welding symbol or in the notes. Figure 10–17 is read (top symbol) V groove weld on the arrow side with a class 3 consumable insert; (below symbol) U groove on the arrow side with a class 2 consumable insert with a convex contour. See Figure 10–17.

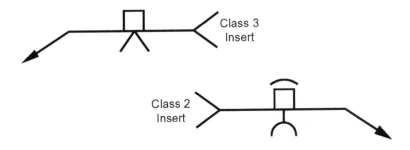

Figure 10–17   Consumable welding insert symbols

What does the contour symbol mean?

Contour symbols describe the desired shape of the completed weld. These contours include flush, convex, and concave. A letter above the contour symbol indicates the method of finishing:

- *C* for chipping
- *G* for grinding
- *H* for hammering
- *M* for machining
- *R* for rolling

The symbols correspond to the actual contour required. See Figure 10–18.

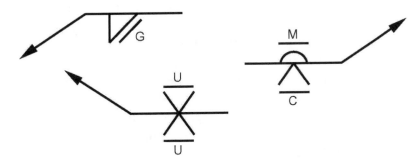

Figure 10–18   Finish contours

How would the welding symbols in Figure 10–18 be read?

The figures above would be read as:

- Top figure is a fillet weld on the arrow side with a ground flush contour
- The middle symbol is a V groove weld on the arrow side with a back or

backing on the other side with the groove weld chipped flush and the back or backing machined flush
- The bottom is a double V groove weld with a flush contour on both sides with the method of contouring being unspecified

What is the difference between a *backing weld* and a *back weld*?
Backing welds are made to the opposite side of a groove weld *before* the groove weld is applied, while back welds are made *after* the groove weld is completed. Back gouging or grinding is required before back welding to ensure complete penetration of weld metal through the joint.

What symbols are used for a *backing weld* and *back weld*?
The same symbol represents both these welds. The order of the welding can be specified in the tail of the welding symbol, or by using multiple reference lines. The backing weld symbol goes on the first line away from the arrow followed by the groove weld symbol on the next line. The symbol always goes on the opposite side of the groove weld symbol. The back weld symbol is placed on the second line from the arrow preceded by the groove weld symbol. See Figure 10–19.

How are the symbols in Figure 10–19 read?
- The first frame is a V groove weld on the arrow side with a back weld on the other side; the only difference between the two symbols is the use of a multiple reference line showing the welding sequence rather than showing the back weld information in the tail
- The second frame should be read a V groove weld on the arrow side with a backing weld on the other side; the only difference between the two symbols, again, is the use of multiple reference lines to show the welding sequence instead of using the tail
- Third frame shows the same weld as the second frame with the only difference being the groove weld preparation dimension 3/4", effective throat dimension 7/8" (weld size) and root spacing 1/8"

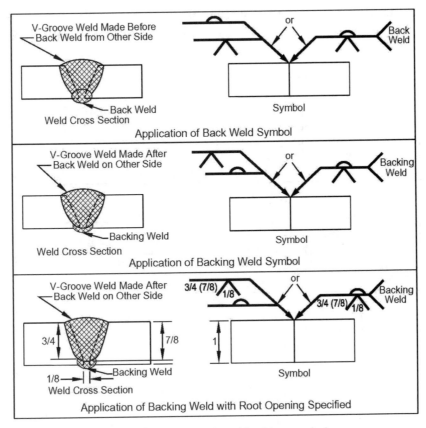

Figure 10–19   Back and backing symbols

Fillet Weld Size

What determines the size of a fillet weld?

The shortest leg of the fillet weld determines the fillet weld size. See Figure 10–20.

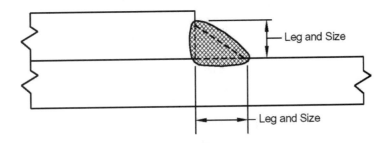

Figure 10–20   Fillet welds of unequal leg size

How are the dimensions of fillet welds shown on the welding symbols?
The leg size is shown on the left side of the fillet weld symbol. See Figure
10–21.

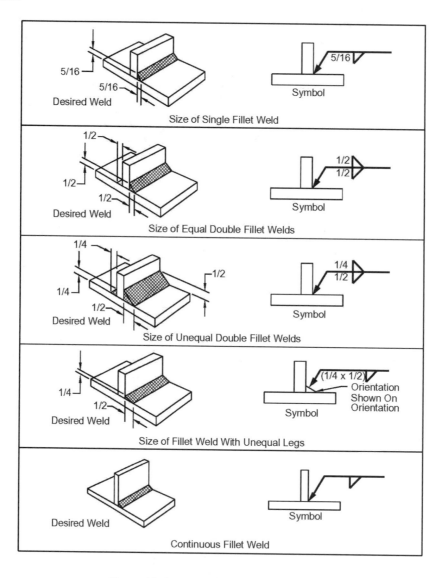

Figure 10–21   Fillet weld measurements

How would the above symbols be read?
- Frame one is a $5/16$" fillet weld on the arrow side
- Frame two is a $1/2$" double fillet weld or $1/2$" fillet weld on both sides

- Frame three is a double fillet weld with a 1/2" fillet weld on the arrow side and a 1/4" fillet weld on the other side
- Frame four is a fillet weld with unequal leg sizes of a 1/4" leg size on the vertical member and a 1/2" leg size on the horizontal member as detailed on the drawing
- Frame five shows a fillet weld with no dimensions but made from one abrupt end of the weld member to the other abrupt end of the member to be welded

What determines the length of a fillet weld?
A fillet weld, unless otherwise dimensioned, is a continuous weld from one abrupt end of the weld member to the other abrupt end. If the weld is not continuous, a single number to the right of the fillet weld symbol tells the welder the length of the weld. See Figure 10–22.

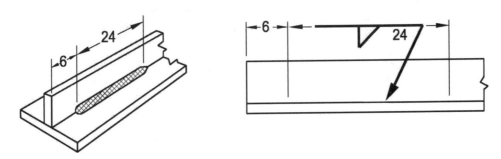

Figure 10–22   Fillet weld length measurements and weld symbol; the length of a fillet weld is always placed to the right of the weld symbol as shown and detailed

Intermittent Welds

What are intermittent welds and what symbols are used to specify them?
An intermittent weld is one that is *not* run continuously from one abrupt end of a joint to the other abrupt end. Intermittent welds are short and spaced. There are two types of intermittent welds, chain intermittent welds and staggered intermittent welds. A *chain intermittent* weld consists of a series of welds made intermittently on both sides of a joint with each increment of weld bead approximately opposite those on the other side of the joint. A *staggered intermittent* weld consists of a series of welds made intermittently on both sides of a joint in which the weld bead increments on one side are alternated with respect to

those on the other side. Refer to Chapter 4, Figure 4–15 for drawings of intermittent chain and staggered fillet welds. See Figure 10–23 to see how these welds are used on welding symbols.

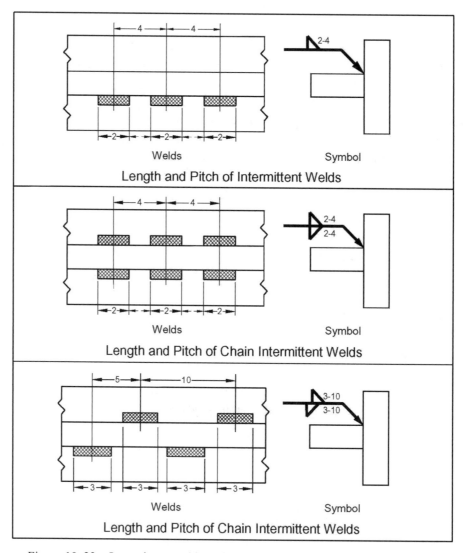

Figure 10–23   Intermittent welds and measurements in a joint (left) and the welding symbols with dimensions (right)

How to read the above welding symbols:

- Frame one is an intermittent fillet weld on the other side the weld is 2" long with a 4" pitch (or spacing between welds)

- Frame two is a chain intermittent fillet weld with 2" welds on 4" pitch so the welds are the same on each side and spaced one across from the other
- Frame three is a staggered intermittent fillet weld (note the fillet weld symbols are also staggered) with the weld being 3" long with a 10" space between the welds on the same side (note the staggering from one side to the other side is 5")

Plug Welds

How are plug welds specified by welding symbols?
Plug welds are specified by weld size, the angle of countersink, and the depth of weld filling. See Figure 10–24.

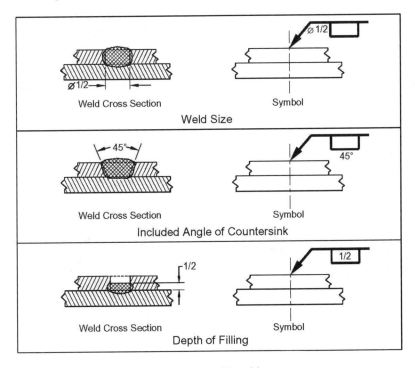

Figure 10–24   Plug welds with measurements

How should the above plug weld symbols read?
- Frame one is a 1/2" plug weld (note the symbol Ø depicting a round hole); the plug weld size is measured at the bottom of the hole where the two members to welded come together (the faying surface)

- Frame two is a plug weld with a 45° number outside the plug weld symbol showing a countersink of 45°
- Frame three is a plug weld with a depth of filling as 1/2 way the structural code states: *The depth of filling of plug or slot welds in metal 5/8 in. [16 mm] thick or less shall be equal to the thickness of the material. In metal over 5/8 in. [16 mm] thick, it shall be at least one-half the thickness of the material, but no less than 5/8 in. [16 mm]*

Slot Welds

What are slot welds?
Slot welds have all the conventions of plug welds except for the Ø symbol is omitted when drawing a slot weld and rather than being round they are elongated holes rounded at the ends having a length and width. See Figure 10–25.

How is the size of slot welds determined?
Slot welds are measured by the width of the slot at the faying surfaces of the members to be joined; the width is the size of the slot weld and on the welding symbol the measurement is to the left of the slot weld symbol. Length measurement of a slot weld is shown on the welding symbol to the right of the weld symbol and spacing or pitch to the right of the length measurement. Plug and slot welds are also discussed in Chapter 4.

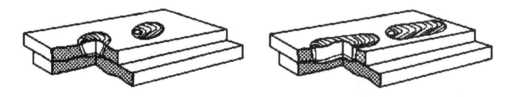

Figure 10–25   Show a plug weld (on the left) and a slot weld (on the right)

Can a single pass be made at the bottom of a plug or slot weld?
Yes, you may make a single pass at the bottom of a plug or slot weld, but that weld is a fillet in a plug or slot weld and not to be called a plug or slot weld. Figure 10–26 shows the symbol to show this.

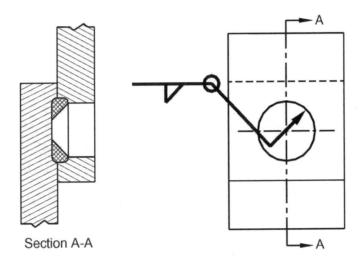

Section A-A

Figure 10–26   Fillet weld in a plug: cross-section (left) and welding symbol (right)
the break in the arrow mean nothing in this case

Surfacing Welds

What are surfacing welds?
Surfacing welds are those applied to the workpiece surface to improve abrasion resistance, increase surface hardness, or replace worn-away metal. Applying filler metal to improve surface hardness or abrasion resistance is called *hard facing*. This is often done to blades and buckets of earthmoving equipment as well as rock crushing equipment or anything used to grind or crush material where the grinding wears away the surface of the metal used to grind, crush, move or dig.

How are surfacing welds measured and shown in weld symbols?
Surfacing welds are measured from the surface of the base metal to the face of the weld bead through the throat, or the height of the weld filler material. Chapter 4 Figures 4–18A and 4–18B show plate and bar surfacing. See Figure 10–27 for dimensions.

What do the frames in Figure 10–27 show about surfacing welds?
• Frame one shows a welding symbol with a ¹/₈" surfacing weld symbol on the edge of a plate with a detail drawing showing where one makes the measurement of the weld

- Frame two shows a welding symbol with a $1/8"$ surfacing weld symbol on a round bar or round material (pipe) the detail drawing shows the weld measurement the weld and the extent of welding

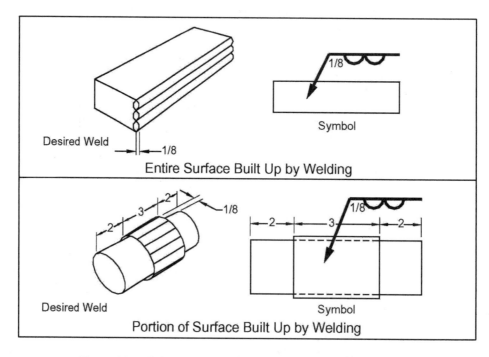

Figure 10–27   Surfacing weld measurements and their symbols

Chapter 11

Inspection & Discontinuities

He that can have patience can have what he will.
Benjamin Franklin

Introduction

Whenever the failure of a weld could lead to financial loss, property damage, personal injury, or death, it is important to know that it will meet its design requirements. These may be set by the designer, fabricator, customer, or law depending on the application. In order to assure compliance with these requirements, a quality control program is needed. A quality control program includes training and qualification of welders, incoming materials testing, record keeping, and weld inspection. This chapter focuses on how welds are inspected and what discontinuities and defects are common in fusion welding processes.

Discontinuities and Defects

How is a discontinuity defined?
In welding a discontinuity is any interruption in the uniform nature of a weld:
- Seams and laps
- Laminations
- Weld undercut
- Cracks (usually considered defects)
- Weld porosity
- Non-metallic inclusions
- Hydrogen inclusions
- Many others

How is a defect defined?
A defect is a non-conforming discontinuity, meaning all discontinuities are not defects but all defects are discontinuities.

What is the difference between a *weld discontinuity* and a *weld defect*?
A discontinuity is some interruption of the weld's typical structure, a location in the weld that is not like the usual parts of the weld. It can be a change in mechanical, metallurgical, or physical characteristics of the weld and its immediately surrounding base metal. A discontinuity only becomes a defect when the size of one discontinuity or the number of separate discontinuities (because of their cumulative effect) prevents the weld from meeting the applicable acceptance standard. A defect makes a weld rejectable. A weld with discontinuities may be acceptable under one code, but be rejectable under another more strict code.

Why are we concerned with discontinuities?
Every discontinuity causes a stress concentration within the weld. Stress levels at concentration points may be many times the stress level of the surrounding material. While a stress concentration may not cause immediate weldment failure, in time stress concentrations lead to cracks—localized points of failure. Cracks tend to grow larger with time and can ultimately lead to complete and catastrophic weld failure.

Why would we want to reject the same weld for one application and accept it for another?
The type and level of stress on the same type weld in different applications varies widely. Obviously, the more lightly stressed weld can tolerate more discontinuities before failing. For economy, the level of weld quality must match the consequences of its failure. It makes little sense spending a lot money on testing a weld whose failure would cost much less than testing cost. Testing extremes vary from testing every weld with several non-destructive methods to testing just a small sample of the total welds. Similarly, a statistical sampling of critical welds may be destructively tested while unimportant welds may never receive regular destructive testing.

Welding Inspectors

How does one become a welding inspector?
Using the AWS designation QC-1 *Certified Welding Inspector* (CWI) as an example, a welding inspector:
- Shall be a high school graduate or have and equivalency diploma
- Shall have, as a minimum requirement, no fewer than five years, verifiable, experience in an occupational function that has a direct relation-

ship to welded assemblies fabricated to a national or international standard and be directly involved in either design, production, construction, examination or repair; there are alternative requirements including education which are also acceptable refer to the *AWS publication QC-1*

- Visual acuity necessary to examine welds as shown by an eye examination
- Each prospective inspector must pass a written examination covering his or her understanding of welding fundamentals, welding codes, and the ability to do hands on inspection

What items does the inspector examine?

The inspector examines every aspect of the welding process to insure the welds meet specifications. Such items might include:

- Reviewing welding plans and drawings
- Witnessing welding procedure qualification and procedure qualification tests
- Verifying welder qualification and certification documents
- Checking the conformance of base metals and welding consumables to specifications
- Verifying welding equipment is appropriate for the process specified
- Checking that welding is performed in accordance with welding procedure qualification, drawings, and specifications
- Performing examinations during the welding and on the completed welds to assure weld conformance to specifications
- Maintaining records of materials, welders, test examination results, and even testing welders

Inspection Methods

How are welds examined?

There are both destructive methods (refer to Chapter 14 which covers Qualification & Certification) and non-destructive methods for testing the quality of welds. We will focus on nondestructive examination and nondestructive testing methods of inspection in this chapter.

How do we define nondestructive examination?

Nondestructive examination (NDE) is the act of determining the suitability of some material or component for its intended purpose utilizing methods which will not have an adversary affect on the serviceability of the finished product.

What is the definition of nondestructive test?
Nondestructive testing (NDT) and nondestructive inspection (NDI) are terms sometimes used interchangeably with nondestructive examination and are considered synonymous. Nondestructive testing and nondestructive inspection are both nonstandard terms for nondestructive examination.

What are some of these tests and how do they work?
The first and foremost method of examination is the visual test; passing a visual test is required before any other testing may continue. Codes set the minimum standard requirements for passing the visual examination.

Visual Test (VT) is the simplest and the first NDE step in examining a weld relying on the detection of acceptability using the human eye often improved by utilizing magnifying glasses to increase the ability of the test method to detect discontinuities. In the case of a welding inspection it is performed before welding, during welding, through all phases of the process and after the welding is complete.

What are the inspection procedures before welding begins?
1. Reviews documents related the welding requirements
2. Review welding procedures and specifications
3. Review welder qualifications
4. Prepare a plan for performing the inspection and recording methods
5. Develop a system for identifying rejects
6. Check the welding equipment for service suitability
7. Check the acceptability of the base and filler materials
8. Verify proper fit-up and alignment of members to be welded
9. Check weld joint quality, accuracy and preparation
10. Check weld zone for cleanliness prior to welding
11. Verify pre-heat post-heat requirements

What are the inspection procedures during welding?
1. Evaluation of the welding technique for compliance with welding procedures specifications
2. Inspection of individual weld passes
3. Check and verify inter-pass cleaning
4. Check and verify inter-pass temperature (when required)
5. Check the placement and sequencing of intermediate welds
6. Verify adequacy of back gouging when required
7. Check for discontinuities in intermediate passes of multiple pass welds

What does an inspector do after welding is complete?
1. Check surface appearance of the completed weld
2. Check the weld for dimensional accuracy
3. Check the weld length and pitch if required
4. Monitor post-weld treatment if required

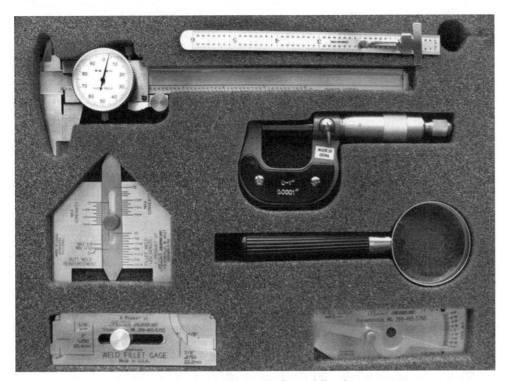

Figure 11–1   Inspection tools for welding inspectors

How does the inspector verify the aforementioned?
The only essential element for visual inspection is white light which may be natural or artificial. Other equipment which may be employed to improve the accuracy, repeatability, reliability and efficiency of the inspection include measuring devices such as:
1. Fillet weld gauges
2. Rulers and tape measures
3. Squares and angle gauges
4. Undercut and reinforcement gauges

See Figure 11–1.

Penetrant testing (PT) reveals surface discontinuities like cracks, undercut, and porosity by bleed-out of a penetrating medium. Penetrant liquid is applied to a cleaned surface of the test piece (by dipping, spraying, or brushing). The penetrant is allowed to remain on the surface for a prescribed time (dwell time); during this time the penetrant will be drawn into any surface openings. Subsequent removal of excess penetrant (by wiping) and application of a developer draws remaining penetrant from discontinuities. The results are indications shown in high contrast and intensify the presence of discontinuities so they may be visually interpreted. See Figure 11–2.

The weld is examined under white light in the case of visible dye (usually red) which produces a vivid red indication against a white developer background when viewed. Fluorescent penetrants produce a greenish, fluorescent indication against a light background when using an ultraviolet light. Since the human eye can more readily perceive fluorescent rather than white light indications, fluorescent penetrants results are a more sensitive test.

Are there different categories of penetrants?
Yes there are different categories referring to the method of removing the penetrant from the test surface. They are classified as either water washable, solvent removable or post-emulsifiable.

How is the water washable penetrant removed?
Water washable contains an emulsifier which allows the oily penetrant to be rinsed off with a low pressure water spray.

How is the solvent removable penetrant removed?
Solvent removable penetrants require a special solvent to remove the dye from the test object; the spray cans usually have specific directions. Solvents must be thoroughly removed before any welding is resumed.

What steps must be taken to remove penetrant with emulsifiers?
Post-emulsifiable penetrants are removed by adding an emulsifier after the dwell time; the application of the emulsifier tends to neutralize the penetrant above the test surface allowing it to be removed with water in a similar manner as the water washable type.

How many types of penetrants are there?
There are six classifications of penetrants and they are:
1. Visible/Water-Washable
2. Visible/Solvent Removable
3. Visible/Post-Emulsifiable
4. Fluorescent/Water-Washable
5. Fluorescent/Solvent Removable
6. Fluorescent/Post-Emulsifiable

Figure 11–2   Examples of typical cleaner-remover, penetrant and developer
spray cans used in dye penetrant examination

What steps must be taken to perform this examination?
1. The test object must be cleaned and free of oil, dirt, rust, paint or anything which will block access to the surface of the test object. Care must be taken when cleaning soft surfaces such as copper or aluminum when using mechanical cleaning devises such as brushes or blasting; aggressive cleaning on theses materials may cause the metal to smear the covering existing surface discontinuities preventing discovery of flaws.

2. Apply the penetrant, small pieces may be dipped in a penetrant bath, larger parts may be sprayed or brushed. Dwell time depends on the manufacturer's recommendations. Capillary action draws the dye into the surface discontinuities.
3. Following the dwell time the penetrant is removed from the surface of the test object and thoroughly cleaned; take care to clean sufficiently to prevent the occurrence of excessive background and other non-relevant indications which could mask real discontinuities. Caution must be taken so the cleaning is not so aggressive as to wash penetrant out of shallow discontinuities.

Magnetic particle testing (MT) detects most surface and some near-surface discontinuities but works locating cracks. MT works only on ferromagnetic materials. Finely divided magnetic particles are applied to the test area, then the test specimen is magnetized either by passing an electric current through it, or by using an external magnetizing coil. If a crack is present, the edges of crack become north-south magnetic poles, attracting magnetic particles, and revealing any cracks by turning the magnetic particles. See Figure 11–3.

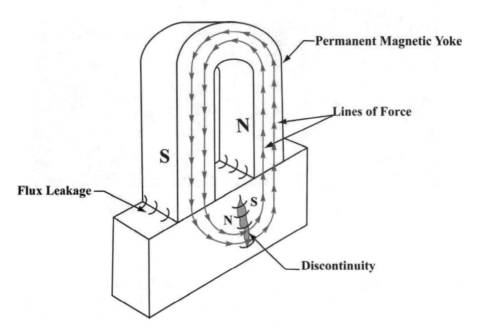

Figure 11–3   U-shaped magnet in contact with ferromagnetic material containing a discontinuity

What can we tell about the magnet in Figure 11–3 and the discontinuity?

By using this U-shaped magnet we are generating a magnetic field in the test piece. There are magnetic lines of force traveling in continuous loops from one pole to the other pole. If magnetic iron particles are sprinkled on the test surface between the legs of the U-shaped magnet these particles will be held in place by this magnet and produce a visual indication of the magnetic field.

The sprinkled-on-particles will be lines up parallel to the magnet loop from pole to pole unless there is a discontinuity. The discontinuity will create an air gap and disturb the magnetic loop, this is called a magnetic flux leak, causing the iron particles between the poles, in the area across the discontinuity, to change their direction and run perpendicular to the north to south poles.

Is magnetic particle testing always done with a U-shaped magnet?

No, most magnetic particle testing is performed using some type of electro-magnet. Electromagnets depend on the principle that there is a magnetic field associated with any electrical conductor.

How does the electromagnet work?

Electricity passing through a conductor, the test material, the magnetic field being developed is oriented perpendicular to the direction of the electricity. There are two general types of magnetic fields created in test objects using electromagnets and their names refer to the direction of the magnetic field generated in the part. Magnetic fields oriented along the axis of the part are referred to as longitudinal magnetism. See Figure 11–4. When the direction of the magnetic field is perpendicular to the axis of the part, it is called circular magnetism. See Figure 11–5.

How does longitudinal magnet particle testing work?

A longitudinal magnetic field created by surrounding the part with a coiled electrical conductor, referred to as a "coil shot" with electricity passing through the conductor a magnetic field is created as shown in Figure 11–4. Flaws lying perpendicular to the lines of force will be easily revealed; those lying at 45° to the magnetic field will also be shown but if the flaw lies parallel to the lines of magnetic flow it will not show.

How does the circular magnetic particles testing work?

A circular magnetic field is created by making the part to be tested an electrical conductor so the indicated magnetic field tends to surround the part perpendicular to the longitudinal axis. On a stationary testing machine this method is referred to as a "head shot." During circular magnetism longitudinal flaws will be revealed but those lying across the long axis of the part will not show while flaws at approximately 45° will show.

What is the main difference between the longitudinal and circular magnetic particle testing?

In circular magnetism the magnetic field is contained within the ferromagnetic material whereas longitudinal magnetic field testing is induced in the part by electric conductor around it. Because of this, the circular magnetic field is generally considered somewhat more powerful making circular magnetism more sensitive for a given amount of electric current.

Are there more portable magnetic particle testing tools?

Yes, one of these is a magnetic yoke which induces a longitudinal field as shown in Figure 11–6. A yoke unit is an electromagnet which uses a winding or coil around a soft magnetic material core. Current flows through the coil inducing a magnetic field which then flows across the test object between the ends of the yoke.

Are there other portable magnetic particle testing tools?

A circular magnetic field unit is the prod technique which uses either alternating or direct current to induce a magnetic field. The magnetic field created by using alternating current is strongest at the surface of the test piece. Alternating current will also provide greater particle mobility on the surface of the part allowing them to move about freely locating flux leakage even on rough and irregular surfaces.

The direct current prod induces magnet fields which have greater penetrating power and can detect near surface discontinuities. A third type of electric current is half wave rectified alternating current which combines both alternating and direct current with this type of power, benefits of both alternating and direct current can be realized. See Figure 11–7.

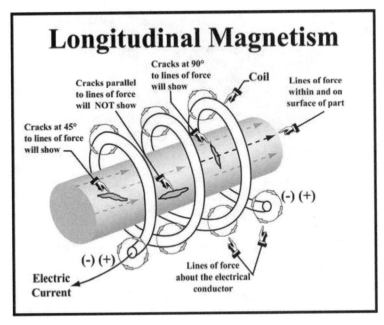

Figure 11–4   Longitudinal magnetism

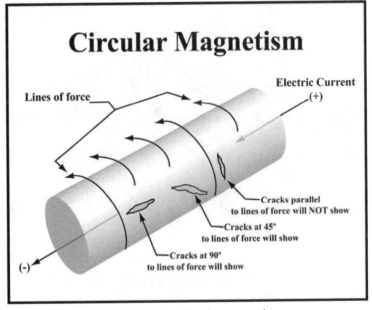

Figure 11–5   Circular magnetism

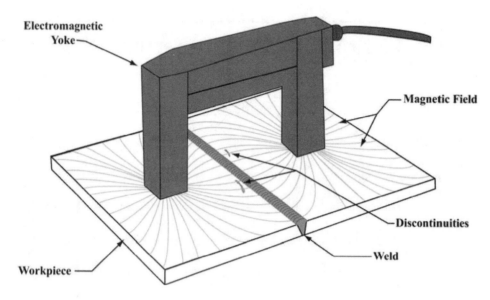

Figure 11–6   Magnetic yoke

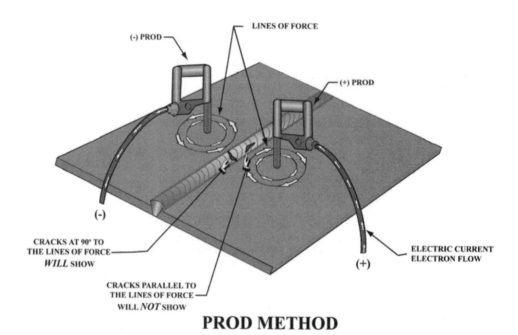

Figure 11–7   Magnetic prod

Radiographic testing (RT) is a nondestructive test method based on the principle of preferential radiation transmission. Areas of reduced thickness or lower density transmit more and absorb less radiation. The radiation passes through the interior of the test specimen, the radiation's strength is modified by the structures inside it and recorded on photographic film, much like a dental x-ray. RT is slow and more expensive than most other methods, but does provide a permanent film record.

How does this show up on film?

Areas of high transmission, low absorption, appear as dark areas on the developed film. Areas of low transmission, high absorption, appear as light areas on the developed film. The thinnest area of the test object produces the darkest area on the film since more radiation is transmitted. The thickest area of the test object produces the lightest area on the film as more radiation is absorbed in the test material so less radiation emission is shown on the film. See Figure 11–8.

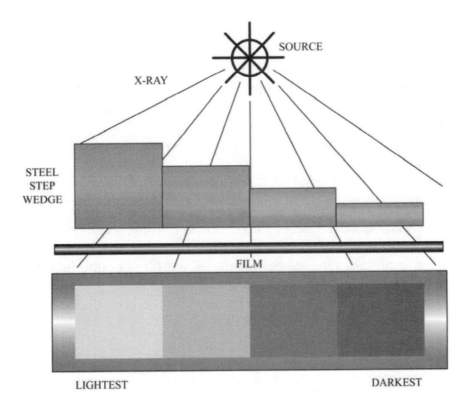

Figure 11–8   Radiographic absorption effect on thin to thick parts

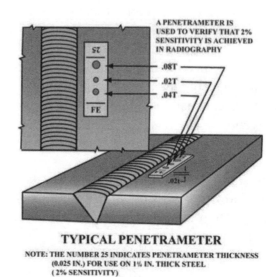

TYPICAL PENETRAMETER

NOTE: THE NUMBER 25 INDICATES PENETRAMETER THICKNESS
(0.025 IN.) FOR USE ON 1¼ IN. THICK STEEL
(2% SENSITIVITY)

Figure 11–9   Penetrameter image quality indicators (IQI) and their placement

How is radiation absorption affected by materials?
Denser materials absorb more radiation and show up lighter on the resulting film; lead has the highest density, followed by copper, steel and then aluminum. See Figure 11–10.

What discontinuities are discovered by radiography?
Subsurface discontinuities readily detected by radiography are voids (porosity), metallic (tungsten inclusions) and nonmetallic (slag) inclusions.

- Porosity, gas pockets, produces a dark area on the film because they represent a significant loss of material density
- Tungsten inclusions produce light areas on the film because tungsten is denser than say aluminum
- Slag inclusions appear light in color because of less absorption

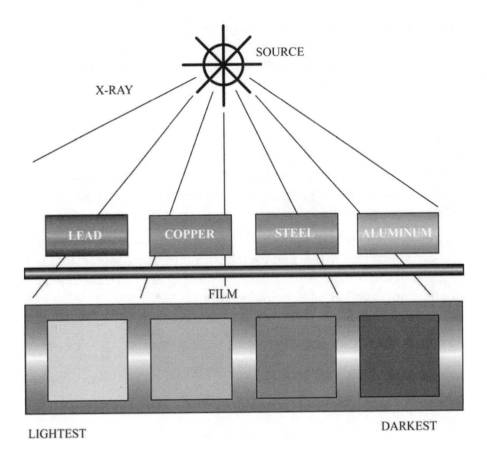

Figure 11–10   Metal absorption

What tools are needed for radiographs?
The equipment needed to perform radiographic testing includes:
- Some source of radiation either an X-ray machine or a radioactive isotope which produces gamma radiation
- Either type X-ray or radioactive isotope require film and a light tight film holder
- Lead letters are used to identify the test object. Lead has high density causing the letters to form light areas on the developed film which can be readily read
- Penetrameters or image quality indicators (IQI) are used to verify the sensitivity of the test. See Figure 11–9.
- Film processing equipment to develop the film
- Film viewers are required for interpretation of the film

What is the advantage of radiographic film?
The major advantage of radiographic film is the detection of surface and sub-surface discontinuities in all common engineering materials. The other advantage is the film can be stored as a permanent record of the testing.

Are there disadvantages?
The biggest disadvantage is the danger of excessive radiation exposure to humans. Technicians are extensively trained in radiation safety to assure safe use for both the radiographic tester and other personnel in the test vicinity. The area where radiographic testing occurs is normally evacuated during the exposure time.

Another disadvantage is the need to have access to both sides of the material to be tested one side for the film and the other for the energy source. Radiographic testing equipment is expensive. Extensive training is required to produce competent operators and interpreters. See Figure 11–11.

Radiographs are not the best method for detecting some more critical flaws including, cracks and incomplete fusion unless the radiation source is preferentially or oriented with respect to the flaw direction.

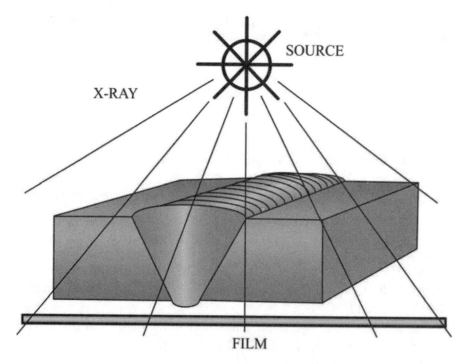

Figure 11–11 Radiation source and film placement

Ultrasonic testing (UT) is a nondestructive examination method which uses high frequency sound waves in the detection of most surface and subsurface weld discontinuities including cracks, slag, and lack of fusion. Ultrasound uses sound energy well above the frequency for the human ear to detect. These sound waves travel in different materials at different speeds but the propagation of these sound waves travel at the same speed through a given material. For instance, one type of sound wave (longitudinal) travels about 13,000 inches per second through air and about 250,000 inches per second through aluminum. See Figure 11–12.

Electrical energy, in the form of applied voltage, is converted to mechanical energy, in the form of sound waves, using a transducer. The transducer works by a phenomenon referred to as the piezoelectric effect. The transducer is a hand held device which directs the high frequency sound beam into the test material on a predictable path, which, upon reflecting back produces a wave that is then displayed on a cathode ray tube (CRT).

There are two basic types of ultrasonic transducers. Longitudinal (straight) beam transducers are used to determine material thickness or the depth of a discontinuity below the surface of materials. Longitudinal transducers send the sound into the part perpendicular to the surface of the part. Shear (angle) beam transducers are used extensively for weld examination because they send the sound into the part at some pre-determined angle; this allows the testing to be accomplished with the need to remove reinforcement. Often the longitudinal beam transducer is attached to a plastic wedge in order to provide the necessary angle.

Usually, the ultrasonic testing is contact testing where the transducer is placed against the surface of the part. High frequency sound is not easily transmitted through air so a liquid is placed between the test object and the transducer; this liquid is called a couplant. Contact testing is convenient and portable.

Another way to use ultrasound testing is immersion testing where the part is placed underwater and sound is transmitted between the transducer and the part through the water.

Ultrasound pulses are coupled into the sample and are used to probe the inside much like a flashlight beam in a dark room. When pulses encounter a discontinuity in the sample, they are reflected back to the transducer that detects them. An experienced UT operator can determine a great deal of information about the dimensions and soundness of the material.

The advantage in ultrasound is that it is a truly volumetric evaluation of material because sound wave will travel through any material no matter how thick and any discontinuity can be detected.

The major disadvantages to ultrasonic examination are:
- A highly skilled operator is required for correct interpretation
- The test object surface must be fairly smooth and a couplant is required for testing
- No permanent record of the cathode ray tube display is generally available
- Testing is generally limited to butt weld inspection of material thicker than 1/4".

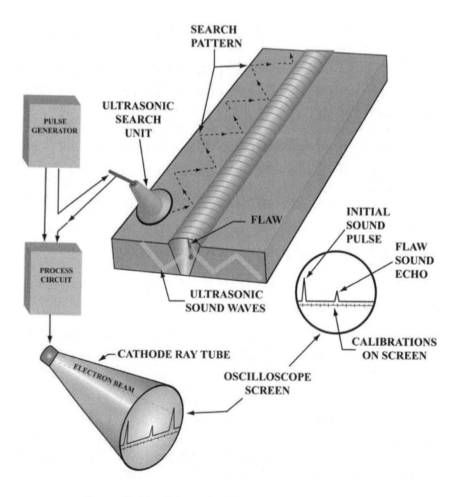

Figure 11–12   Schematic of pulse echo flaw detector

Discontinuities—Definitions, Causes, and Remedies

Porosity

What is *porosity*?
Porosity is the existence of cavities or voids within the weld metal. They are generally spherical, but may also be elongated.

Are there different types of porosity?
Yes, there are four main types:
- Uniformly scattered porosity, see Figure 11–13.
- Cluster porosity, see Figure 11–14
- Linear porosity, see Figure 11–15.
- Piping porosity describes elongated gas pores. In fillet welds they extend from the root of the weld to the face. If the gas bubbles make it to the weld surface, they are visible to the eye. These visible bubbles suggest the presence of other porosity below the surface and require additional inspection with UT or RT.

What causes porosity?
Porosity is the result of gas becoming entrapped in solidifying weld metal. Hydrogen is the major cause of porosity; it may come from many sources including the atmosphere, the disassociation of water, electrode coverings, base metal surface contaminants, residual lubrication on filler wire, or being dissolved in the filler metal itself. Nitrogen from the atmosphere may cause porosity. Oxygen from the atmosphere, from oxides on the filler wire, base metal, or electrode coverings may also cause porosity.

Why be concerned about porosity?
In general it is an indication that the welding process is out of control and needs adjustment. In small numbers and sizes, porosity can remain a discontinuity rather than a defect. Welding codes provide guidelines for their permissible size and number. In general, porosity at levels below 3% of weld volume will not cause significant changes in weld strength or properties. Porosity is also considered a less serious defect than cracks since voids and cavities have rounded ends and will not propagate.

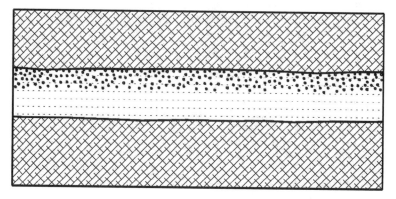

Figure 11–13   Uniformly scattered porosity

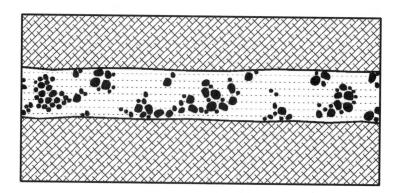

Figure 11–14   Cluster porosity

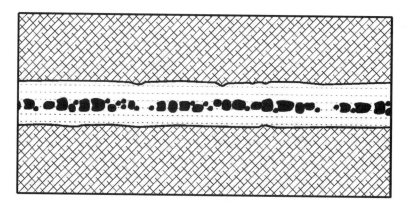

Figure 11–15   Linear porosity

How can porosity be remedied?

See Table 11–1.

Causes	Remedies
Excessive hydrogen, nitrogen, or oxygen in welding atmosphere	Use low-hydrogen welding process; filler metals high in deoxidizers; increase shielding gas flow
High solidification rate	Use preheat or increase heat input
Dirty base metal	Clean joint faces and adjacent surfaces
Dirty filler wire	Use specially cleaned and packaged filler wire; store it in a clean area
Improper arc length, welding current, or electrode manipulation	Change welding conditions and techniques
Volatilazation of zinc from brass	Use copper-silicon filler metal; reduce heat input
Galvanized steel	Use E6010 electrodes and manipulate the arc heat to volatilize the zinc ahead of the molten weld pool
Excessive moisture in electrode covering or on joint surfaces	Use recommended procedures for baking and storing electrodes
High-sulfur base metal	Use electrodes with basic slagging reactions

Table 11–1   Common causes and remedies of porosity

Slag Inclusions

What are *slag inclusions*?

Slag inclusions are nonmetallic solid materials trapped inside the weld metal. Inclusions occur in arc welding processes. See Figure 11–16.

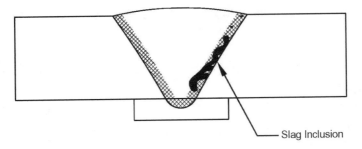

Slag Inclusion

Figure 11–16   Slag inclusions

What are the causes and cures for slag inclusions?
See Table 11–2.

Cause	Remedies
Failure to remove slag	Clean surface of any previous weld bead.
Entrapment of refractory oxides	Power wire-brush the previous weld bead.
Improper joint design	Increase groove angle of joint.
Oxide inclusions	Provide proper gas shielding.
Slag flooding ahead of welding arc	Reposition work to prevent loss of slag control.
Entrapped pieces of electrode covering	Use undamaged electrodes.

Table 11–2   Common causes and remedies of slag inclusions

How serious are slag inclusions?
Because slag inclusions are generally rounded and have no sharp corners like cracks, they are not considered as serious as cracks since they are less likely to propagate. The applicable codes indicate how much slag is allowed.

Tungsten Inclusions

What are *tungsten inclusions*?
Tungsten inclusions are small pieces of the GTAW or PAW tungsten electrode embedded inside the weld.

What causes tungsten inclusions?
If the tungsten electrode is dipped into the molten weld metal, or the welding current is too high and causes melting and transfer of tungsten droplets into the weld metal, tungsten inclusions can result.

What makes tungsten inclusions look different from other metallic inclusions on weld X-rays?
Because tungsten absorbs more radiation than other metals, tungsten inclusions appear as lighter areas on the radiograph. Almost all other inclusions show up as dark areas. See Figure 11–17.

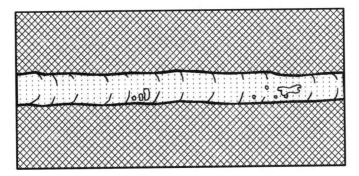

Figure 11–17   Tungsten inclusions as seen on radiograph

Incomplete Fusion

What is *incomplete fusion*?
Incomplete fusion exists when fusion did not occur between weld metal and fusion faces or the adjoining weld beads.

What are the causes of incomplete fusion?
There are:
- Insufficient welding current
- Lack of access to all weld joint surfaces requiring fusion
- Inadequate pre-weld cleaning of joint surfaces

Inadequate Joint Penetration

What is *inadequate joint penetration*?
Inadequate joint penetration is the discontinuity seen at the root of a joint where the root penetration of a weld is less than that specified in the design or code. See Figure 11–18.

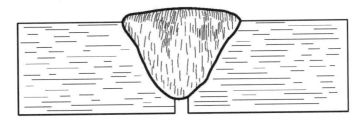

Figure 11–18   Weld cross-section showing inadequate joint penetration

What are the causes of inadequate joint penetration?
- Insufficient welding heat
- Improper joint design—too much metal for the welding arc to penetrate
- Incorrect bevel angle
- Poor control of the welding arc

Cracks

How are discontinuities called *cracks* defined?
Cracks are fracture-type discontinuities characterized by a sharp tip and a high ratio of length to width at its opening.

When do cracks form?
- *Hot cracks*—Form *during* metal cooling.
- *Cold cracks*—Form *after* metal has cooled.

How are cracks categorized?
Cracks are categorized with respect to the position to the weld—longitudinal, transverse, or under-bead. See Figure 11–19 for an overview of crack types and where they occur.

- *Crater or star cracks* occur in weld craters when a bead terminates. They develop as the weld pool cools, shrinks and solidifies. Since they form in the last metal to freeze, they are subject to the restraint of the surrounding weld metal. They are minimized by gradually reducing the arc at the end of a bead and allowing the crater to fill and cool gradually. Some GMAW power supplies have an automatic crater-fill control function that automatically slowly decreases the arc when the welder turns it off. See ① in Figure 11–19.

- *Face cracks* occur on the weld surface and can be transverse across the weld, or longitudinal—the length of the weld. Longitudinal face cracks usually result from transverse shrinkage of welding or service conditions. Transverse face cracks are generally caused by longitudinal shrinkage stress of welding acting on weld or base metals of low ductility. See ② and ⑤ in Figure 11–19 and Figure 11–20.

- *Heat-affected zone cracks* occur in the base metal whose microstructure has been altered by the heat of welding. The less ductile microstructure of these areas is more likely to crack under the stress of weld shrinkage. See ③ in Figure 11–19.

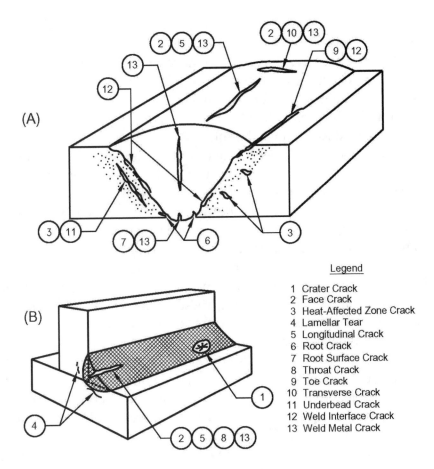

(A)

(B)

Legend

1 Crater Crack
2 Face Crack
3 Heat-Affected Zone Crack
4 Lamellar Tear
5 Longitudinal Crack
6 Root Crack
7 Root Surface Crack
8 Throat Crack
9 Toe Crack
10 Transverse Crack
11 Underbead Crack
12 Weld Interface Crack
13 Weld Metal Crack

Figure 11–19   Cracks in and around a weld

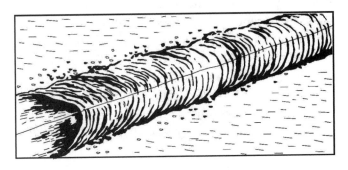

Figure 11–20   Longitudinal face crack

- *Lamellar tears* are parallel to and under the base metal's surface. They are the result of thin layers of non-metallic materials included in the metal during the rolling process. Because they have lower ductility than the metal itself, failure occurs along these layers. When the weld is put under load these tears may open up becoming laminated cracks. They are usually small and not noticeable unless they extend to the plate edges. Redesign of the weld joint to reduce strain is the solution. See ④ in Figure 11–19.

- *Root cracks* initiate at the weld root or root surface. Root cracks are usually longitudinal, but propagation may be in either the weld or the base metal. Root cracks are generally related to the existence of shrinkage stresses from welding. These cracks are considered to be hot cracks and are often the result of improper fit up. Large root openings may cause stress concentration resulting in root cracks. See ⑥ and ⑦ in Figure 11–19.

- *Throat cracks* extend through the weld along the weld throat, the shortest path through the weld's cross-section. Throat cracks are longitudinal and are usually considered hot cracks. These cracks can be observed on the weld face and are called centerline cracks. Throat cracks are longitudinal cracks in a fillet weld. See ⑧ in Figure 11–19.

- *Toe cracks* are base metal cracks propagating from the weld toe. Weld convexity and changes in the HAZ base metal may provide a stress concentration at the weld toes leading to cracking. This type of crack is a cold crack. Toe cracks occurring in service are usually the result of metal fatigue at a stress concentration. See ⑨ in Figure 11–19.

- *Under-bead cracks* occur in the HAZ and typically lie directly adjacent to the weld fusion line. In cross-section, under-bead cracks appear to run parallel to the fusion line of the weld bead. Under-bead cracks are usually subsurface making them difficult to detect and are particularly troublesome because they may not propagate until many hours after the welding is completed. For materials susceptible to under-bead cracking, final inspection should be delayed for 48 to 72 hours after weld completion. They are also called *delayed cracks* and *cold cracks*. They result from hydrogen in the weld zone that can come from the base material, filler metal, surrounding atmosphere or organic surface contamination. To avoid under-bead cracking, use electrodes from the low-hydrogen group (especially in multiple pass welding) and store them in ovens at the proper temperature to avoid water pickup. Do not weld on materials with moisture present. Having the ambient temperature and weldment at 70°F is also helpful. See Figure 11–21.

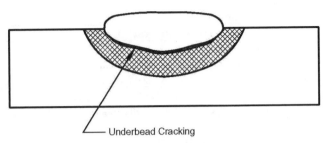

Figure 11–21   Under-bead cracking

- *Weld interface cracks* may result from a lack of fusion between the weld and base material. Incomplete fusion may be a result of an improper welding electrode, improper filler material manipulation or the presence of *non-metallic inclusions*. Non-metallic inclusions are usually a result of improper material cleaning before welding. Slag inclusions are an example of improper cleaning. See Figure 11–19.

What is the significance of weld cracks?
They are the most critical discontinuities because their sharp ends cause stress concentrations that can lead to crack growth and ultimately sudden weld failure. Welds subject to impact loading and low temperatures are particularly vulnerable to cracks. Regardless of size, they are not normally permitted in welds governed by welding codes. Cracks must be removed and replaced with sound weld metal.

What are the causes and remedies for weld cracking?
See Table 11–3.

Arc Strikes

What are *arc strikes*?
Arc strikes are small, localized points of remelted metal, heat-affected metal, or changes in surface profile away from the joint caused by an arc.

What is the remedy for existing arc strikes?
Avoid arc strikes outside the weld area if possible; should you incur arc strikes they should be ground to a smooth contour and checked for soundness.

Causes	Remedies
Weld Metal Cracking	
Highly rigid joint	Use pre-heat. Relieve residual stress mechanically. Minimize shrinkage stresses using back-step welding sequence.
Excessive dilution	Change welding current and travel speed. Weld with covered electrode negative; butter the joint faces prior to welding.
Defective electrodes	Change to new electrodes. Bake electrodes to remove moisture.
Poor fit-up	Reduce root opening; build up the edges with weld metal.
Small weld bead	Increase electrode size, raise welding current; reduce travel speed.
High-sulfur base metal	Use filler metal low in sulfur.
Angular distortion	Change to balanced welding on both sides of joint.
Crater cracking	Fill crater before extinguishing arc; use a welding current decay device when terminating weld bead.
Heat-Affected Zone Cracking	
Hydrogen in welding atmosphere	Use low-hydrogen welding process; preheat and hold for 2 hours after welding or post-weld heat treat immediately.
Hot cracking	Use low heat input, deposit thin layers; change base metal.
Low ductility	Use preheat; anneal the base metal.
High residual stress	Redesign the weldment change the welding sequence; apply intermediate stress-relief heat treatment
High hardenability	Preheat; increase heat input; heat treat without cooling to room temperature.
Brittle phases in microstructure	Solution heat-treat prior to welding.

Table 11–3   Common causes and remedies of cracking

Overlap

What is *overlap*?

Overlap occurs when weld metal deposits are larger than the joint can accept and the filler metal flows over the base metal surface without fusing to it.

Undercut

What is *undercut*?

Undercut occurs when the welding arc plasma removes more metal from the joint face than it replaces with filler metal. See Figure 11–22.

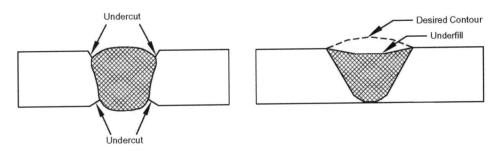

Figure 11–22   Cross section of weld showing undercut and under-fill

Under-fill

What is *underfill*?
Underfill is the depression on the weld face or root surface extending below the adjacent surface of the base metal. See Figure 11–22.

Laminations

What are *laminations*?
Laminations are flattened discontinuities that run generally parallel to the surfaces of rolled products like plates and shapes. They may be completely internal and invisible except with ultrasonic inspection.

Delaminations

What are *delaminations*?
Delaminations are the opening up of laminations and failure to carry load when subject to stress across the laminations.

Incorrect Weld Size and Profiles

When do weld size and profile constitute a defect?
Welds sizes and profiles constitute a defect whenever they fail the standards for size and profile in the applicable specifications. For example, the *ANSI/AWS D1.1 Structural Welding Code–Steel* provides accept/reject size and profile criteria with the drawings in Figure 11–23.

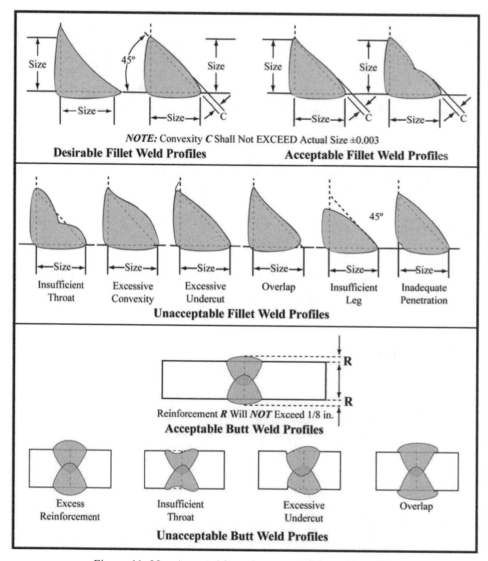

Figure 11–23   Acceptable and unacceptable weld profiles

Other Weld Discontinuities

What are other discontinuities that may require the weld to be classified as defective?

- Excessive warpage of the weldment
- Joint misalignment
- Incorrect joint preparation

Chapter 12

Welding Metallurgy

*Generally the theories we believe, we call facts and
the facts we disbelieve, we call theories.*
Felix Cohen

Introduction

Because welding exposes metal to high temperatures, welders need to understand how metals react to heat. To do this, we examine the properties of metals and how they are measured. Then we investigate the crystalline nature of metals and how heat can cause changes in their crystal structure and properties. We study the effects of alloying elements, the iron-carbon diagram and explain the hardening, tempering, and heat treatment of steel. We also look at steel classification systems and consider the different types of cast iron. With this background, we then look at ways to reduce the negative effects of welding heat on metals.

Fundamentals

What is the difference between *metallurgy* and *welding metallurgy*?
Metallurgy is the overall field of extracting and applying metals. Welding metallurgy is a subdivision concerning the behavior of metals during welding, and the effects of welding on the metal's properties.

Why are welders interested in metallurgy and welding metallurgy?
To solve routine fabrication and repair problems, welders must understand:
- Effects of welding on the properties of metals.
- How to correct or minimize the negative effects of welding on metals.
- Hardening, annealing, and tempering of metals, particularly steels.

Mechanical Properties of Metals

What are the four basic types of stresses?
See Figure 12–1.
- Tension
- Compression
- Shear
- Torsion

All stress, however complex, can be described by a combination of two or more of these basic types.

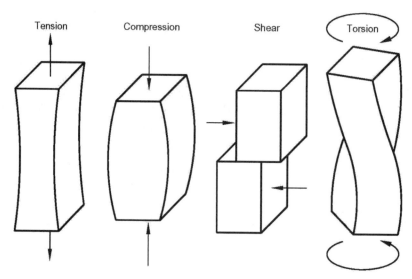

Figure 12–1   Four types of stress

What is *tensile strength*?
It is the ability of a test specimen to resist being pulled apart. Tensile strength is calculated by dividing the load by the cross-sectional area of the item under tension. This measurement is given in pounds/inch^2 or MPa.

How is tensile strength determined?
A testing machine that subjects the specimen to increasing tension measures tensile strength. See Figure 12–2. The tension, or pull, on the specimen and its resulting changes in length are recorded simultaneously. The plot of this data is called the stress-strain curve for the specimen. See Figure 12–3.

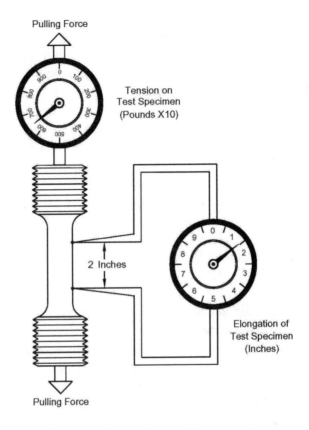

Figure 12–2   Tensile testing machine

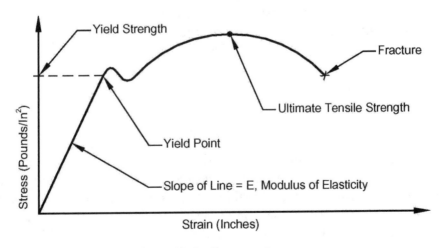

Figure 12–3   Stress-strain curve

What additional information does the stress-strain curve provide and what do these parameters tell us?

- *Yield strength*—the maximum stress that can be applied without permanent deformation of the specimen. We usually want to keep our design loads well below yield strength so structures are not permanently deformed.
- *Yield point*—the point on the stress strain curve beyond which stress and strain are no longer proportional having a straight-line relationship.
- *Ultimate tensile strength*—the maximum stress value obtained on a stress-strain curve.
- *Stiffness or modulus of elasticity*—the straight line relationship between stress and strain for non-permanent changes in length. This is the slope of the stress-strain curve below the elastic limit. It is also called Young's modulus and is represented by E in equations. The more vertical the linear portion of the stress-strain curve, the more rigid and stiff the material. See Chart 12–1.

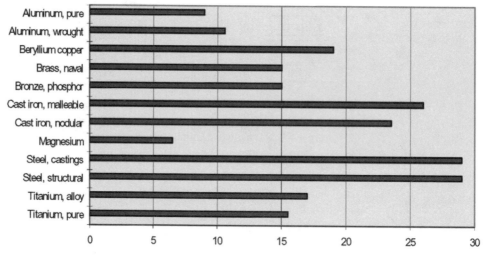

Chart 12–1　　Modulus of elasticity for some common metals $\times 10^6$ psi

How is *ductility* related to elasticity?

Elasticity concerns the relationship between stress and strain below the point of non-permanent deformation. Ductility measures how much a material can be permanently deformed without breaking. For example, aluminum and copper are quite ductile and for this reason are easily drawn into wire. Although all steels have about the same elasticity, ductility varies greatly depending on composition: low-carbon steels are generally ductile, high carbon steels are not.

How is ductility measured?

It is expressed in either of two ways:
- Percent elongation in two inches
- Percent reduction in cross-sectional area

How is the *hardness* of a metal defined?

Hardness is the ability of a material to resist indentation or penetration. It is an excellent indicator of a material's strength and is easily performed without destroying the material tested. Portable hardness testers are often used to check the hardness of welds in the field. Conversion tables relate tensile strength and hardness. See Figure 12–4.

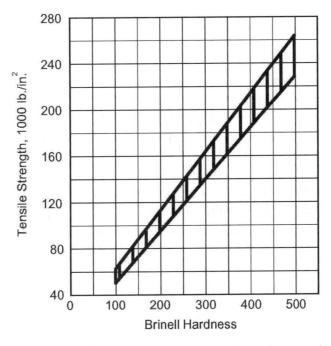

Figure 12–4   Conversion of hardness to tensile strength

What are the most important hardness measuring scales for metals?
- Rockwell—Uses a 10 mm steel or tungsten carbide ball indenter
- Brinell—Uses a cone-shaped diamond indenter

How is *metal hardness* measured?

A very hard indenter usually of diamond or tungsten carbide is pressed into the specimen under a given weight. How far the indenter penetrates the metal indicates the specimen's hardness. See Figure 12–5.

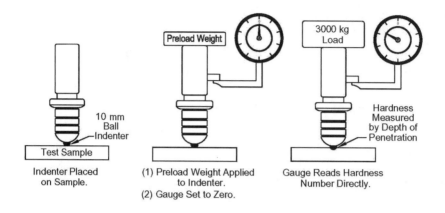

Figure 12–5   Rockwell hardness measurement

What about testing *compressive strength* in metals?

Compressive strength, the ability of a metal to resist a gradually applied squeezing force, is not usually important for metals. This is because metals have at least as much strength in compression as in tension, so compressive strength is seldom important.

What is *fatigue strength* and how is it determined?

Fatigue strength is the ability of a material to withstand repeated applications of stress, or the repeated application and reversal of stress. Figure 12–6 shows a fatigue-testing machine.

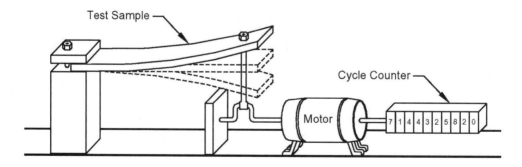

Figure 12–6   Fatigue testing machine

Why is *fatigue strength* important?
Under many stress cycles, failure may occur through fatigue even though the maximum stress is considerably *below* the maximum stress at which failure would normally occur if the stress were constant. Knowing fatigue strength is important in products subjected to cyclic stress and vibration such as engines, airplanes and bridges. But just knowing the maximum yield stress of a metal is not enough.

What are *impact strength* and *fracture toughness*?
These terms apply to the ability of metals to withstand large shock forces applied suddenly. Impact strength is the energy needed to fracture a notched specimen and is usually determined by the *Charpy V-notch* or *Izod* impact tests. Fracture toughness is the ability of a metal to resist crack propagation when a crack already exists and a large force is suddenly applied.

How is impact *strength* measured?
A notched test specimen is subjected to the impact from a pendulum. The height of the pendulum swing after breaking the specimen, shows how much energy was used breaking the sample. See Figure 12–7.

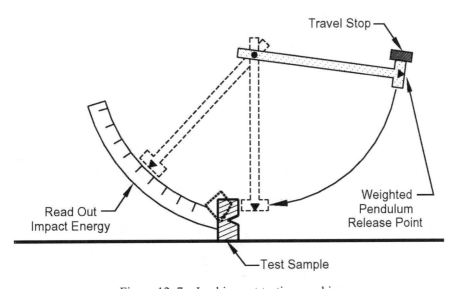

Figure 12–7   Izod impact testing machine

Non-physical Properties of Metals

What are the important non-physical properties of metals?

- *Oxidation resistance*—the ability of a metal to resist oxide formation in air and to continue oxide formation once it has begun.
- *Corrosion resistance*—the ability of metals not to rust in water or react with specific chemicals, usually in solution.
- *Magnetic properties*—the ability of a material to become magnetized, or resist magnetization.
- *Electrical conduction*—the ability of the metal to conduct electricity without generating heat.

Note: The same metal or alloy can have different non-physical properties depending on its cold-work and heat history.

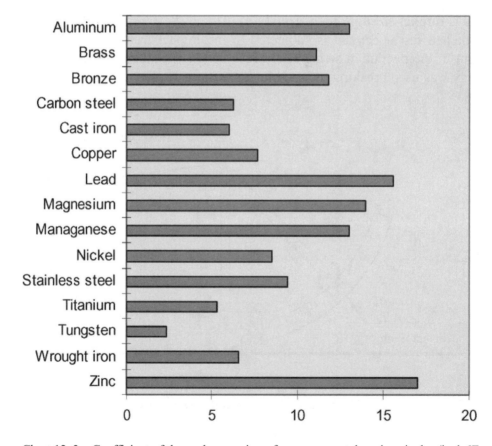

Chart 12–2   Coefficient of thermal expansion of common metals, micro-inches/inch-°F

Thermal Expansion

What is the *coefficient of thermal expansion* and why is it important to welders?

This coefficient indicates how much change in length per unit of length a material has with one degree of temperature change. English measurement units are microinches/inch-°F; metric units are micrometers/meter-°C.

The formula for calculating change in length (ΔL) with change in temperature (ΔT) over a part of length L is:

$$\Delta L = \text{coefficient of expansion} \times L \times \Delta T$$

Here are two examples where the coefficient of thermal expansion can cause problems if not taken into account:

A machinist turns a part on a lathe to 6.000 inches (150 mm) diameter, but when he measures the same part after cooling to room temperature, a 100°F (56°C) drop, the part measures nearly 0.004 inch (0.1 mm) undersize.

A steel bridge in Chicago is 300 feet (90 m) long. In winter temperatures of −25°F (−32°C) occur; summer temperatures reach 100°F (37°C), so the bridge steel sees a temperature change of 125°F (70°C). How much change in length must the expansion joints provide for? The answer is just under 3 inches (76 mm) change due to temperature changes. Failure to provide for this movement can cause buckling of metal or weld, or foundation failure. Any large structure, even a steel pipe fence, can get into serious trouble without expansion joints.

Structure of Metals

What is the difference between the structure of metal when molten and solid?

In the hot liquid state, metals have no particular structure meaning there is no orderly, defined structure among the metal atoms. However, at lower temperatures, the atoms have lower energy, move less rapidly and atomic forces tend to arrange them into particular structures or patterns. These arrangements are called crystals. All metals and alloys are crystalline solids.

What are the crystalline structures seen in common metals?
- Face-centered cubic (FCC)
- Body-centered cubic (BCC)
- Hexagonal close-packed (HCP)

These three-dimensional structures, also called space lattices, are shown in Figure 12–8.

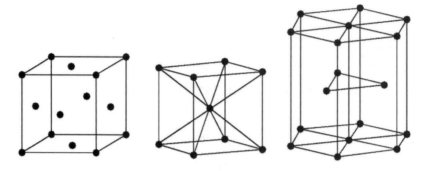

Figure 12–8   The three most common crystal structures in metals and alloys:
Face-centered cubic (left), body-centered cubic (center) and
hexagonal close- packed (right)

Metal	Crystal Structure
Aluminum	FCC
Chromium	BCC
Cobalt	HCP
Copper	FCC
Gold	FCC
Iron (alpha)	FCC
Iron (gamma)	BCC
Iron (delta)	BCC
Lead	FCC
Magnesium	HCP
Nickel	FCC
Silver	FCC
Titanium	HCP
Tungsten	BCC
Zinc	HCP

Table 12–1   Crystal structure of metals

Are there materials without crystal structure at room temperature?
Yes, common examples are glass and silicon.

How do crystalline solids form?

As molten metal cools to its solidification temperature, solid particles called nuclei form. Because these nuclei already have the crystal structure characteristic of the particular metal (or alloy), additional atoms add on to the nuclei repeating and enlarging this structure into larger solid particles called grains. As grains continue to grow, they eventually meet on irregular and disordered grain boundaries, so a grain is a crystal with an irregular atomic boundary. Each grain has the same atomic structure and spacing internally, but at the grain boundary this regularity is interrupted. Grains and grain boundaries are visible under a microscope at 50–1000× magnifications. This cooling process takes place whenever a metal cools, regardless of its heat source. Because molten metal in a weld solidifies on cooling, we can regard the interior of a fusion weld as a miniature casting. See Figure 12–9.

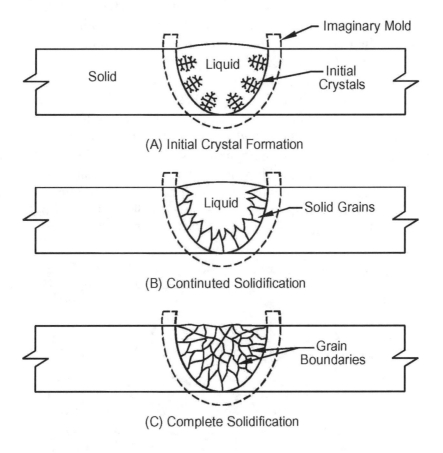

(A) Initial Crystal Formation

(B) Continuted Solidification

(C) Complete Solidification

Figure 12–9   Steps in the solidification of molten metal

When a second metal (or other element) is added to a pure metal and an alloy formed, what happens in the formation of crystals of the new alloy? How new atoms replace atoms of the parent metal in the crystal lattice have a profound effect on the properties of the alloy. There are two ways the new atoms combine with the base metal:

- *Direct substitution*—when the new metal is similar to the parent metal, it can fit into the position in the lattice formerly occupied by an atom of the parent metal. The new metal is dissolved in the base metal in a solid solution, Figure 12–10. Examples are copper in nickel, gold in silver, carbon in iron as ferrite. See Figure 12–11.

- *Interstitial solid solution*—if the atomic size of the added metal is small in comparison with the atomic size of the base metal, it can fit into the parent metal lattice without displacing any of the original atoms. See Figure 12–12. Because the "fit" is imperfect, strain is created in the crystal structure: tensile strength increases, elongation decreases, Figure 12–13. This alloy is called an interstitial solid solution. An example of an interstitial solid solution is small amounts of copper in aluminum.

- *Multiphase system*—if neither direct substitution, nor interstitial solid solution can completely dissolve the added metal atoms, mixed atomic groupings result within the new alloy. That is, each grouping has its own atomic crystalline structure and is called a phase of the alloy and the alloy is a *mixture* of these groupings or multiphase system. Pearlite, Figure 12–14, is an important example of a multiphase system within the carbon-iron family. The way these grains, boundaries and phases exist in an alloy is called its *microstructure* and collectively they are responsible for the physical and mechanical properties of the alloy.

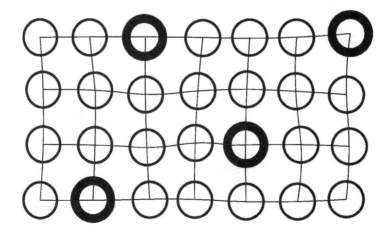

Figure 12–10   Diagram of substitutional solid solution

Figure 12–11   Photomicrograph at 500× of ferrite, a solid solution of carbon in iron

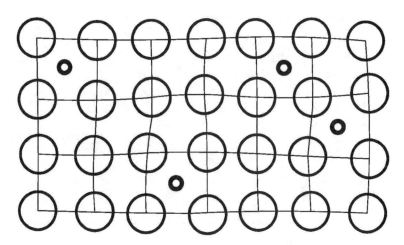

Figure 12–12   Diagram of interstitial solid solution

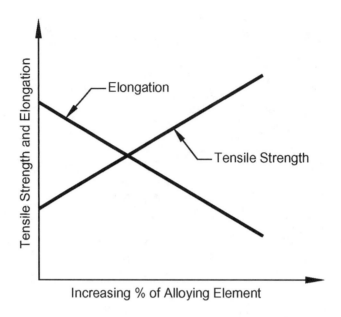

Figure 12–13   How adding an alloying element affects tensile strength and elongation

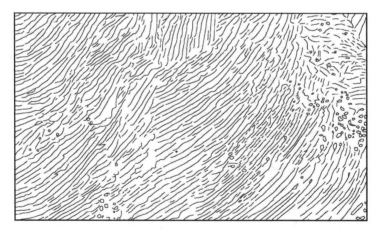

Figure 12–14   Photomicrograph of pearlite at 1500×, a multiphase system
                with a lamellar structure

Why are we interested in the properties of alloys?

Most engineering metals are alloys consisting of one major metal *and* one or
more alloying elements. They have better mechanical properties for engineering applications than pure metals and are used more often.

Multiphase systems that occur only in alloys permit us to modify the physical properties of alloys to meet engineering requirements that neither alloy metal could meet when used by itself. This is especially true for steels.

What determines the grain size?
Grain size depends on the rate of growth, which in turn depends on the rate of cooling. A fast cooling rate forms small grains rapidly and a slow cooling rate produces large grains.

Why are grain size and structure important?
Grain size, grain boundary structure, and alloy phases present determine an alloy's properties—and its usefulness. Distortion of the microstructure by cold working hardens it. The alloying elements—foreign atoms in the lattice—distort the structure and harden it too. Atoms of the alloying element dissolved in the base metal and then precipitated out also distort the microstructure and harden the alloy.

What is peculiar about the atoms at the grain boundaries?
Grain boundaries contain lower melting point atoms than those within the grains, since they are the last areas to freeze as the alloy cools. These foreign elements cause distortion of the microstructure and harden the alloy at room temperature. However, they reduce the alloy strength at elevated temperatures as the grain boundary foreign atoms begin to melt and allow slippage between grains.

Because the atomic structure is irregular and may be larger than normal at the grain boundary, foreign or odd-sized atoms tend to congregate there. This may lead to formation of undesirable phases that reduce ductility and lead to cracking when subjected to welding.

How does grain size affect a metal's strength?
In general, fine-grained metals have better properties at room temperature than coarse-grained ones. Grain structure is classified as small grain (fine-grained), large grain (coarse-grained) or mixture of both (mixed-grained).

Phase Transitions and Diagrams

What are the two types of phase changes?
- When a metal or alloy goes from solid to liquid or the reverse, we say it has had a phase change.

- Some metals change crystalline structure with temperature while remaining solid and this too is called a phase change. Iron, for example:
  - From room temperature to 1670°F (910°C) is BCC.
  - From 1670°F (910°C) to 2535°F (1390°C) is FCC.
  - From 2535°F to its melting point of 2800°F (1538°C) is BCC.

Phase transformation is also known as *allotropic transformation*.

What is the best way to show how phase changes relate alloy composition and temperature?

Phase diagrams (also called *constitution* or *equilibrium diagrams*) graphically display these relationships. For example, pure metals have a single melting point, while most alloys melt and freeze over a range of temperatures. Phase diagrams show this. Figure 12–15 shows the copper-nickel phase diagram. Because these metals are equally soluble in each other, they show no allotropic behavior. The diagram shows the single melting point for pure copper on the left (0% nickel) and the single melting point for nickel on the right (100% nickel). It also shows how at all alloy mixtures between these points, there is not a single melting point, but a transition zone consisting of a mixture of solid and liquid.

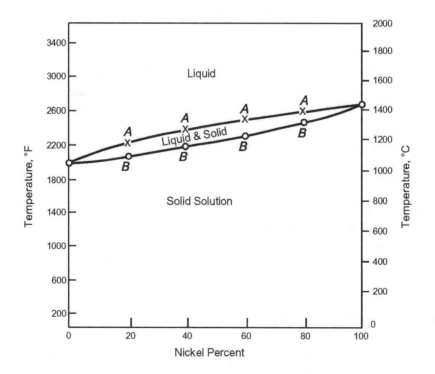

Figure 12–15   Copper-nickel phase diagram

What are the major limitations of phase diagrams?
- Phase diagrams show only what happens in equilibrium. An equilibrium state means the alloy is in its final state for a given set of temperatures and happens only after slow heating (or cooling) conditions and long hold times. Phase diagrams fail to show what intermediate or transitional compositions or phases exist when temperature changes too rapidly for the various atoms to go from one set of combinations at one temperature and to settle into their final, or steady-state combinations at another. These intermediate structures are *not* shown by the phase diagram. This is especially true for steel where an intermediate structure may be "frozen" or maintained if we change the mixture's temperature rapidly enough. Since temperature changes very rapidly during welding processes, phase diagrams fail to display many possible events.
- Phase diagrams show only the relationship between *two* metals. Phase diagrams of more than two elements are very complicated.

Carbon Content of Steel and Cast Iron

What carbon levels define major iron-containing products?
- *Iron* (as in wrought iron) contains almost no carbon (<0.008%).
- *Steel* runs from just above 0.008 to 2.1% carbon; however most steels contain less than 1% carbon. They may be subdivided into four sub-families:
  - Low-carbon steels contain less than 0.30% carbon. They are also called *mild* steels. They are the most frequently used grades; they are more ductile and can be machined and welded more readily than the higher carbon grades.
  - Medium-carbon steels range from 0.30 to 0.45% carbon. With increasing carbon content hardness and tensile strength increases, ductility decreases and machining becomes more difficult.
  - High-carbon steels have from 0.45 to 0.75% carbon. Welding these alloys is difficult and heat-treating before, during and after welding must be done to achieve good welds and mechanical properties.
  - Very high-carbon steels have up to 1.50% carbon. These steels are used for springs and metal cutting tools.
- *Cast iron* contains over 2.1% carbon.
- *Alloy steels* contain more than 1.65% manganese or more than 0.60% copper or a guaranteed minimum of any other metal. The metals most frequently found in alloy steels are nickel, chromium, molybdenum, and vanadium.

How is the percentage of carbon sometimes described for carbon steel?
Carbon content is often described in *points* with each point being 0.01% carbon.

What is the carbon percentage range in which steel can be hardened by a heat-quench-temper cycle?
Steels between 0.35 and 1.86% carbon content can be hardened.

What are the four types of cast iron and what are their properties?
- *Gray cast iron*—relatively soft and readily machinable and weldable. Frequently used for engine cylinder blocks, machine tool structures, and cast iron pipe.
- *White cast iron*—very hard and brittle, used for high abrasion applications, but more often used as basis for malleable cast iron. Not weldable.
- *Malleable cast iron*—somewhat ductile, made by annealing white cast iron, weldable. Good strength, machinability, shock resistance, and ductility.
- *Ductile cast iron*—also called *nodular* or *spheroidal graphite cast iron*; comes from heat-treated malleable cast iron, so it has the same chemical composition. Its carbon content is in the form of small spheres instead of flakes that makes it malleable and improves ductility. It is weldable.

Steel Classification Systems

What are two common numbering and classification systems for steels?
The Society of Automotive Engineers (SAE) and the American Iron and Steel Institute (AISI) have nearly identical systems. They are based on a four-digit numbering system. The first digit indicates the type of steel; the second, the percent of that alloying element; and the last two digits, the carbon content in hundredths of a percent. Hence, 2315 would identify nickel steel, with about 3% nickel and 0.15% carbon. See Table 12–2 for this steel classification system.

Iron-Carbon Diagram

Why is the Iron-Carbon diagram important?
Steel is mainly an alloy of iron and carbon. Figure 12–16 shows the microstructures of different carbon steels at various temperatures. These microstructures, in turn, determine the physical properties of the alloy. It also shows how welding thermal cycles and heat treatments affect steel's properties.

Code	Steel Classification
1XXX	Carbon steels
12XX	Special sulfur-carbon steels with free-cutting properties
12XX	Phosphorus-carbon steels
13XX	Manganese steels
2XXX	Nickel steels
3XXX	Nickel-chromium steels
4XXX	Molybdenum steels
5XXX	Chromium steels
6XXX	Chromium-vanadium steels
7XXX	Tungsten steels
9XXX	Silicon-manganese steels

Table 12–2   SAE/AISI steel classification codes

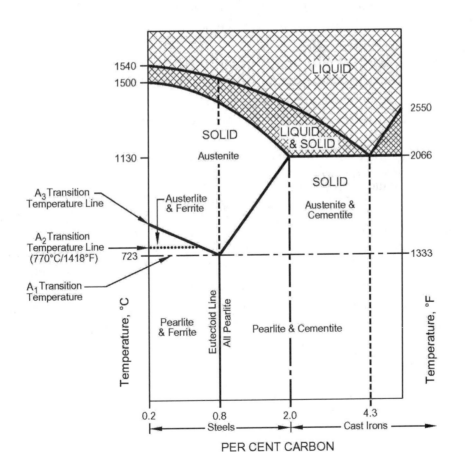

Figure 12–16   Simplified Iron-Carbon diagram

Why does the Iron-Carbon diagram stop at 5% carbon?

Iron-carbon alloys of more than 5% carbon are cast irons, not steel alloys. They do not exhibit the important heating/quenching/tempering properties of low-carbon content steels, so they cannot be hardened. Hence the phase diagram in the higher percentages of carbon is *not* particularly useful.

What microstructures appear in the iron-carbon diagram, Figure 12–16?

- *Ferrite*—a solid solution of small amounts of carbon (>0.02%) in alpha iron (BCC). It is magnetic
- *Cementite*—Iron carbide, Fe_3C, contains about 6.69% carbon. Because our iron-carbon diagram cuts off at about 5% carbon, no region of pure cementite appears. Only structures containing cementite-pearlite and cementite-austenite mixtures, appear.
- *Pearlite*—alternating layers of ferrite and cementite make up pearlite. Pearlite *always* contains 0.77% carbon. Steels and cast irons with more than 0.77% carbon will form mixtures of cementite and pearlite. Steels with pearlite are usually ductile.
- *Austenite*—a solid solution of carbon in gamma iron (FCC); it can dissolve up to 2% carbon. While austenite is never stable below 727°F (386°C) in carbon steel, additional alloying elements can make austenite stable at room temperature. It is strong, ductile, nonmagnetic, and easily work hardened.

What microstructures of iron-carbon appear when an alloy is cooled rapidly and what are their properties?

- *Martensite*: The quenched structure is generally too hard and brittle for most engineering applications. Tempering restores some ductility without significantly reducing its strength. It is very useful for metal cutting tools, punches, dies, and many other applications. This structure is what makes hardening steel possible, Figure 12–17. Because it is a non-equilibrium structure, it does not appear on the iron-carbon diagram.
- *Bainite*: This structure of fine carbide needles in a ferrite matrix, which forms from austenite at cooling rates below those needed to form martensite. It is hard and has low ductility. Also a non-equilibrium structure, like martensite.

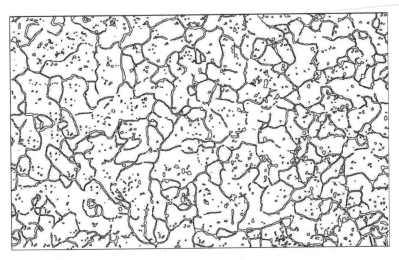

Figure 12–17   Microphotograph of martensite at 1500×

Based on the iron-carbon diagram, what can be said about alloys of exactly 0.80% carbon, slowly cooled to room temperature from 2000°F?
A vertical straight line at 0.80% carbon content represents this transition point, called the eutectoid. Here are the three important cases:

- Steels with *below* this carbon content will convert to a mixture of both pearlite and ferrite (called *hypo-eutectoid* steels).
- Steels with *exactly* this carbon content will convert *all* its austenite to pearlite.
- Steels *above* 0.80% carbon content will convert to mixtures of pearlite and cementite (called *hypereutectoid* steels).

Techniques to Strengthen Metals

What are the four ways to increase the strength of metals?

- *Cold working* the metal to deform and stress its crystal structure induces *work hardening*. For example, steel mills repeatedly run steel back and forth through rollers at temperatures below a plastic state. This process distorts the metal's grain structure. This deformation increases its tensile strength and hardness, and decreases ductility. Cold working can be performed on many other metals and alloys.
- *Solid solution hardening* is done by adding alloying metals that do not easily fit into the base metal's crystal lattice; they cause stress within the metal's crystal structure increasing tensile strength and decreasing ductility.

- *Transformation hardening* (heat/quench/tempering heat treatment cycle) produces the combination of strength and ductility desired in iron-carbon alloys. With this cycle we can adjust the strength and ductility needed for a specific application.
- *Precipitation hardening* is mainly done with aluminum alloys and is detailed later.

Transformation Hardening of Steel

How is transformation hardening of carbon steel done?
There are three separate steps:
1. Heat the steel 50 to 100°F (27 to 55°C) above its A_3-A_{cm} transformation temperature so it becomes entirely austenitic. This process is called *austenitizing* and is shown in the iron-carbon diagram in Figure 12–18.

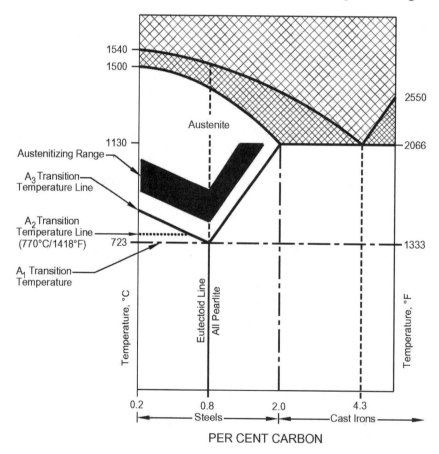

Figure 12–18   Shaded area shows austenitizing temperature range

2. Cool the steel so rapidly (called *quenching*) that the equilibrium products of pearlite + ferrite (or pearlite + cementite) cannot form. This leaves the transitional structure martensite which is very hard and brittle. We attempt to form 100% martensite at this step.

3. Temper the martensitic steel to reduce its brittleness, by further heating at temperatures below A_1. This is typically between 400 and 1300°F (204 and 704°C) and allows some of the martensite to convert to pearlite and cementite. The work is then allowed to cool slowly in air. The complete transformation hardening cycle is shown in Figure 12–19.

The more martensite, the steel becomes harder and less ductile; the more pearlite and cementite, the steel becomes more ductile and less brittle. By selecting steel with the proper carbon content and proper heat treatment, we can get just about any degree of hardness and ductility needed.

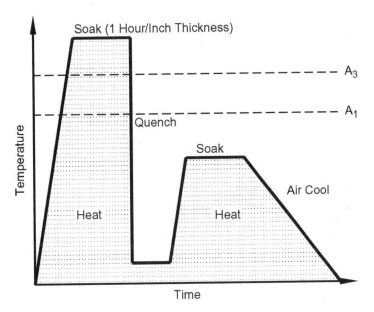

Figure 12–19   Transformation hardening of steel: austenitizing, quenching, and tempering

What are the two other names for a *time-temperature-transformation (TTT)* diagram?
They are also called *isothermal transformation (IT) diagrams* and *S-Bain curves*.

How is a TTT diagram determined for a particular alloy of steel?

The objective is to determine when an austenitized steel sample will begin and end transformation when quickly cooled and held at temperature below A_1.

For example, following from Figure 12–20:

1. Austenitize the six samples by holding them 50 to 100°F (27 to 55°C) above their A_1 temperature.
2. Immerse the six samples in a heated salt bath to hold them at the 1200°F (650°C) isothermal temperature.
3. Withdraw the samples one by one after fixed time intervals shown, A through F, quench sample in iced brine, then use microscopic examination to determine extent of transformation. The sample will show martensite if transformation has not begun and other transformation products, like pearlite, if it has begun.
4. Samples A and B show only martensite, so no transformation has begun.

   Sample C shows first non-martensite structure, so a transformation has begun, mark point C's time and 1200°F as start of S-curve.

   Sample D shows increasing amount of non-martinsitic products, but still some martensite, so transformation is not complete; data point D is *inside* the S-curve.

   Sample E shows no martensite, transformation is complete, so mark end of transformation curve at point D's time and 1200°F.
5. Repeat this process at other temperatures to determine the shape of the complete transformation curve.

What does a TTT diagram show?

When our objective is to devise a quenching cycle for hardening steel to form martensite and avoid formation of other microstructures, the TTT diagram shows us cooling paths to do the job.

This task is complicated because:

- Austenite is unstable at temperatures below A_1 and will transform into softer, non-martensitic microstructures.
- The time interval before the unstable austenite begins to transform depends on the temperature at which it is held.
- These microstructures, once formed, will not form martensite.
- Austenite can only begin to transform into martensite at temperatures below about 400°F (204°C).

Putting these phenomena together means there is a competition between the time-dependent formation of non-austenitic, soft microstructures and the temperature-dependent formation of martensite. The TTT shows all these relationships and what cooling paths will produce martensite.

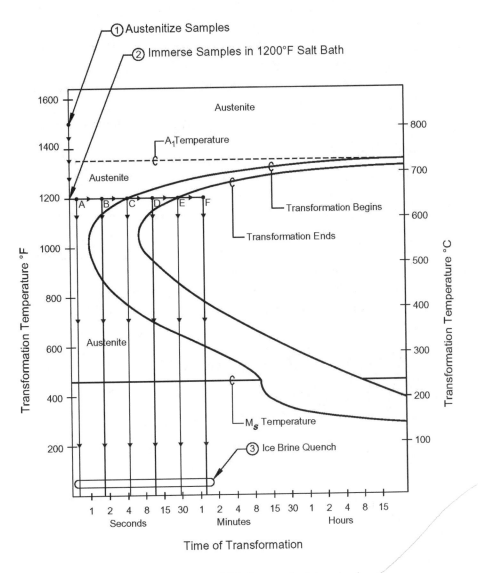

Figure 12–20   How a TTT diagram is determined

What cooling paths does the TTT diagram in Figure 12–21 display and what are their end microstructures?

There are four cooling paths in this TTT diagram and all begin at 1470°F (800°C) where the entire sample is in the austenite phase:

- Path B-D represents very rapid quenching, dropping more than 1300°F (704°C) in one second. As the steel's temperature drops to about 480°F (248°C), it reaches the M_S (martensite starts forming) temperature and when it reaches about 460°F (238°C), it reaches the M_f (martensite finished forming) temperature: all austenite has transformed to martensite. The steel is extremely hard and brittle and is ready to be tempered. Such a rapid quench is only possible with thin, small parts. Larger parts will require more time to have heat removed from their interiors and their temperatures reduced.
- Path B-K-E is the *critical cooling path* and is the slowest quenching rate, that converts *all* austenite to martensite. This is called the *critical cooling rate*. Note that because the cooling path just misses the nose of the S-curve, the austenite reaches and passes through the M_s and M_f and zones: all austenite converts to martensite. *No* pearlite or bainite forms because the cooling path does not enter the S-curve region. This quenching took four seconds.
- Path B-L-M-F does enter the S-curve region. Some pearlite and bainite forms from the austenite. Some martensite is formed too. This quenching took nearly 20 seconds.
- Path B-G represents very slow cooling as would happen in a furnace cooling operation. As the steel enters the S-curve region austenite is transformed to pearlite and then to bainite. These microstructures are stable end products, so no martensite forms and the final result is a combination of bainite and needle-like bainite.

Different cooling paths produce different microstructures with different properties.

In what ways are metals quenched?

From slowest to fastest cooling rates:
- Slow furnace cooling
- Cooling in still air
- Oil bath quench
- Water bath quench
- Salt brine quench

See Figure 12–22.

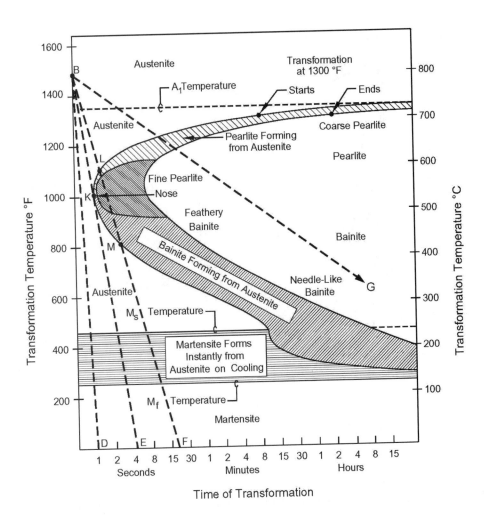

Figure 12–21   Time-temperature-transformation curve

What is the effect of alloying elements on the iron-carbon diagram?

In general they shift the nose of the TTT diagram to the right, transforming austenite to martensite at *lower* quenching rates than unalloyed steels. This permits larger parts to form martensite even though their quench rates are slower because of their size and heat transfer limitations. Some alloys have their S-curves shifted to the right so far that they can form martensite with air cooling.

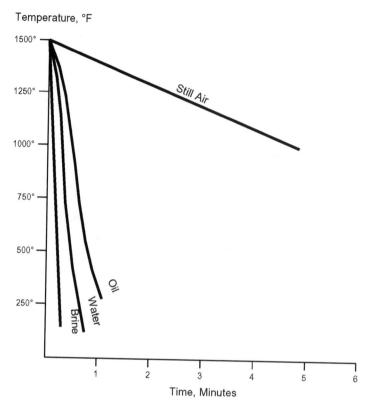

Figure 12–22   Different cooling methods produce different cooling rates

What factors determine the amount and location of martensite formed in a part?

- Non-iron components of the steel, its chemistry
- Grain size
- Size of the part as it affects heat removal rate
- Shape of the part as it affects heat removal rate
- Previous heat treatment/heat history of the part
- Quench rate and beginning/end temperatures

Thin parts are easier to cool because of their greater ratio of surface area to volume and are therefore easier to harden. Thick and/or large parts may prevent rapid heat quenching and so may not be hardenable unless made of an alloy which will harden at slower quench rates.

What is the difference between hardness and hardenability?

Hardness is the resistance to penetration while hardenability is the *ability* to be quench hardened. Steels and alloys with high hardenability require lower cooling or quench rates to achieve a given hardness than steels with lower hardenability. It is readily possible to have a steel with very high hardenability to be soft in the unquenched state.

What methods can be used to determine temperatures for the various heat-treating processes?

There are several ways:

- Electrical instruments called pyrometers can accurately determine temperature. Furnaces are usually equipped with them and hand held instruments are also available.
- Temperature indicating crayons that turn color on reaching specified temperature.
- Temperature indicating cones and pellets that melt on reaching temperature.
- Color change of the clean metal surface in dim light requires skill and judgment, but has been used for hundreds of years. See Table 12–3.
- Testing the work with a magnet to determine when it becomes non-magnetic is a simple way to tell when A_2, the *Curie temperature*, occurs. See horizontal line in Figure 12–16. This is a crude but effective method.

Degrees Fahrenheit	Degrees Centigrade	Steel Color
430	221	Very pale yellow
440	227	Light yellow
450	232	Pale straw yellow
460	238	Straw yellow
470	243	Deep straw yellow
480	249	Dark yellow
490	254	Yellow-brown
500	260	Brown-yellow
510	266	Spotted red-brown
520	271	Brown-purple
530	277	Light purple
540	282	Full purple
550	288	Dark purple
560	293	Full blue
570	299	Dark blue
640	338	Light blue

Table 12–3   Temperatures indicated by the color of plain carbon steel

Alloying Elements for Steel

What are the common alloy elements and what is their purpose?

- Carbon is steel's most important alloying element. Although some steels contain up to 2% carbon, most welded steels have less than 0.35%. The more carbon, the harder, stronger and brittle the steel; more carbon also improves hardenability. Higher carbon content reduces weldability because of the tendency to form martensite.
- Sulfur causes brittleness and reduced weldability at levels above 0.05%. Sometimes added to increase machinability.
- Phosphorous at levels above 0.04% causes brittleness.
- Silicon is added as a deoxidizer in rolled steel. It is also found in castings from 0.35-1.00%. By dissolving in iron, silicon strengthens it.
- Manganese, like silicon, is soluble in iron and increases hardenability. Also used as a scavenger for sulfur. More than 1.0% reduces weldability.
- Chromium at levels to 9% increases oxidation resistance, hardenability, and high-temperature strength. At higher levels of 12% steels become stainless steels with high oxidation resistance.
- Molybdenum increases hardenability and high-temperature strength.
- Nickel in low-alloy steels increases toughness and hardenability at levels of 3.5%. High alloy and some stainless steels contain as much as 35%.
- Aluminum in very low levels is added as a deoxidizer and to increase toughness.
- Dissolved gases of hydrogen (H_2), oxygen (O_2) and nitrogen (N_2) are serious contaminants, since they readily dissolve in steel and cause embrittlement if not removed.

What is the difference between *rimmed* and *killed steels*?

The terms rimmed, semi-killed and killed in low-carbon steels indicate whether steps have been taken to remove oxygen in solution with the steel. When rimmed steel is welded, carbon and oxygen in the steel combine to form carbon monoxide. This makes bubbles and leads to porosity and weakness in the completed weld. Silicone added to the steel before being poured into ingots makes semi-killed steel. Aluminum added produces killed steel with nearly all the oxygen removed. Silicone and oxygen form compounds that chemically tie up oxygen. This prevents the oxygen from coming out of solid solution as bubbles.

Precipitation Hardening

How does *precipitation hardening (age hardening)* work?
The essential requirements for precipitation hardening are:
- Alloy has at least two phases, B, the base metal, and H, the hardening metal.
- Structure of the B-H alloy phase diagram such that increasing amount of B is soluble in A with increasing temperature, Figure 12–23.

The Process:
- Metals B and H are combined and heated to make a single solid solution in the *solutionizing* treatment temperature range where all B goes into solid solution with A.
- Quenching the alloy rapidly enough to form a super saturated solution with H in B; in other words, there is not enough time for the equilibrium conditions of (B+H) to be formed, so the super saturated solid solution of H in B is created.
- When the alloy is reheated with careful control of time and temperature, H precipitates as very finely divided, uniformly distributed particles *inside* of B's crystals. This stresses and hardens B's structure and is known as precipitation or age hardening.

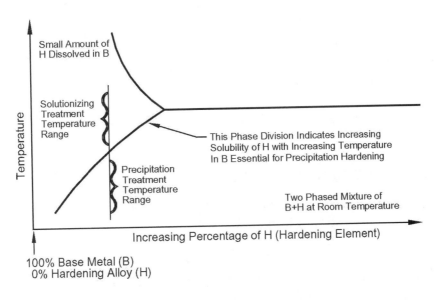

Figure 12–23   Precipitation hardening

What happens if the precipitation hardened alloy is reheated into the solutionizing temperature range?

If heated again into the precipitation temperature range, it will revert to equilibrium conditions and will become softer; its yield strength and hardness will fall. This process is called overaging, Figure 12–24.

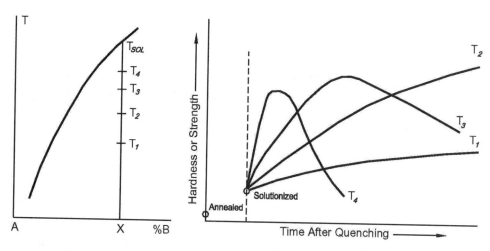

Figure 12–24   Aging temperatures for a given alloy (left). Aging response for this alloy (right); note that hardness in the annealed state as shown on the left axis is lower than the hardness in the solutionized state as shown on the dashed axis

On what metals is precipitation hardening used?
- Aluminum is alloyed with copper
- Nickel with aluminum or titanium
- Copper with beryllium

How Welding Affects Metals

What are the negative effects of welding heat on metals?
- Warping and distortion from high residual stresses caused by localized and uneven heating.
- Cracking from loss of ductility and increasing hardness in the HAZ.
- Reducing joint toughness, particularly in the heat-affected zone.
- Destroying the favorable effects of heat treatment and work hardening done to the metal *before* welding.

What metallurgical processes occur in welding?

Many possible processes, some at the same time:

- Melting
- Solidification
- Gas-metal chemical reactions
- Slag-metal chemical reactions
- Surface phenomena
- Solid state reactions

These reactions occur very rapidly when compared with most other metallurgical reactions in metal refining, forming, casting, and heat treatment. Welding metallurgy studies rapidly changing temperature conditions not often seen in normal metallurgical processes.

How do we usually classify the parts of a weld according to their level of heat exposure during welding?

- Weld metal: This metal has been melted and resolidified and is composed of base metal, filler metal or a mixture of both.
- Heat-affected zone (HAZ): This base metal adjacent to the weld metal has not been melted, but has been exposed to such high temperatures as to have its pre-weld properties modified. The region is not defined by having reached a specific temperature, but by having a *measurable* change in it properties.
- Unaffected base metal, sometimes called parent metal: This area is the remaining original structure that has not seen enough welding heat to change its pre-weld properties.

See Figure 12-25.

What is the nature of weld heat seen by metals?

There is very rapid heating followed by a slow cooling, and in general:

- Peak temperature reached decreased at increasing distance from the weld.
- Time to reach peak temperature increases with distance from the weld.
- Heating and cooling rates decrease with distance from the weld.

See Figure 12–26.

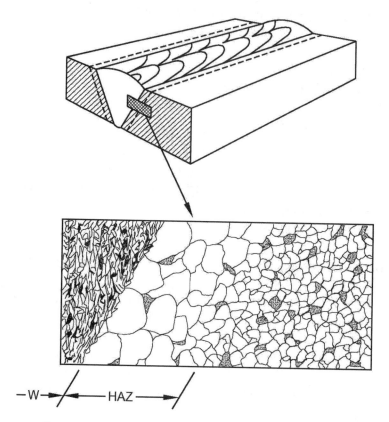

Figure 12–25   Groove weld showing area magnified to see
weld metal, heat-affected zone, and base metal structure

How can we estimate the *heat input* to a weld?

While the actual heat input to the weld is influenced by many complex factors (radiation of heat from the arc not going into the weld, cooling effect of metal, slag and gas leaving the weld area and others) and is hard to precisely calculate, we can get a *rough* numerical idea of the heat input by the following formula:

$$H(joules/inch) \cong \frac{E(volts) \times I(amperes) \times 60}{S(inches/\min)}$$

In this formula S is the travel speed along the weld in inches. Using this formula we can compare the energy input of different welding setups and different processes.

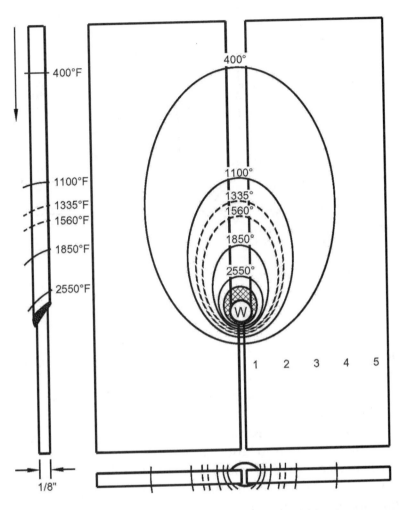

Figure 12–26   Temperature distribution around a weld as it progresses

If we increase the energy input to a weld, how does it affect heating and cooling in the weldment?

Increasing the energy input to the weld slows heating and cooling rates everywhere and heats more of the weldment to a higher temperature. This leads to a larger weld and larger HAZ, Figure 12–27. The advantage to a higher heat input and a larger HAZ, is that the larger HAZ's area can support greater stress than a smaller one can.

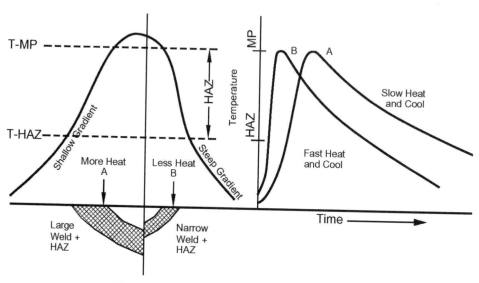

Figure 12–27   How heat input affects HAZ size

Beside energy input, what other factors affect heating and cooling conditions in the weld vicinity?

- Preheat has the same effect as increasing the energy input. It increases the HAZ and retards cooling.
- Increasing plate thickness reduces HAZ width because the larger the plate, the more material there is to absorb heat.
- Thermal characteristics of the metal, called diffusivity, determines how fast heat flows away from the heat source and into the surrounding material. High diffusivity metals like copper and aluminum permit heat to flow more rapidly than iron and nickel. In fact, copper has such a high diffusivity, that a low-energy GTAW arc can be struck without melting the base metal.

Weld-Induced Distortion and How to Minimize It

What causes weld-induced distortion and related stress?

They are caused by nonuniform heating of the weld and base metal. Consider the steel bar of Figure 12–28:

- A. Undistorted bar prior to heating.
- B. Weld arc heat begins to raise metal temperature.
- C. Metal surrounding weld pool heats and expands bending metal downward.

D. Heat continues to flow from weld and soften surrounding steel.
E. As the top portion of the bar begins to cool, it contracts and as it shrinks, bends the bar upward.
F. Residual stresses keep the bar permanently deformed after cooling.

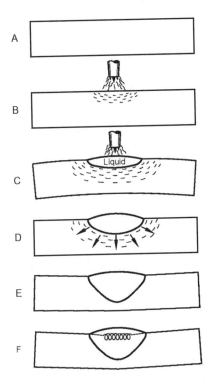

Figure 12–28   Welding causes uneven heating, distortion, and residual stress

What are some of the factors affecting distortion and residual stress?
- Coefficient of thermal expansion
- Melting point
- Phase changes
- Difference in thermal expansion between two metals
- Energy input
- Preheat
- Heat treatment history
- Weld application methods such as *backstepping* and weaving
- External restraint during welding

What are some methods to reduce residual stress?

- *Peening*: The stretching or deforming of the weld surface material by some form of hammering. This approach is difficult to control and may induce tensile stress where there initially was compressive stress. The peening operation may itself cause distortion.
- *Heat treatment*: Much more controllable and effective than peening.

Will one time-temperature heat treatment stress-relief cycle work for all metals?

Absolutely not, each metal and alloy has its own requirements. Consult the metal supplier's literature. Control of hold times and cooling rates is critical to effective heat treatment. There are few applications where a torch can be used for the stress relief of welds because temperature, heat distribution, and heating/cooling rates cannot be controlled. Also, hold times can be measured in hours making torch use impractical for this reason alone.

Work Hardened Metals

How is the HAZ affected in metals or alloys that have been work or strain hardened to increase their strength and what can be done to minimize chances of weld joint failure?

Work or strain hardened metals will recrystallize in the HAZ and soften considerably with the local, intense welding heat. With proper filler metal selection, recrystallization occurs only in the HAZ and is the only area weakened; when correctly chosen the as welded filler metal is as strong or stronger than the base metal when by design.

Joint design must be strengthened to compensate for the weakness in the HAZ when such metals are welded and put under stress. This is important for:

- Wrought iron
- Cold rolled steel
- Drawn or rolled parts of copper, brass, and aluminum

Some brazing processes may even have enough heat to cause weakening in cold worked materials. See Figure 12–29.

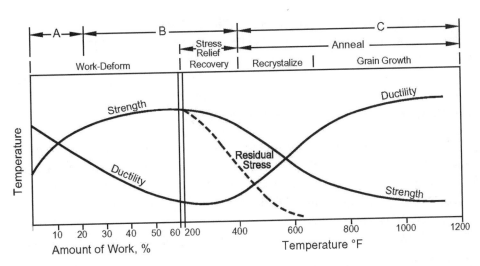

Figure 12–29   The effect of work hardening on a metal (left) and the result of heating this material as in welding appears in zone C, particularly its effect on joint strength (right)

Precipitation Hardened Metals

How is the HAZ affected in metals or alloys that have been precipitation hardened to increase their strength and what can be done to reduce the chance of weld joint failure?

The changes in precipitation-hardened alloys are more complex than work-hardened metal, but the result is similar, metal in the HAZ undergoes an annealing cycle and is softened.

With heat and time, the size and distribution of the precipitate that gives the alloy its strength begins to grow and agglomerate. The effect of precipitation hardening diminishes. The higher the temperature generated by the weld, the shorter the time it takes to reach the weakened, *over-aged* state.

Careful selection of filler material so the deposited material matches the base metal aging characteristic is essential for post-weld heat treatment.

Solid Solution Hardened Metals

How is the HAZ affected in metals or alloys that have been solid solution hardened to increase their strength and what can be done to reduce the chance of weld joint failure?

Solid solution hardened metal shows the least changes in the HAZ compared with the three other hardening methods. There will be grain growth next to the fusion line, but usually only a few grains in width. This will have little effect on the metal's properties.

Transformation Hardened Metals

How is the HAZ affected in metals or alloys that have been transformation hardened (heat-quench-tempering-heat treatment cycle) to increase their strength and what can be done to reduce the chance of weld joint failure? Transformation hardened alloys that have formed martensite in previous heat treatments or have enough hardenability to form it if they have not been previously heat treated, respond in much the same way as solid solution hardened metals.

Figure 12–30 shows four regions: Region 1, closest to the weld, shows austenitic grain growth. This region can readily form martensite on cooling. Region 2 may not show much grain growth, but may still transform to martensite if the cooling rate is fast enough. In region 3, some austenite grains may form, but they are very fine. Region 4's ferrite grains will be softened by the welding heat.

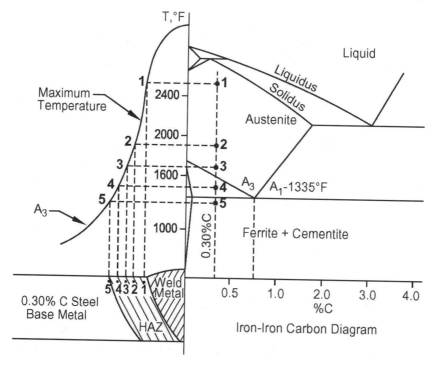

Figure 12–30   Temperature profile in transformation hardened metal

Heat input determines the width of the HAZ as well as the width of each region within it. Higher heat input creates a wider HAZ than lower heat input. Also, higher heat inputs lead to slower cooling rates which are less likely to produce martensitic regions. Preheating the weld to reduce cooling rate will reduce martensite formation and weld brittleness. Post weld heat treatment may also be needed.

The hardness of the HAZ is a good indicator of the martensitic content and likelihood of cracking. At hardnesses below 250HB, cracking rarely occurs, but is common at 450HB and above unless precautions are taken.

Why is post-weld heat treatment helpful in maintaining weld joint strength?
First, it will soften or temper any martensite or bainite formed in the HAZ, and second, it relieves stress in the weldment which may lead to future crack formation.

Heat Treatment of Metals

What are the purposes of heat treatment?
- Alter balance between ductility, hardness, toughness, or tensile strength
- Change grain size
- Improve machinebility
- Improve magnetic or electrical properties
- Modify chemical composition and properties of the surface (case hardening)
- Re-crystallize cold worked metals
- Relieve stress

What are the principal factors in heat treatment?
- Temperature
- Time
- Rate of cooling
- Chemical nature of surrounding materials

What methods are used to heat metals for heat treatment?
- Torches (oxyfuel or fuel-air)
- Furnace heating
- Induction heating
- Electrical resistance heating
- Salt baths heated by electricity or natural gas
- Molten metal baths heated by electricity or natural gas

What methods are used for cooling metals in heat treatment?
- Gradual cooling in furnace
- Cooling in still air
- Fan cooling of part
- Water spray on part
- Water cooling of part
- Cooling buried in sand

What are some heat treatments of metals and how are they done?
- *Annealing* is done to soften metal before cold working it to make it more machinable or to relieve internal stress and strain developed in welding or forming.

    Steps for annealing are:
    1. Heat the metal 50 to 100°F (27 to 55°C) above its A3 temperature.
    2. Hold at this temperature one hour/inch of thickness.
    3. Slowly cool to 50°F (28°C) below its A1 temperature.
    4. Cool to room temperature.
- *Normalizing* makes the metal's internal structure more uniform, improves ductility and reduces internal stresses, but does not soften it as much as a full annealing. It is often performed to prepare the metal to respond better to later heat treatments.
- *Thermal stress relieving* consists of heating the part below the lower transformation temperature and holding the temperature while its internal locked-up stresses are relieved. The part is slowly cooled to room temperature.
- *Spheroidizing* is any process that produces a rounded or globular form of carbide and is used to improve machinability in continuous cutting operations such as those performed by lathes and screw machines.
- *Flame hardening* first uses an oxyacetylene flame to heat the part, then uses a rapid quench to produce a hardened surface. This provides a hardened outer surface to resist wear and a tough unhardened interior to resist shock loads. This process is often used on gear teeth and threads.
- *Case hardening* is the two-step process of carburizing followed by quench hardening. Used on low-carbon steels to harden the outer few thousandths of an inch of the part, it leaves a tough, resilient interior in the unhardened state. The part may be carburized by furnace heating in a carbon monoxide atmosphere, heating the part while packed in a carbon compound, or dipping the part in a commercial product and then heating it. Carbon is absorbed by the steel in the outer skin of the part and makes the steel hardenable with subsequent quenching.

Carbon Equivalent Formula

What is the carbon equivalent formula and what does it tell us?
This formula will produce a carbon equivalent percentage (CE) that indicates weldability—and how much the weld affects the HAZ—of a specific alloy.

From this information, we can adjust the welding process. Here is the formula:

$$CE\ (\%) = \%C + \left[\frac{\%Mn}{6}\right] + \left[\frac{\%Mo}{4}\right] + \left[\frac{\%Cr}{5}\right] + \left[\frac{\%Ni}{15}\right] + \left[\frac{\%Cu}{15}\right] + \left[\frac{\%P}{3}\right]$$

Use Table 12–4 to interpret the carbon equivalent results

Carbon Equivalent %	Adjustments to Welding Process
0.40 or less	No special requirements.
0.40 to 0.60	Use low-hydrogen electrodes.
0.60 or more	Use low-hydrogen electrodes, increase welding heat inputs, pre-heat, post-heat, or slow cooling rates.

Table 12–4   Weld process adjustments based on carbon equivalent

Identification of Metals

Why is proper identification of metals important?
Metal must be identified to select the right welding process and filler metal to make a repair weld or identify unlabeled new material.

What are the main tests by which we can identify metals?
While there are accurate tests available with analytical laboratory instruments like spectrometers, listed below are some rapid, inexpensive means to identify metals in the field.

- Appearance test—The metal's visual appearance often will offer clues to its identity:
  - Many metals have a particular color or luster.
  - Often the shape or application of a part will suggest its material: castings for pump housings, drawn steel for oil pans, copper and brass for heat exchangers.

- – Some castings will show mold marks or grinding where they were removed.
- Chip test—removing a small chip from the edge of the sample with a hammer and chisel, will indicate hardness and ductility.
- Fracture test to examine the structure of failure—Structure revealed by the fracture shows brittleness; grain structure may show; freshly cut metal may have a distinctive color.
- Magnetism test—a small magnet will identify most ferrous metals which range from strongly magnetic to mildly magnetic in the case of some stainless steels.
- Melt test using oxyacetylene torch.
  - – Watch how rapidly the unknown metal melts when compared with known material.
  - – Consider the color of the metal when it melts with known metals.
  - – Look at the surface of the cooled metal pool.
- Ring of metal upon being struck.
- Spark test using a grinding wheel—the color, spark volume and the pattern of spark branching can pinpoint a metal. There is extensive literature with comparative photos of spark patterns. You can compare the pattern of an unknown metal with that of a known metal.
- Specific gravity—density comparison provides more clues to a metal's identity.
- Hardness comparison.
  - – Test with a file and compare with known samples.
  - – Strike a sample of known hardness against the unknown; if the unknown was dented, it is the softer metal.
- Chemical spot test—commercial chemical test kits are available.

While these tests can be helpful in quickly narrowing down the possible composition of a sample, experience is needed. Assembling a kit of samples of known metals for comparison with the unknown can be very helpful.

Chapter 13

Power Supplies & Electrical Safety

*As scarce as the truth is, the supply has always
been in excess of demand.*
Josh Billings

Introduction

All, electrically based, welding processes require a power supply. Their electrical characteristics must match the needs of each process. After reviewing the basics of electricity, we will look at the electrical components of a power supply and then study the difference between constant-voltage and constant-current power supplies. Next, we will follow the development of power supply designs from motor-generator sets to transformer-based supplies to inverters. Finally, we will look at the importance of equipment grounding and ground fault interrupters for personnel safety.

Electricity Fundamentals

What is an *electric current*?
When a group of electrons moves from one point to another, we say there has been flow of electric current or simply, a current.

What is an *electric circuit*?
It is a path along which an electric current can flow. Current flows through a circuit much like water flows through a pipe. The principal requirement of any circuit is to form a complete and unbroken path from one side of the voltage source (for example, a battery or generator) to the other. Hence, current only flows when the switch is closed and completes the circuit path, Figure 13–1.

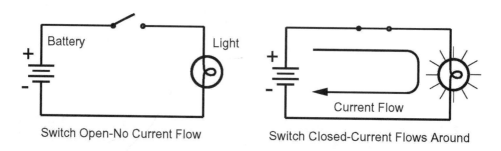

Figure 13–1   Electric circuit

What makes current flow around a circuit?

A *source* provides the electric pressure, called *voltage*, which is measured in volts to push the current around the circuit. The abbreviation for volts is *V*. Voltage is also referred to as *potential* or *electromotive force (EMF)*. Common voltage sources are generators, batteries, and the electric utility company's network. The letter *E* is used to represent voltage in equations.

What units are used to measure electric current?

The unit *amperes*, often called *amps* or *current*, is the unit of current. One ampere represents a very large number of electrons (about 6×10^{18}) moving along the current path every second. The abbreviation for amperes is *A*. We use the letter *I* to represent current in equations. In welding, the setting of a given number of amperes is called *amperage*.

Why does current flow around the circuit and not some other path?

We set up a circuit so it consists of good *conductors* of electricity. This provides an easier or more desirable path for the current than *insulators*. At the same time, we surround the conductors with insulators, poor conductors of electricity. The insulators act like fences confining the current to the desired path. An example of this is an insulated wire: the copper interior is the circuit path; the exterior plastic or rubber insulation is the fence.

What causes the difference between an *insulator* and a *conductor*?

An insulator does not readily conduct electricity because its electrons are not easily stripped from their orbits around the nucleus. If electrons cannot move about easily when a voltage is imposed across a conductor, current cannot flow, and we say that the material has a high resistance to current flow. We can think of resistance just like a constriction in a water pipe creating a reduction in water flow.

A conductor's electrons are readily removed from their outer atomic orbits and can easily move around inside the conductor's atomic structure. Another electron will shortly fill the place of the electron that left. Such materials are said to have a low *resistance*. In the water analogy, a good conductor can be thought of as a large diameter pipe with a smooth-walled interior.

In what units is *resistance* measured?

Resistance is measured in the unit *ohms*, abbreviated by the Greek letter omega, Ω, and symbolized by the letter R in equations. The electrical resistance of conductors and insulators, sometimes called non-conductors, can differ by very large ratios. In fact, a conductor can conduct 10^{20} (that is, 10 followed by 20 zeros) times more current for a given voltage than an insulator of the same size and shape.

In a conductor, what effect does resistance have on the flowing electrons?

In materials with a high resistance, electrons experience friction as they make their way through it. This friction converts electrical power to heat. Lowering the resistance of the conductor lessens heat generated by the same current flow and lessens heat loss. We provide large conductors when large currents flow and we want to minimize heat generation as the current goes from the power source to the load. For this reason, we provide large welding cables with low resistance, so power on its way to the arc is not wasted in heat before it gets there.

In what direction does a current flow?

Using the water flow analogy, electric current moves from a point of more voltage (higher electric pressure) to a point in the circuit of lower voltage (lower electric pressure). Electric current can be thought of as flowing from *plus* (positive), a point of more voltage, to *minus* (negative), a point of lower voltage. Note that sources of DC voltage such as batteries, generators, and welding power supplies have terminals marked with plus and minus signs, Figure 13–2.

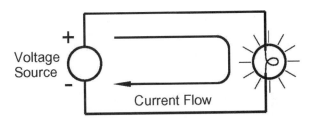

Figure 13–2   Electric current can be thought of as flowing from plus to minus

Why does an electric current go around a circuit in one direction and the flow of electrons in the opposite direction?

Before electrons were identified as minute particles of *negative* charge, scientists arbitrarily designated *electric current* as consisting of *positive* charges flowing from plus to minus terminals of the voltage source. This is commonly called *conventional current flow*.

We now know that electrons are negative particles and flow from minus to plus terminals of the voltage source. See Figure 13–3. Electron and current flow are two views of the same action; however, you must remember that current flows from plus to minus and electron flow goes from minus to plus. Usually, we will be concerned with current flow, but when we explain the movement of positive ions in an arc plasma transferring metal from the electrode to the weld, we must deal with electron flow.

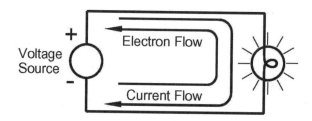

Figure 13–3 Current and electron flow directions

How are voltage, current, and resistance related?

They are related by Ohm's Law:

$$I = E/R$$

This equation tells us that current is proportional to voltage—the more voltage, the more current flows. Also, that at a given voltage, the current is inversely proportional to the circuit resistance: the more the circuit resistance, the lower the current will be. See Figure 13–4.

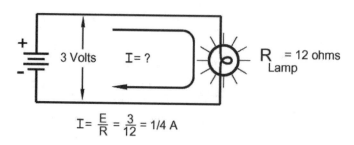

Figure 13–4   Ohm's law in a simple circuit

What unit is used to measure electrical power?

The unit of electrical power is the *watt*, abbreviated *W*. In equations the power in watts is represented by the letter *P*. Power is also called *wattage*. In a resistor, power can be calculated by multiplying the amperes flowing through times the voltage across the load:

$$P = E \times I$$

Also for a resistive load:

$$P = I^2 \times R$$

And:

$$P = V^2 / R$$

Show that the three formulas given for power calculation in a resistor give the same results for the circuit in Figure 13–5.

Using $P = E \times I$:

$$P = 3 \times 1/4$$
$$P = 3/4 \text{ watt}$$

Using $P = I^2 \times R$:

$$P = (1/4)^2 \times 12$$
$$P = (12/16) = 3/4 \text{ watt}$$

Using $P = V^2 / R$:

$$P = (3)^2 / 12$$
$$P = 9/12 = 3/4 \text{ watt}$$

Conclusion: In a resistive circuit any two of the three variables—current, voltage, and resistance—enable us to calculate power used by the resistor. See Figure 13–5. Because welding arcs are resistances we can readily compute power into the arc.

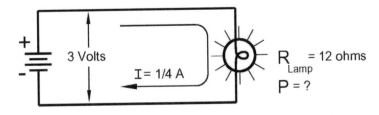

Figure 13–5   Power in a resistor

Now that we have defined the units of voltage, current, resistance, and power, what are some examples of the values of these units we would measure in everyday items during welding operations?

In everyday items:

- Voltage from a car battery: about 13 volts.
- Current used by 100 watt light bulb on 120 volts: 0.8 amperes.
- Resistance of 620 feet (190 m) of number #12 (2 mm) copper wire used in house wiring: 1.0 ohm.
- Power consumed by a 19 inch (48 cm) color TV set: 300 watts.

In welding operations:

- Voltage across GMAW on ¼ inch (6 mm) steel plate: 20 volts.
- Current used by SMAW on ¼ inch (6 mm) steel plate: 200 amperes.
- Resistance of 100 feet (30 m) of 0000 AWG size, also written as #4/0, (about 13 mm diameter) welding cable: 0.0054 ohms.
- Power consumed by FCAW welding ½ inch (13 mm) steel plate: 5500 watts.

What is meant by *polarity*?

In general, polarity refers to the way a circuit or load is hooked to the plus and minus terminals of a DC power supply. See Figure 13–6. To reverse a circuit's polarity is to swap the positive and negative power cables.

In welding, polarity means *electrode* polarity: whether the work or the electrode is connected to the positive terminal of the welding power supply. Many welding-power supply machines have internal switches to reverse polarity. If the machine lacks a polarity-changing switch, the welder need only reverse the electrode and work cables at the power supply. Polarity applies only to DC power sources. AC power sources change polarity continuously at twice the rate of their frequency—120 polarity changes/second for a 60 Hz power source, 100 polarity changes/second for 50 Hz sources.

What is the difference between a *direct current* (DC) and an *alternating current* (AC) source of voltage?

A source of direct current, like a battery, maintains a single output polarity. That is, the positive output terminal *always* remains positive, and the negative terminal *always* remains negative. Since the source polarity remains constant, the direction of current flow does too.

A source of alternating current, like the power company's lines, constantly changes polarity. Every second the AC source reverses polarity 120 times, producing 60 positive half cycles and 60 negative half cycles. One complete positive and one complete negative half-cycle make up a complete cycle of 1/60-second duration. We can say the frequency of the power is 60 Hertz (Hz), meaning 60 cycles per second. Note that power sources of all frequencies can be made. Some welding power supplies generate a high frequency output voltage. See Figure 13–7.

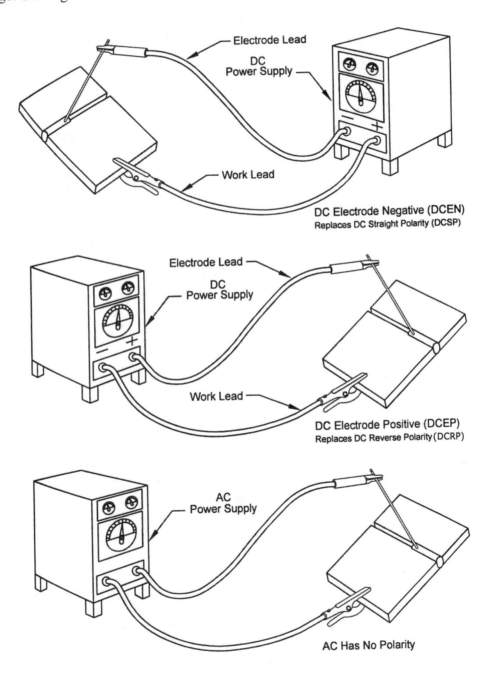

Figure 13–6   Welding polarity designations

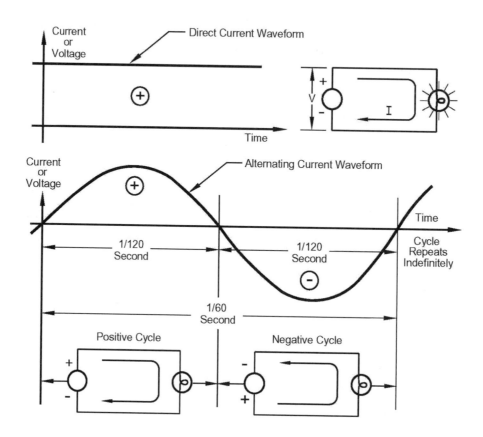

Figure 13–7   DC and AC voltage sources and their voltage waveforms

What is an *ion*?

An ion is an atom which has had electrons stripped from it (leaving it a positive ion), or added to it (leaving it a negative ion). In welding, we frequently see normally neutral metals stripped of some electrons to become positive ions. Metal ions traveling through a welding arc deposit the metal of the welding electrode.

What is *plasma*?

When an electric current passes through a gas in an arc and heats it to high temperatures (8000 to 11,000°F, or 4430 to 6100°C), electrons are stripped away from the gas atoms and surrounding materials creating ions. These ions create a stream of electrons traveling in one direction (remember electrons travel in the *opposite* direction of conventional current) and ions stripped of electrons travel in the other. These ions are positively charged as they were neutral atoms which have lost electrons, and they behave as if they were electric particles flowing from positive to negative polarity.

Electrical Components of Welding Power Supplies

How are AC generators constructed and how do they work?

AC generators are designed so a set of coils, called *windings*, is subjected to a *changing* magnetic field. When the magnetic field cutting through the windings fluctuates as the generator shaft is turned, a voltage is induced in the generator's windings. Since the field passes through a maximum strength in one direction, through zero strength and then a maximum strength in the reverse direction, an alternating current is generated. Whether the coil is fixed and the magnetic field rotates, or the magnetic field is fixed and the coil rotates between the magnet's poles, the result is the same: a voltage appears at the generator's output. See Figure 13–8.

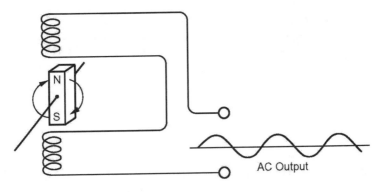

Figure 13–8   Simplified AC generator

How do practical AC generators differ from these simplified models?

Practical AC generators have many more windings and turns of wire in each winding. They usually use additional windings to generate the magnetic field, rather than permanent magnets. Frequently, the field winding, the one making the magnetic field, will have some interconnection or circuitry to the output winding to produce the desired output characteristics.

How are DC generators made and how do they work?

If we want to make a DC generator, we need to introduce a rotating switch, called a *commutator* into the generator. The commutator insures that the voltage and current always flow in one direction by a switching action. Changing the direction of generator rotation changes the polarity of its output. See Figure 13–9.

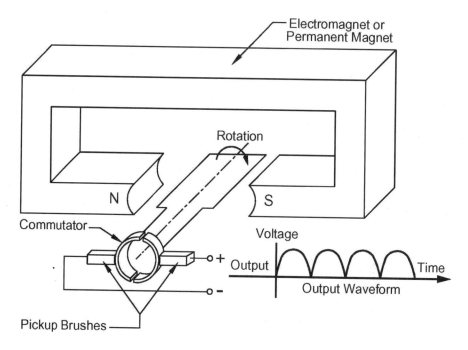

Figure 13–9   Simplified DC generator

How does a practical DC generator differ from the simplified model pictured in Figure 13–9?

Practical DC generators have many coils (20 to 50) with many wire turns on each coil. For each coil there is a pair of commutator bars, but the principle remains the same. There may also be a smaller generator (called an exciter) to supply DC to the field windings on the main generator.

How is a transformer made?

Transformers consist of three components:

- Iron alloy core that is easily and temporarily magnetized by a magnetic field.
- Primary winding consisting of multiple turns of insulated wire wound over the core, and which connects to the power source (AC power lines or engine-driven alternator).
- Secondary winding connected across the load. See Figure 13–10.

The terms *winding* and *coil* are interchangeable. Also, transformers often have more than one secondary winding. Transformers have only one primary winding as the primary is always the winding connected to the source of power.

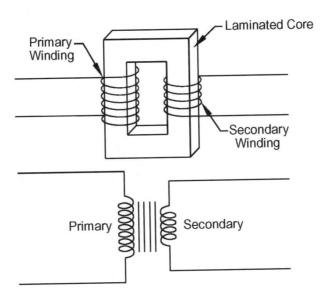

Figure 13–10   Transformer construction (top) and its electrical symbol (bottom)

How do transformers work?

The principle is simple: the voltage relationship between input and output coils is directly proportional to the *ratio* of turns between the secondary and primary coils. More turns on the secondary coil means more magnetic field lines cut it as the AC runs through the coil producing more voltage. Put another way:

$$\frac{N_{primary}}{N_{secondary}} = \frac{V_{primary}}{V_{secondary}}$$

$N_{primary} / N_{secondary}$ is known as the *transformer turns ratio* and directly indicates the ratio between primary and secondary voltages.

This means that if the:

- Secondary winding has fewer turns than the primary, the secondary winding voltage will be lower than the primary winding voltage, and we have a *step-down* transformer.
- Secondary and primary winding have the same number of turns, the secondary will have the same voltage as the primary; also called a *one-to-one* transformer.

- Secondary winding has more turns than the primary, the secondary winding voltage will be greater than the primary winding voltage, and we have a *step-up* transformer.
- Because transformers link the primary and secondary by the constantly reversing magnetic field through the iron core, they only work with alternating current.

Looking at a specific example in Figure 13–11, we see that we can readily calculate the input or primary winding current, since the power into the primary equals the power out of the secondary:

$$P_{primary} = P_{secondary}$$

$$E_{primary} \times I_{primary} = E_{secondary} \times I_{secondary}$$

$$400 \times I_{primary} = 30 \times 100$$

$$I_{primary} = 7.5 \text{ amperes}$$

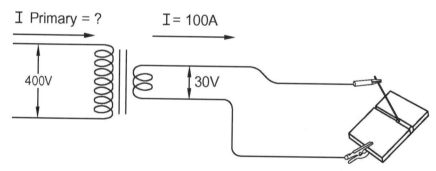

Figure 13–11   Relationships between primary and secondary windings

How does the number of turns in the primary and secondary windings determine the output voltage?

At 60 Hz, the AC voltage to the primary reverses every 1/120 second. As this voltage flows back and forth through the primary winding, it generates a magnetic field in the iron core. This changing field in the core induces (or generates) a voltage across the secondary winding. The critical factor is the *ratio* between primary winding turns (loops of wire around the core of the primary winding) and secondary winding turns (loops of wire around the core of the secondary winding). The fewer turns on the secondary versus the primary, the more current and lower voltage the secondary can deliver. This is exactly what we want for many SMAW operations.

What is a *resistor*, how is it made, and what does it do?

A resistor is an electrical device inserted in a circuit to restrict the flow of current through its portion of the circuit by its property of *resistance*. See Figure 13–12.

Small wattage resistors typically consist of a carbon compound encapsulated in a plastic case. Larger high-power resistors consist of wire with a high electrical resistance wound on a ceramic form.

If a resistor had a value of zero ohms, it would offer no electrical resistance and be the same as a huge wire placed between its terminals (a *short* in electrical terminology). It would also no longer really be a resistor. If it had an infinite, or extremely high resistance, it would look like an open circuit. Between these two extremes lie the resistance values of useful resistors: they will pass some current, but not as much as a short. They are the most common electrical components and useful for limiting the flow of current through a circuit without stopping it completely. In the water analogy, they would be a smaller pipe causing a constriction and reduction in flow between two larger diameter pipes.

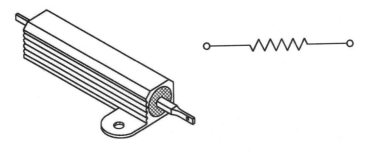

Figure 13–12   Resistor and its electrical symbol

What is an *inductor* and what does it do?

An inductor is built like a transformer with only one winding as in Figure 13–13. It exhibits the electrical property of *inductance* that means it resists only *changes* in the flow of current through it. If no current flows through an inductor, it resists a current flow; if current is flowing, the inductor attempts to keep current flowing when the circuit is interrupted. They are also called *chokes*, *dynamic reactors*, *reactors*, and *weld stabilizers*. They are used to fine tune the

power supply to the job by adjusting the *dynamic response,* the shape of the
waveform when the arc is struck and the time it takes to reach a stable or
steady-state level. Adding more inductance to an SMAW power supply will
reduce the harshness of the arc and reduce spatter.

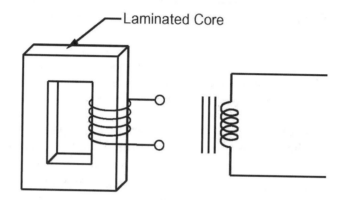

Figure13–13   Inductor and its electrical symbol

What is a *variable inductor* and what does it do?

Variable inductors are made the same way as transformers with adjustable outputs with taps, switches, movable coils, and movable iron cores. See Figure
13–14.

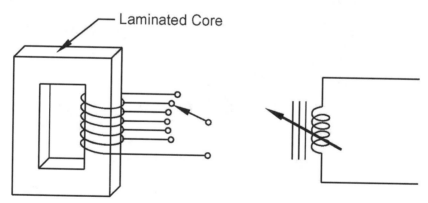

Figure 13–14   A variable inductor or reactor and its electrical symbol

What is a *capacitor* and what does it do?

A capacitor is built so large areas of metal or metallic plates that are insulated
from each other are in close proximity. They are just the opposite of inductance: they oppose *changes* in voltage across their terminals. They will allow
AC to pass through themselves, but not DC. In the water analogy they are like

a fluid shock absorber. Most often they are used to "smooth" the output from a rectifier that is "lumpy" and make it look more like DC to the arc. See Figure 13–15.

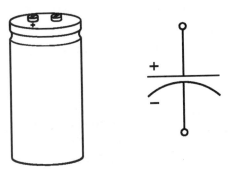

Figure 13–15   Capacitor and its electrical symbol

What is a *diode* and what does it do?

Diodes are check valves for electricity. They will pass current only in one direction, from plus to minus. They block current flow when hooked "backwards" preventing current flow. Rectifiers control only the *direction*, not the amount of current flowing in a circuit. Note the anode (positive) and cathode (negative) designations on the symbol shown below. Current can flow only in the direction of the arrow on the symbol. See Figure 13–16.

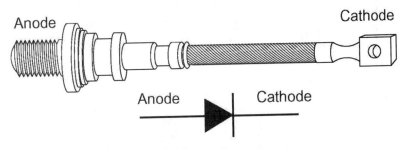

Figure 13–16   Semiconductor diode and its electrical symbol

What is *silicon controlled rectifier* (*SCR*)?

Although an SCR permits current to flow through it in only one direction like a diode, it will not permit any current flow until it is triggered—until a small positive voltage appears across its gate or trigger lead. See Figure 13–17. At that point it will conduct current positive to negative, and will continue to conduct current until the voltage across it is zero. Typically between 1/100 and 1/1000 of the current it switches is needed to trigger the SCR. This means that if an SCR is placed in series with a welding transformer secondary and the welding cables, the SCR can not only turn AC into DC, in conjunction with additional electronics it can control the amount of current available to the welding process. They have only two states, off and on. They can only regulate by turning on at different points in the AC cycle. See Figure 13–18. The earlier in the cycle the SCR fires, the more power that is fed to the load.

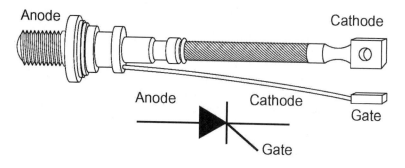

Figure 13–17   SCR and its electrical symbol

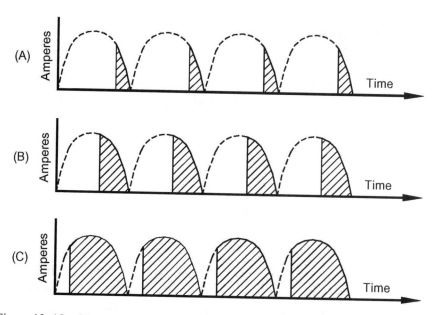

Figure 13–18   SCR's firing at different points on AC cycle. Drawings show SCR firing late in cycle (A), about half way through cycle in (B) and early in cycle in (C)

What is a *transistor* and what does it do?

Like an SCR, a transistor, see Figure 13–19, can also control the flow of electricity through itself from a small control signal, but with the following important differences:

- Transistors have a full range of adjustment from full off to full on and an infinite range of intermediate steps in between.
- Unlike SCRs, transistors pass current only while a control signal is applied. They do not remain on once turned on and can be turned off by cutting off the control signal.
- Transistors can switch—turn off and on—at very high frequencies, much more rapidly than SCRs and so can generate the high frequency pulses and signals used in GTAW. High frequency switching makes possible smaller and lighter power supplies for several processes as the iron core in a transformer decreases with increasing frequency, making weight reductions in transformer size of 70% possible.

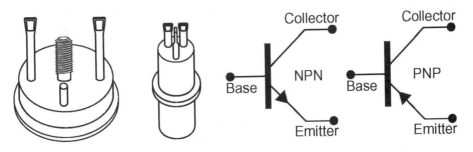

Figure 13–19   Two transistors and the electrical symbols for NPN and PNP transistors

Electrical Utility Service

How is electricity supplied by the utility company?

- Three-wire single-phase service
- Three-phase service

What is the difference between these two types of service and what are their advantages?

Three-wire single-phase electricity service, Figure 13–20, shows the most common means of supplying power to homes, offices, and light industry. This hook-up provides both 110 and 220 volts.

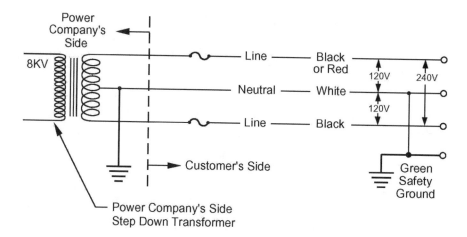

Figure 13–20   Three-wire singe-phase power

Three-phase service is preferred by industry and especially welding operations. Both the electric power company and the user enjoy benefits from this hook-up. In fact, both the generators at the powerhouse and the electric power distribution system *are* three-phase. See Figure 13–21.

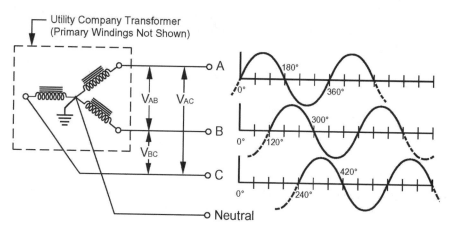

Figure 13–21   Three-phase power

What are the advantages of three-phase electric service over single-phase service?

There are lower losses when transmitting three-phase power than single-phase both on the distribution system and within the user's facility.

- Motors run with three-phase power are more efficient than single-phase motors of the same size.

- Heavy loads on a single-phase system would tend to unbalance the power distribution system, but heavy loads run on three-phase systems do not.
- Most important to welders, when three-phase power is converted to direct current, the result is a much smoother, even output which is better for welding (more on this later).

What does the term *power factor* mean and why is it important?
To the power company a welding transformer load "looks" like both a resistor and an inductive load. Although inductors do not consume power themselves, the power company must supply the current they use and return on each cycle. This charging current that does not get measured on a watt-hour meter, *does* consume expensive power as it goes to and from the power plant through the power distribution network to the customer's inductive load each half AC cycle. In effect, the power company is supplying a charging current at no cost, but must pay for the heat losses incurred in transmission. Power factor measures how much of the load is resistive and how much is inductive. A power factor of exactly one means the load is purely resistive. A power factor less than one indicates the load has inductive currents. Because the power company charges extra fees to industrial users with low power factors consider the power factor of a welding power supply in evaluating competitive models. Power factor can be increased (corrected) by adding capacitors inside the welding machine.

Welding Power Supplies

Why is there a need for welding power supplies?
Welding power supplies are needed for three reasons:
- The voltage supplied by power companies, usually 120 to 575 volts is too high for welding processes that require voltages in the 20 to 80 volt range. Welding power supplies reduce utility voltage to within this range and provide the ability to adjust, or fine tune the power supply's output to the welding process. Were we to try to weld directly from the power line, excessive current would flow, blowing fuses, meting welding cables and making our experiment very short!
- A welding power supply's output characteristics (voltage-versus-current curve) provide arc stability when matched to its respective welding process. This means that the arc is maintained even when the arc length changes as the electrode to work distance fluctuates during the progress of the weld. With the wrong output characteristics, small changes in the electrode-to-work distance would extinguish the arc.

- Also, welding power supplies frequently convert 60 Hz utility-supplied alternating current to direct current, square wave, or to high frequency pulses. All these modifications either make a specific process possible or improve its properties. Modern welding power supplies sometimes modify their output rapidly under microprocessor control to simplify weld starting, weld completion, or sustain the arc in response to changes "seen" at the weld by the machine's electronics. Some welding machines remember the machine settings from a successful job storing them for when they are needed again.

Welding Power Supply Characteristics

What is *open circuit* or *no load voltage*?
This is the voltage measured between the electrode and work when no current is being drawn, typically 50-80 volts for SMAW supplies. It is also the voltage on the vertical axis in a voltage versus current plot of welding machine characteristics. The higher the open circuit voltage, the easier it is to strike an arc; but there is also more risk of electric shock.

What is *closed circuit* or *operating voltage*?
This is the voltage measured across the arc during welding. It will vary depending on the type of electrode, polarity, arc length, and current type. Typically it runs between 17 and 40 volts. Not including the voltage drop of the welding cables, the closed circuit voltage is also the voltage across the power supply output terminals during welding. If we ignore the voltage drop of the welding cables, the closed circuit voltage is also the voltage across the power supply output terminals during welding.

What are the two main classifications of welding power supply output characteristics (voltage-versus-current characteristics)?
- Constant-current (CC)
- Constant-voltage (CV)

How do we determine what the CC and CV curves look like?
We use the test set up with a variable resistive load to simulate the welding arc. By gradually decreasing the load resistance from an open circuit (or infinite ohms) to a short circuit (zero ohms) and recording the voltage and current (as shown in the Table 13–1), we determine the characteristics of the welding machine. These are the *static* characteristics, because we are changing the load resistance slowly. The *dynamic* characteristics determine how power supply

voltage and current respond to rapid changes in the load or arc. These too are important and will be discussed later.

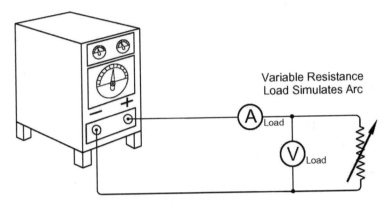

Figure 13–22   Test setup to determine static voltage-versus-current output characteristic curves for a welding power supply

R_{Load}	V_{Load}	A_{Load}
$R_{Load} =$: (open circuit)	Maximum Voltage	0
↓	↓	↓
Decreasing R_{Load} values	Decreasing V_{Load} values	Increasing A_{Load} values
↓	↓	↓
$R_{Load} = 0$ (short circuit)	0	Maximum Current

Table 13–1   Data format for voltage-versus-current output characteristic curves taken with Figure 13–22 equipment

What do the ideal CC and CV output voltage versus current curves look like? See Figure 13–23.

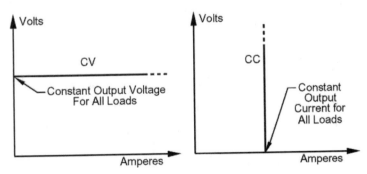

Figure 13–23   Ideal CV (left) and CC (right) output voltage-versus-current curves

What do these curves tell us?
Ideal constant-current (CC) welding power supplies have voltage-versus-current characteristic curves and that maintain a *constant current* to the arc despite changing arc length.

Ideal constant-voltage (CV) welding power supplies have voltage-versus- current characteristic curves and that maintain a *constant voltage* to the arc despite changes in arc length.

What do the output curves of practical CC and CV welding power supplies look like? See Figure 13–24.

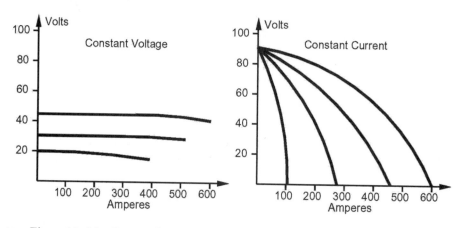

Figure 13–24   Output characteristics of practical CV (left) and CC (right) welding power supplies

How can we weld satisfactorily when the real characteristics differ so much from the ideal characteristics?
Because the slope of the CV curve is almost flat, nearly constant voltage is maintained on the arc with big changes in current caused by changing arc length. The real CV characteristic curve comes very close to the ideal CV one and acts very much like an ideal power supply over its rated range of welding current. Because of fundamental physical laws, it is much easier to build a nearly perfect CV power supply than it is to build a nearly perfect CC one.

Unlike the CV curves, the CC characteristic curves of the ideal and practical power supplies differ greatly. However, since the slope of the CC curve is steep, and changes in arc length create much larger changes in voltage for a given change in current, the power supply acts very much like—but not exactly like—

a CC power source *within the range of current and voltage where welding occurs*. Because of the shape of characteristic curves for CC machines, they are sometimes called *droopers*. See Figure 13–25.

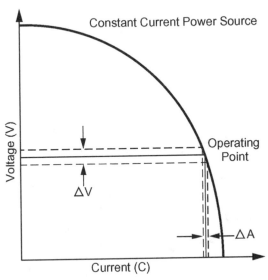

Figure 13–25   How a practical CC power supply approximates an ideal supply within the range of current and voltage where welding occurs

What is the relationship between the type of welding process and the type of output characteristics of its respective power supply?

The issue is which type of power supply for a given electrode feed system provides a stable arc over a wide range of electrode-to-work distances. For welding processes with a manual electrode feed system such as SMAW, constant-current power supplies provide maximum stability and operator control of the weld pool size. And for welding processes with a continuous feed electrode system like GMAW and FCAW, constant-voltage supplies work best because they are *self-regulating*.

When the electrode-to-work distance changes, how does an SMAW arc change with a constant-current (CC) power supply?

As the arc length increases, the weld current is slightly reduced. This causes the arc to spread out over a greater area and the weld pool freezes more quickly. Conversely, with a shorter arc length, the welding current increases and the weld area is reduced making the weld pool freeze more slowly, as shown in Figure 13–26. This adjustment of arc length gives the welder control of the weld pool and is especially useful when welding in the vertical and overhead positions.

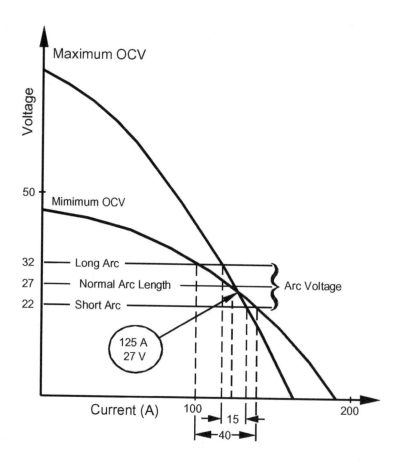

Figure 13–26   SMAW stability with CC power supply

When the stickout distance changes, how does a GMAW or FCAW arc achieve self-regulation with a constant-voltage (CV) power supply?

A typical CV power supply output curve in the welding current and voltage range has a slope of –2 volts per hundred amperes. An increase in output current of 100 amperes causes the output voltage to *drop* 2 volts.

Increasing the stick-out distance (work-to-torch distance) increases the resistance of the arc as seen by the welding machine. In a GMAW or FCAW process, the electrode wire is fed into the weld at a constant speed. For the arc to be stable, the electrode wire feed rate must exactly equal the wire burn-off rate. Here is what happens when the arc length *increases* from a stable arc length at a given wire feed rate:

When the stick-out distance increases, Figure 13–27, point A, arc resistance seen by the welding machine increases. Making the current drop from I_A to I_B

and increasing the voltage from V_A to V_B. A reduced current through the arc slows the wire burn-off rate and allows the constantly feeding wire electrode to shorten the arc and return to the equilibrium at point A. Therefore a decrease in stickout distance causes a decrease in arc resistance, and an increase in arc current causes an increased wire burn-off rate and a re-establishment of the equilibrium condition at point A. This process is called *self-regulation.*

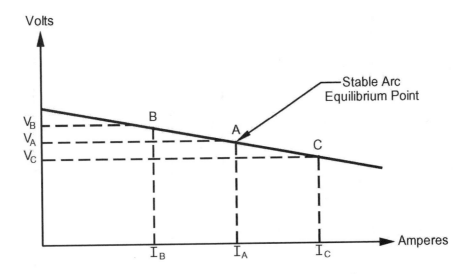

Figure 13–27   GMAW stability with CV power supply

We have talked about one of the two factors that control arc stability, static power supply characteristics, what is the other one?

Dynamic power supply characteristics are the second factor in arc stability. While static characteristics are measured with a fixed (or very slowly changing) load on the power supply, dynamic characteristic determine how the arc current and voltage recover after the supply has been shorted (when the arc is struck). The shape and speed of this recovery curve is usually in the millisecond range and determines how smoothly a process runs. Too rapid a recovery will generate spatter, while too slow recovery will give a cold arc.

Welding power supplies have evolved from quite simple to very complex devices. We will follow the path of this development in time as we review the general classes of power supplies. This table provides an overview of welding power supplies. Note the column reference numbers along the bottom row of Table 13-2 on the following page.

Power Source	Gasoline or Diesel Engine			Electric Utility Power Lines							
Conversion Starts with:	AC from Generator	DC from Generator		AC from Utility Lines @ 120/240/480 volts — Single-phase for light & medium welding machines, portable and fixed — Three-phase for heavy industrial, mostly fixed welding machines							
Means of Conversion	None Required	None Required		Transformer	Transformer + Rectifier		Transformer + Rectifier + SCRs and/or Transistors				AC Motor Drives DC Generator
Output Waveform	AC	DC		AC	AC or DC	DC	AC or DC	DC	Inverter Designs HF Pulses		DC
Output Characteristics	CC	CC	CV	CC	CC	CV	CC	CV	CC	CV	CC
Process(es) Supported											
SMAW	•	•		•	•		•		•		•
GMAW			•			•		•		•	
FCAW						•		•	•	•	
GTAW	•	•			•		•		•		
PAC		•			•		•		•		
PAW					•		•		•		
Column Reference	❶	❷		❸	❹	❺	❻	❼	❽	❾	❿

Table 13–2   Welding power supply classifications.

Welding Power Supplies

What were the first commercial welding power supplies?

The first commercial welding power supplies were motor-generator combinations, sometimes called M-G sets. See ❿ in Table 13–2. These were three-phase AC motors driven by electric utility power; these motors drove DC generators. They worked well for SMAW and were widely used, but are no longer made in the U.S. Most have been replaced by more efficient transformer-based power supplies, which have no moving parts (except perhaps a fan). Today many of these motor-generator sets are tucked away in maintenance departments where their low (electrical conversion) efficiency is not a consideration since they are not in full time use.

What welding power supplies are used for field service where no electric utility power is available?

Gasoline or diesel engines drive AC or DC generators. The simplest arrangement is to have an engine-driven DC generator for SMAW, traditionally the most popular field process (❶ in Table 13–2.). With the increasing popularity of GMAW and FCAW, constant-voltage engine-driven welding power supplies supporting these processes became available (❷). While older engine-driven welding power supplies did not contain semiconductors, most modern ones do. By using semiconductors they can provide AC and DC in constant-current and DC in constant-voltage outputs as well as 110/220 VAC for auxiliary site power. For portable or field use these engine-driven generators are the only choice and are widely used in construction, cross-country pipelining, and oil refinery operations. Progressing from lightest duty to heaviest duty service the engines used are: air-cooled gasoline engines, water-cooled gasoline engines, and diesel engines.

What is the least complicated welding power supply running on utility power?

A transformer-only welding power supply is the least complicated least expensive. It provides constant current AC for SMAW. They are used in automotive shops, farm repair, and hobbyist applications (❸). See Figure 13–28.

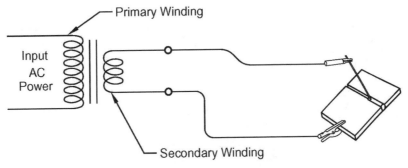

Figure 13–28   Transformer-based AC welding supply

What does a welding machine transformer do?

The transformer converts an AC power source of high voltage and low current input power to a source of high current and low voltage AC power, a *step down*. Welding power supplies use transformers to step down incoming voltage from 120, 240 or 480 volts to between 18 and 80 volts.

How is the output of a welding transformer adjusted?

There are many methods of adjusting the output. Here are some of the most common:

1. Using *taps* on transformer windings, Figure 13–29. Taps are connections to transformer windings in addition to the two at the extreme ends of the winding. Taps effectively change the number of turns on a winding that changes the turns ratio between the primary and secondary windings and thus the output voltage. Taps may be adjusted by a series of plugs, by a switch, or even a continuous slider on the exposed secondary winding. In SMAW supplies taps are on the secondary winding, while in GMAW they may be in the primary winding.

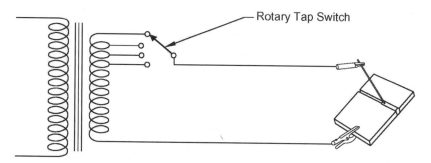

Figure 13–29   Taps on transformer

2. Adjusting the distance separating primary and secondary windings, usually with a crank or lever, changes the magnetic linkage between them that changes the output current. See Figure 13–30

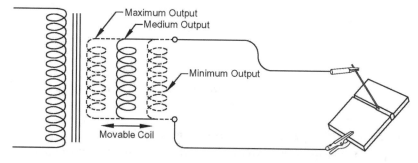

Figure 13–30   Changing the output by changing the separation
and magnetic coupling between the primary and secondary
windings changes the output current

3. By altering the properties of the magnetic path linking the primary and secondary windings by moving parts of the iron core, Figure 13–31.

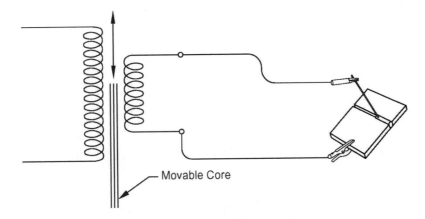

Movable Core

Figure 13–31   Method of electrically altering the iron core
magnetic path and the transformer current output

In what other ways can the output of a welding power supply be controlled?
The welding power supply can be controlled by adding resistance in series with the load as in Figure 13–32. This is very wasteful of power, but there are occasions for its use.

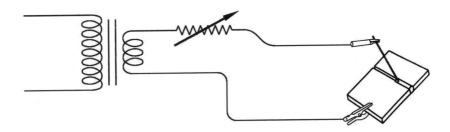

Figure 13–32   Resistance in series with the load

Adding a variable reactor, Figure 13–33

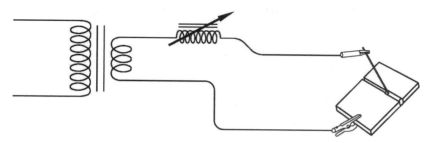

Figure 13–33   Variable reactor added

Note that both modification of transformer output and adding devices in series or parallel with the load are used in all types of welding power supplies with transformers. By adjusting the output of the transformer or the devices in series or parallel with it, we can adjust the output of all types of power supplies whether AC or DC output. These devices cannot only alter the arc properties, but can also change the supply's output characteristics between CC and CV.

What is the next step up in complexity from the transformer-only welding power supply?
The next step is adding DC capability for SMAW. Semiconductor diodes rectify the transformer's AC output to DC (❹). All require 220 VAC or higher service voltage. The larger versions of these designs (250-350 A output and 60% or higher duty cycle) are the workhorses of SMAW industrial welding. See Figure 13–34.

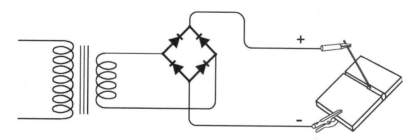

Figure 13–34   Single-phase transformer and rectifier power supply

How do we extend this transformer-rectifier power supply design to three-phase supply? See Figure 13–35.

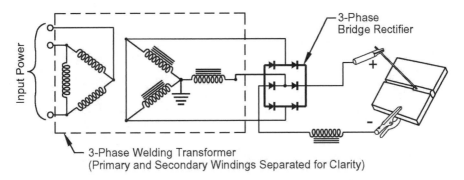

Figure 13–35   Three-phase transformer and rectifier power supply

Why is a three-phase rectifier output better for SMAW than a single-phase one?

Three-phase power provides a smoother, more even output which looks more like DC output to the arc than the rectified output of a single-phase rectifier, Figure 13–36. Engineers say the output has less *ripple*.

Now that we have DC, how do we create a power supply with constant-voltage output for GMAW and FCAW?

In the simpler power supplies with no electronics, we reduce the total *impedance* (the total of AC and DC resistance) in series with the load, usually by reducing the smoothing inductance and other components.

How are SCRs and transistors used in (non-inverter design) welding power supplies?

By changing the point on the AC input power cycle when the SCR turns on, or fires, we can control the output of the welding power supply. In fact, we can eliminate many of the tap, moving coil, and variable inductor schemes used to control output. With electronics, we can sense the voltage or current to the arc and use semiconductors to control the welding machine's output to maintain either CC (❻) or CV (❼) characteristics. With little difficulty we can incorporate both CC and CV characteristics in a single power supply. See Figures 10–37 and 10–38.

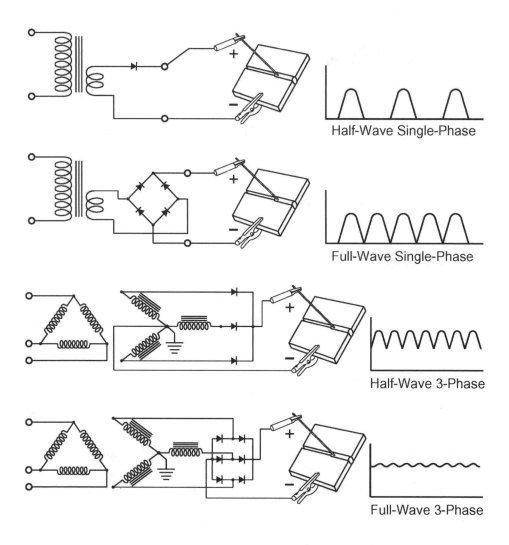

Half-Wave Single-Phase

Full-Wave Single-Phase

Half-Wave 3-Phase

Full-Wave 3-Phase

Figure 13–36   Single-phase and three-phase transformer rectifiers and their outputs

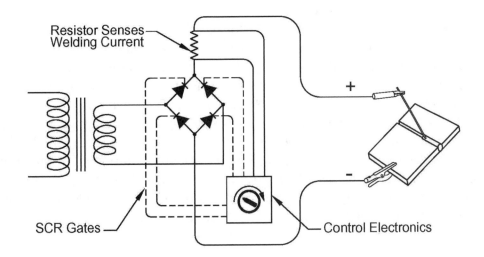

Figure 13–37   SCR-based CC power supply. Control electronics senses
current to load and adjusts SCR firing to maintain CC output

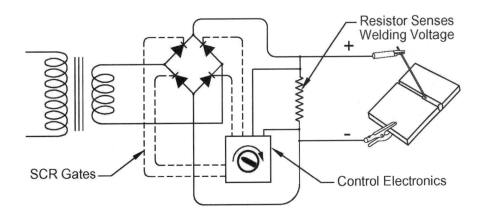

Figure 13–38   SCR-based CV power supply control electronics senses
voltage across load and adjusts SCR firing to maintain CV output

How does a typical inverter power supply design look?

Figure 13–39 shows a typical transistor-based inverter design. Depending on the switching frequency either SCRs (lower frequencies) or transistors (higher frequencies) are used. The incoming AC power feeds a bridge rectifier. The rectifier's output is then "chopped" by a transistor bridge circuit to provide the high frequency AC to drive the transformer. This step-down transformer reduces the voltage and increases the current output. Its output goes to a second high frequency rectifier so all output pulses have the same polarity. The control determines when and how long the transistors turn on. Sensors at the output provide feedback to the control electronics and maintain CC or CV output characteristics. By adding a microprocessor to the control electronics, special waveforms can be added which enhance the welding process.

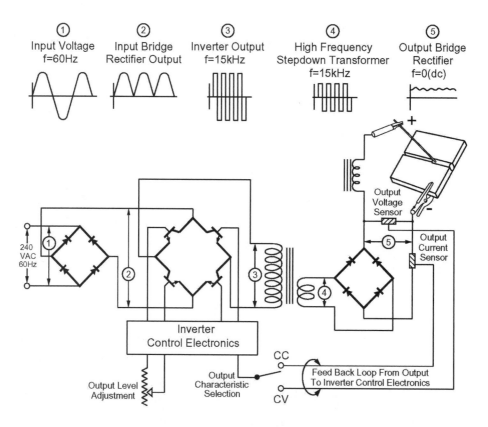

Figure 13–39   Basic transistor inverter with CC and CV outputs

What advantages do inverter-based welding power supplies have over older power supply designs?

- A reduction of weight and volume can be up to 70% over 60 Hz transformer-based designs.
- Control electronics provide multi-process capability in a single unit.
- Increased conversion efficiency and higher power factor than 60 Hz transformer designs.
- Increased portability permits shorter welding cables.
- With microprocessor controls added, it provides rapid adjustment of output to changing conditions at the weld as well as special waveforms.

What is a welding machine's *duty cycle*?

In the US it is the percent of arc time to total time in any ten-minute period that a welding machine can be operated continually at its rated output. In Europe a five-minute period is used. For example, a machine delivering 50 amperes for seven of the ten-minute interval would have a 70% duty cycle. Because a manual welder does not weld all the time, a 60% duty cycle is typically recommended. The other 40% of the time, a manual welder is loading or unloading parts, chipping slag, inspecting work, or changing electrodes. Welding power supplies for automatic welding operations must have 100% duty cycles as they work nearly continuously and may have very long intervals between work stoppages. The factor that determines duty cycle is the speed heat generated inside the power supply can be removed. Excessive internal temperature will destroy critical components.

Can a power supply deliver more output current than the maximum current output at its maximum duty cycle?

Yes, it can deliver more output current than its maximum for its rated duty cycle but for a shorter time. Charts showing this for a particular machine are available from supply manufacturers. Figure 13–40 shows the output curves for welding machines of three different current ratings. All three machines are rated at a 60% duty cycle. These three curves show that each machine can supply *more* output current for a shorter duty cycle than their nameplate ratings. Each can also supply less current for a longer duty cycle than its nameplate duty cycle. In general, each machine can provide the combinations of current and duty cycle represented by their duty cycle curve.

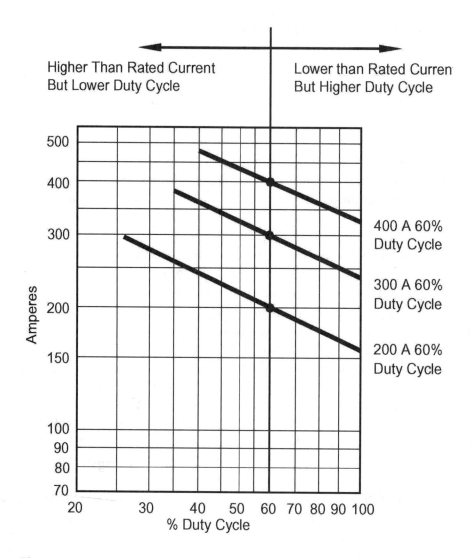

Figure 13–40   Duty cycle versus rated current curve for three welding machines

What are the common duty cycle ratings and who sets them?

The National Electrical Manufacturers' Association (NEMA) has established three classes for welding machines based on duty cycle percentages:

- Class I: rated output at 60%, 80%, or 100%.
- Class II: rated output at 30%, 40%, or 50% duty cycle.
- Class III: rated output at 20% duty cycle.

Class I machines are for heavy industrial use, Class II for automotive mainte-nance repair shops, but not heavy production use, and Class III machines for farm repair and hobbyist use.

How are the input voltages described by NEMA in the three classes of welding power supplies?
- Class I and Class II:
  - @ 60Hz 200, 230, 460, and 575 volts
  - @ 50 Hz 220, 380, and 440 volts
- Class III:
  - 110, 220 volts

Most welding power supplies have taps on the transformer's primary winding to permit operation at two or more input voltages.

Besides the source of power, means of power conversion, output waveform, output characteristics, welding process and duty cycle, in what other ways are welding power sources specified?
Welding power sources are specified by:
- Maximum output current rating—The maximum current within the duty cycle.
- Input voltage and current requirements, single-phase and three-phase.
- Rated load voltage—The voltage across the arc when welding.

What special power supply designs are used for gas tungsten arc welding (GTAW)?
Two separate power supplies are connected in series, Figure 13–41. The AC constant-current supply (high-current, low-voltage) provides the welding power and the spark gap oscillator (high-voltage, low-current) provides a source of high voltage spikes (about 3000 volts) to assist in plasma reformation when the AC cycle reverses and the colder workpiece becomes the cathode. Without these high voltage spikes, to initiate the arc, as the AC waveform goes through zero voltage, the arc formation would be delayed (and the process would run rough) or fail completely on the reverse (DCEP) part of the AC cycle. Note the different scales on voltage for the AC output and the spark out-put. This is discussed further in the GTAW chapter.

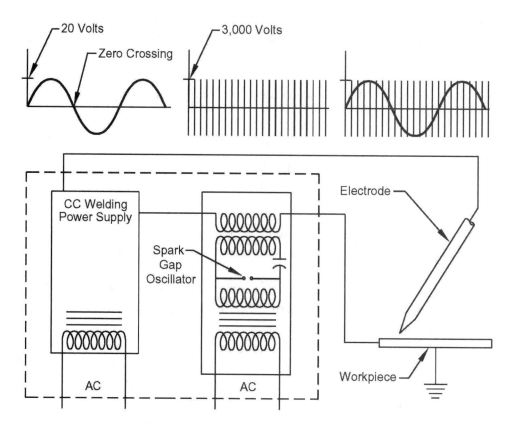

Figure 13–41　GTAW AC power supply: a constant-current supply plus
a spark gap oscillator

What other output waveforms are commonly used in welding ?

- Pulses plus DC waveform—GTAW uses pulses from 2 to 20/second. The idea is to have the metal transfer at high enough current (during the peak pulses) to achieve a forceful arc without overheating the work. GMAW uses a similar waveform to achieve spray transfer during the pulse peaks without overheating the work. The background current provides enough energy to keep the plasma alive, yet prevents the transfer of metal and allows the weld pool to cool between cycles. See Figure 13–42.

- High frequency switched DC (Figure 13–43): GTAW uses pulses of 20 kHz to increase the force of the arc and reduce both arc blow and the effect of wind on shielding gas.

- Square wave + DC—GTAW uses this waveform to provide cleaning action on the DCEP portion of the cycle. See Figure 13–44.

These waveforms and many others are readily produced by inverter power supplies under microprocessor control.

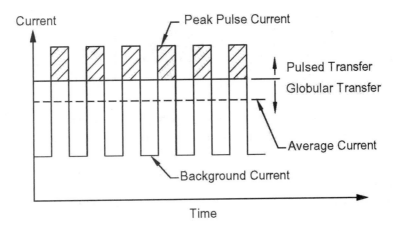

Figure 13–42   Pulsed DC waveform used by GTAW and GMAW

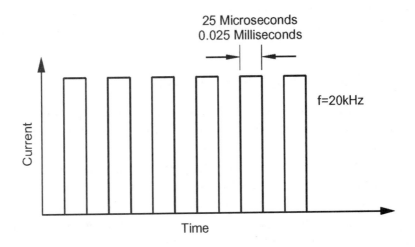

Figure 13–43   20 kHz pulses maintain arc force

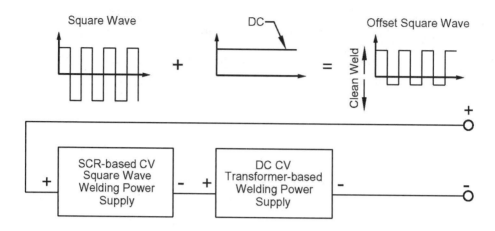

Figure 13–44   Square wave plus DC to provide cleaning action when
welding aluminum with GTAW

Why does welding with square wave (usually high frequency) power supplies have a different feel than 60 Hz powered equipment?

The time interval when the voltage across the plasma is near zero is dramatically reduced when using square wave power supplies. In a 60 Hz sine wave power supply, there is a comparatively long time at every time the voltage reverses when the plasma is not being kept alive by a high voltage, while with square wave supplies the slope of the waveform as it crosses the zero voltage line is very steep, and the change from positive maximum voltage to negative maximum voltage (and back) is very rapid. This means the plasma is energized nearly all the time and so has little time to loose its ionized state. See Figure 13–45.

Welding Power Supply Controls

In an SMAW (CC) power supply, what controls will appear on the welding machine and what effect will they have on the characteristic curves and arc properties?

Most transformer/rectifier CC power supplies will have one adjustment for arc current. This adjustment controls the electron volume of the arc. Increasing the current shifts from curve A, at the low setting, through curves B and C, to curve D, at the highest setting, Figure 13–46. On machines with tapped transformers, each tap point provides a separate current curve. There will be as many curves as there are taps.

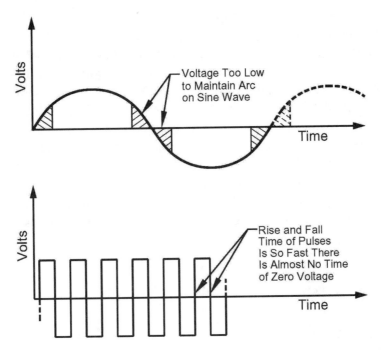

Figure 13–45   Why zero crossing is more rapid with square wave power
supplies than with 60 Hz ones

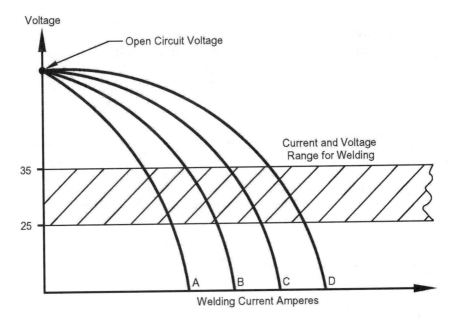

Figure 13–46   Effect of increasing output current on single-control supply

Some machines have two controls: a coarse current control, corresponding to the current control described above, and a fine current control (sometimes labeled *voltage control*). The fine current control (a poor choice of label) really controls the open circuit voltage, and moves point A in Figure 13–47 vertically on the voltage curve and the coarse current control moves point B on the current axis. With these two adjustments, we can set an infinite number of characteristics, one of which will put the operating point at the proper current and with the proper degree of slope at the operating point to get the desired weld and degree of weld pool control. A steeply sloping curve at the operating point will give the operator the ability to control weld pool by arc length adjustment, while a shallow curve, one that provides less change in current with change in arc length provides less control. This may appeal to the beginner as it is easier to control.

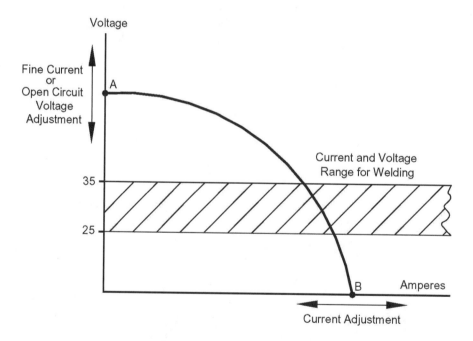

Figure 13–47   Adjustment affects of coarse and fine current adjustments settings on a constant current shielded metal arc welding power supply. Using both these controls permit the welding anywhere within the desirable range of current/voltage range

In a GMAW or FCAW (CV) power supply, what controls will appear on the welding machine and what effect will they have on the characteristic curves and arc properties?

Most machines will have a control for electrode wire feed speed and another for output voltage. There may be an additional control marked *slope*. Figure 13–48 shows how slope control affects output characteristics. For GMAW with non-ferrous electrodes and inert gas or FCAW in large electrode sizes slopes of 21 1/2 to 22 volts per hundred amperes work best; for GMAW with CO_2 and smaller FCAW electrodes, slopes of minus 22 to 23 volts per hundred amperes are preferred. For GMAW and FCAW with short-circuiting transfer, 23 to 24 volts per hundred amperes is optimum. See Figure 13–48.

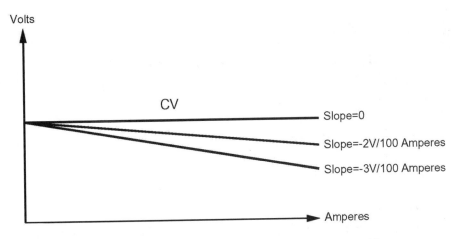

Figure 13–48   Slope control applies only to CV power supplies. Slopes are
exaggerated for clarity

Welding Cables

What voltage appears across welding cables?
Typically 14 to 80 volts. The voltage is measured from one cable to the other.

What is *lead voltage drop* and its typical value in proper application when welding?
Welding cable voltage drop results from the welding current flowing through the resistance of the cables. Remember that although the cables are relatively heavy, the welding current is substantial. This voltage drop is usually in the

range of 4 volts. Higher cable voltage drops will result in excessive power losses, power that should be in the weld, not radiated by the cables.

How are welding cables constructed?
- The conductor is usually copper, sometimes aluminum.
- 400–2500 fine wires make up the cable's conductors and give it flexibility and to reduce welder strain positioning the electrode.
- The conductor bundle is covered with several layers of insulation with an internal fabric or cordage to provide strength and an outer coating of neoprene as a further insulation and wear surface. The entire package is called a cable.

Electrical Safety

What factors determine the severity of electric shock?
- The magnitude of the current flowing through the body
- The duration of the electrical shock
- DC or AC, frequency if AC
- Point in the heart's electrical cycle when shock begins
- Skins resistance
- The currents path through the body

What are some effects of electric shock on humans?
- As more current flows the severity of the shock and injury increases:
- At the lowest current levels, a slight tingling sensation occurs. (1 mA = 1/1000 A)
- A slight shock is annoying but not painful (5 mA).
- Painful shock causes loss of muscle control, inability to let go, burns (6–30 mA)
- Severe shock causes pain, respiratory arrest, severe muscle contractions, deep burns, possibly death; injury possible from shock induced falls (50–150 mA)
- Possible ventricular fibrillation (the heart's electrical rhythm is disrupted, it fires rapidly, but since its chambers pump out of synchronization it cannot pump blood), possible death (1000–4300 mA)
- Can cause cardiac arrest, severe burns, death very probable (10,000 mA)

What happens when a circuit contains a *short*?

A short, or short circuit, permits current to flow through the short and diverts it from flowing through the rest of the circuit where we normally intended it to flow. This is because the short provides an easier—lower resistance—path between the power source terminals than the rest of the original circuit. In welding, shorts can be a serious hazard and even fatal if a human *is* the short and the voltage is sufficiently high. Shorts can also start fires, ruin equipment, trip circuit breakers and blow fuses.

What is the purpose of electrical grounding and how is it accomplished?

Grounding is done to protect men and machines from the effects of short circuits. One side of the power supply is connected to "ground." An earth ground can be obtained by:

- Driving a metal pipe into the earth and connecting the ground line on the equipment to it
- Making a solid electrical ground to the structural steel of a building that is itself properly grounded (proper structural steel grounding is usually required by building codes)
- By connecting the welding machine to the grounding connection of a building's electrical system, usually a green wire, included in all wiring. This wire is usually connected to a grounded water pipe at the point the power lines enter the building
- Grounding connections provide a safe, low resistance path to earth, so that personnel will not be exposed to a short's high voltage. When properly grounded equipment has a short, so much current will flow from the short to ground that a fuse will blow, or circuit breaker will trip shutting down the circuit and alerting the user to a problem. Figures 13–49 and 13–51

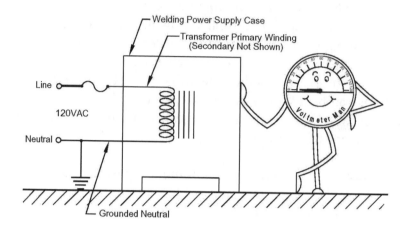

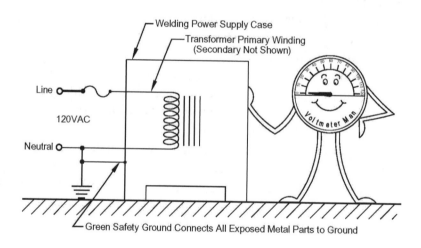

Figure 13–49   Properly fused 120 VAC welding transformer without safety
ground on (above) and with safety ground (below). Shock hazard
exists should transformer insulation fail

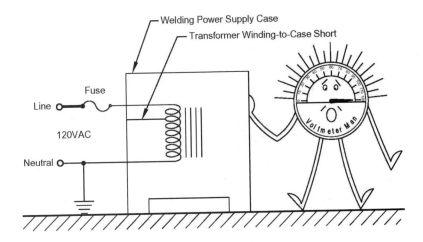

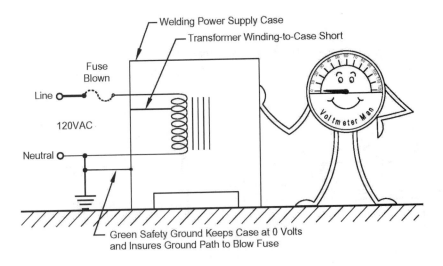

Figure 13–50   Properly fused 120 VAC welding transformer with winding-to-case
short without safety ground (above) and with safety ground (below).
Grounding holds case at ground voltage and blows fuse

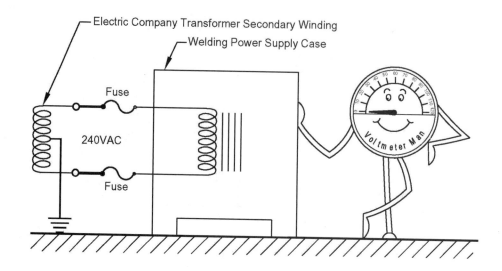

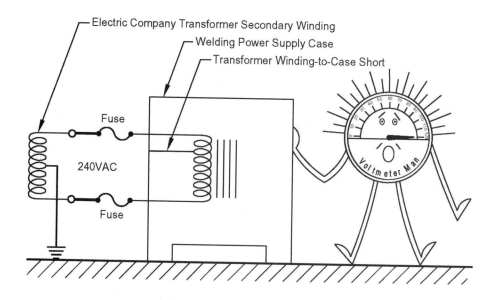

Figure 13–51   Properly fused welding transformer with 240 VAC input
without winding-to-case short without safety ground (above)
and with winding-to-case short (below)

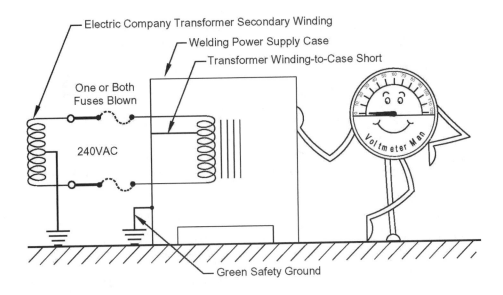

Figure 13–52   Properly fused welding transformer with 240 VAC input with
safety ground and winding shorted to case. Grounding of case holds
case voltage to ground and blow one or both fuses

What equipment is used to protect men and equipment when they are in wet areas such as a rain-flooded construction site and how does this equipment work?

Ground Fault Interrupters (GFIs) are an inexpensive way to protect men and equipment on a job site, where both the men and equipment may become part of the ground path; GFIs measure the current entering the equipment through one side of the power line and compare it with the current leaving the equipment on the other side of the power line. If there is a difference between these two currents, the GFI breaks the circuit to the load in just a few milliseconds. The assumption is that current imbalance to and from the load is due to a short to ground. GFIs interrupt power to the load so rapidly that the current through the short to ground does not have a chance to injure or kill personnel. Note that GFIs will *not* protect a person who is not grounded who gets *across* the power terminals; there will be no current imbalance and the GFI will not detect the problem. GFIs, good equipment grounding, and proper fusing should all be used to provide maximum protection. Note in Figure 13–53 that even though the welding machine was not properly grounded, the GFI provided protection. See Figure 13–52.

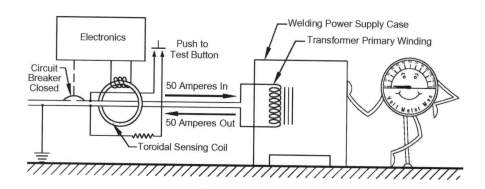

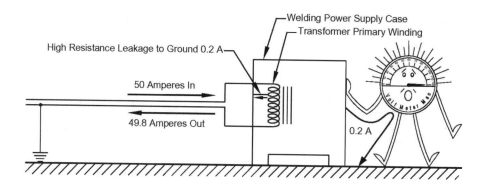

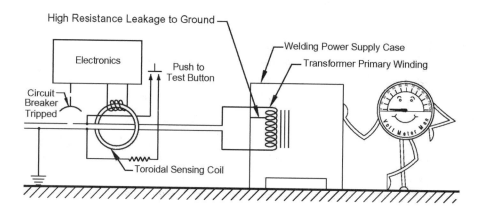

Figure 13–53   Ground fault interrupter leaves circuit breaker closed with equal current into and out of load (top), machine without GFI with high resistance ground to case does not blow fuse, but poses an electrocution hazard (middle)and GFI detects small leakage to ground through man and trips out load before harm is done (bottom)

Why do GFIs provide workers better protection than fuses alone?
It is possible to have a high resistance short to ground so that the case or frame of the machine is capable of supplying enough current to electrocute a man *without* enough to blow a fuse. It takes much less current to kill or injure a person than it takes to blow a fuse or circuit breaker.

Safety

What are the major electrical safety points to remember?
- Make sure all equipment has a safety ground and that it is connected at the power feed point
- Do not weld during thunderstorms, especially when welding on high steel. A lightning strike on welding cables (or power cables) could be fatal.
- Use GFIs on all outside equipment
- Keep all power and welding cables in good condition.
- Do not stand on wet ground or concrete when welding, working on welding power supplies or operating switches
- Be careful in rescuing a person who has been electrocuted so that you too do not become a victim. If you are unsure what to do, wait for the rescue service
- Know the location of breakers and fuse boxes for emergencies.
- Turn off external power to welding power supplies before attempting to service them

Chapter 14

Qualification & Certification

Ignorance once dispelled is difficult to reestablish.
Laurence J. Peter

Introduction

Most welding codes, standards, and specifications include a plan to assure production welds provide the quality required for the product. These plans usually include methods to verify that the welders who will apply the welds have the skill needed; this is done by welder qualification and certification documents. We will consider why welding codes are used, where they come from, which ones are most common, and where they are required. We will present typical qualification test specimens, weld test methods, and weld acceptance criteria. We will discuss the need for developed and documented Welding Procedure Specifications (WPS).

Welding Codes

Why are welding codes (including standards and specifications) used?
Since codes have evolved over many years, they contain a body of collective experience on how to make sound welds. Codes are arranged as a systematic and comprehensive set of rules and standards for welding applications which are mandatory where the public interest is involved. By following established codes, weld reliability is greatly increased. Codes are updated continually as processes and product requirements change.

Where do welding codes come from?
Many codes are issued by professional organizations such as the American Welding Society (AWS) and the American Society of Mechanical Engineers (ASME) and trade associations like the American Petroleum Institute (API).

They are known as consensus standards. Committees of senior engineers and scientists within these organizations establish and update these codes.

Frequently government organizations adopt consensus codes outright giving them the force of law. Sometimes governments use them as a basis for their laws and adding their own modifications.

Who determines which code to use for an application?

Federal, state, city, and provincial laws mandate codes for many applications. When the code is not determined by law, the contract between the manufacturer and buyer may specify the code that applies. Typical examples of welded products where welding codes are a matter of law are:

- Aircraft
- Construction equipment
- Industrial machinery
- Nuclear reactors
- Ordinance
- Pressure vessels
- Railroad rolling stock
- Ships, barges, drilling rigs
- Storage tanks
- Structural steel for buildings
- Bridge construction

What are some of the most common welding codes?

Here are the three most popular codes in the US:

- *AWS D1.1, Structural Welding Code—Steel*
- *ASME Boiler and Pressure Vessel Code, Section IX (Welding Qualifications)*
- *API STD 1104, Standard for Welding Pipelines and Related Facilities*

Like most codes, these require qualification and certification of welders.

Welding Procedure Specification

Where does the qualification and certification cycle begin?

Before we can evaluate the skill of welders to make a joint, we must first define the joint itself and the process to make it. To do this we must develop a *welding procedure specification* (WPS). This document provides detailed information on welding or variables for a specific application to assure repeatability by properly trained welders or welding operators and technicians.

What welding variables are described in the WPS?

The welding variables in the WPS are:

- Welding process
- Base material
- Base material thickness
- Pipe diameter and schedule
- Filler metal
- Electrical characteristics
  (polarity, current, voltage,
  travel speed, wire feed speed,
  mode of metal transfer,
  electrode size)
- Weld type (groove, fillet)
- Welding position(s)
- Shielding gas
- Preheat conditions
- Post-weld heat treatment
- Welding progression
  (upward or downward)
- Backing (metal, weld metal,
  flux, gas)

What happens now that a WPS exists?

The Welding Procedure Specification must be qualified (proven) to show that joints made with it meet the prescribed requirements. This is done by recording the actual welding conditions used to make acceptable test joints and the results of the tests on the weldment in a *Procedure Qualification Record* (PQR). On this basis an approved WPS is issued.

How can a PQR be developed?

Development of a Procedure Qualification Record may be accomplished by following the guidelines, of a code. Codes will list the essential variables necessary to develop a successful welding procedure. Following these guidelines welds are produced and tested until it is proven the variables developed will produce a sound weld consistently. These variables include such things as:

- Joint design
- Base metals
- Filler metal
- Welding positions
- Welding technique
- Electrode diameter (in GTAW electrode type)
- Electrical characteristics (polarity DC+ DC- AC)
- Electrical parameters
- Shielding gas or gases

Qualification and Certification

Now that an approved WPS exists, are we ready to test welders?

Yes, given the WPS welders are required to demonstrate that they can produce the joints it describes. This testing process is called *welder performance qualification.*

What do these documents look like?

The ASME section IV and API have examples of qualification documents as does the American Welding Society's D1.1 structural code appendix. The documents include examples of Procedure Qualification Records (PQR), Welding Procedure Specifications (WPS) and Welder, Welding Operator, or Tack Welder Qualification Test Record. See these forms at the end of this chapter.

Can a common qualification and its welding procedure specification test be described?

A common qualification is the American Welding Society unlimited structural welding test. This is a pre-qualified complete joint penetration groove welded joint as detailed in the *AWS, D1.1 Structural Welding Code—Steel*, single V-groove weld butt joint. Figure 14–1 shows this joint and also the locations from which the side-bend test specimens are to be taken. This test can be welded with SMAW, GMAW, or FCAW.

Where can one find pre-qualified welds?

The AWS code books have pre-qualified weld joints in the D1.1 structural code these joints may be found in figures 3.3 and 3.4. The AWS and other codes have a standard for writing PQR's and WPS's. See the forms at the end of this chapter.

What positions may be used to produce the weld in Figure 14-1?

The welding can be performed in any of the four welding positions or combinations of positions. The welding process and the weld position(s) used in the test determine the qualification if successfully welded in:

- Flat position—the welder's qualifications are limited to flat position and horizontal, complete, and partial joint penetration, groove welds, and fillet in the flat and horizontal positions.
- Horizontal position—the welder is qualified for flat and horizontal complete, and partial joint penetration grooves and fillets
- Vertical position—the welder is qualified for flat, horizontal, and vertical complete and partial joint penetration grooves and fillets.

- Overhead position—the welder is qualified for flat, and overhead complete and partial joint penetration groove and fillet welds
- Both vertical and overhead—the welder is qualified to weld grooves and fillets in all positions

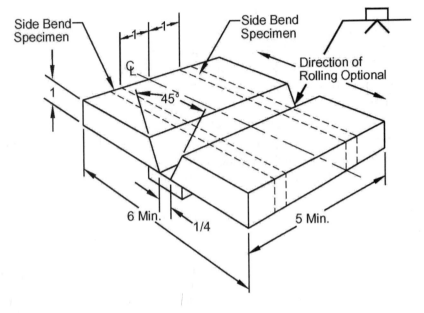

Figure 14–1   Test plate dimensions

Testing Qualification Welds

What type of weld testing is required for qualification?

All codes define test procedures to determine whether qualification welds meet their requirements. For groove welds, guided bend test specimens are cut from specific locations in the welded plates and bent in specified jigs. Because fillet welds do not readily lend themselves to guided bend tests, fillet welds are usually subjected to weld break tests or macro-etch tests or both. In most cases, testing includes one or more of the following:

- Visual inspection
- Guided bend tests
- Tensile tests
- Fracture tests
- Macro-etch tests
- Micro tests
- Radiographic tests

What are the visual examination's criteria?

After welding is complete, the welds must be visually inspected in accordance with the *AWS, D1.1, Structural Welding Code—Steel, Section 4.8.1.* Visual inspection for acceptable qualification requires welds:

- Must be free of cracks.
- Have all craters filled to the full cross-section of the weld.
- Have the face of the weld flush with the surface of the base metal.
- Undercut shall not exceed $1/32$ inch (1 mm).
- Weld reinforcement shall not exceed $1/8$ inch (3 mm).
- The root of the weld shall be inspected, and there shall be no evidence of cracks, incomplete fusion, or inadequate joint penetration. A concave root surface is permitted within the limits shown below, provided the total thickness is equal to or greater than that of the base metal.
- Maximum root surface concavity shall be $1/16$ inch (1.6 mm) and the maximum melt-through shall be $1/8$ inch (3 mm).

If the visual examination is good, what other testing is needed?

We may also test the integrity of the weld by non-destructive methods such as x-rays or ultrasound. If these tests are performed and the specimen meets acceptance criteria according to code, it may be used. These test criteria are found in the *AWS, D1.1, Section 6, Part C.*

What mechanical testing is used to evaluate weld integrity?

Mechanical testing is a common laboratory examination of a weld. This method is destructive, so the weld is not useable after testing. Mechanical testing is performed as follows: Root, face, and/or side bend specimens are prepared, Figure 14–2. Any convenient means may be used to move the plunger member with relation to the die member. See Figure 14–3.

- The specimen is placed on the die member of the jig with the weld at mid-span. Face bend specimens shall be placed with the face of the weld directed toward the gap. Root bend and fillet weld soundness specimens are placed with the root of the weld directed toward the gap. Side bend specimens are placed with that side showing the greater discontinuity, if any, directed toward the gap.
- The plunger forces the specimen into the die until the specimen becomes U-shaped. The weld and heat-affected zones must be centered and completely within the bent portion of the specimen after testing.

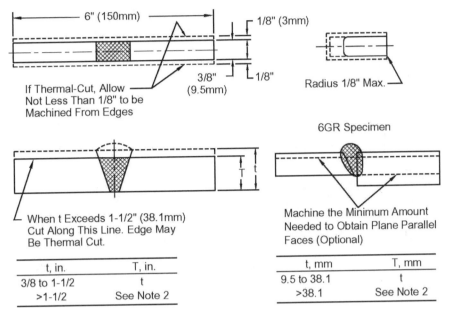

Figure 14–2   Bend test preparation.

t, in.	T, in.
3/8 to 1-1/2	t
>1-1/2	See Note 2

t, mm	T, mm
9.5 to 38.1	t
>38.1	See Note 2

Notes:

1. A longer specimen length may be necessary when using a wraparound-type bending fixture or when testing steel with a yield strength of 90 ksi (620 MPa) or more.
2. For plates over 1-1/2" (38.1mm) thick, cut the specimen into approximately equal strips with T between 3/4" (19.0mm) and 1-1/2" and test each strip.
3. t = plate or pipe thickness.

After the destructive tests are complete what determines acceptability?

The convex surface of the bend test specimen is visually examined for surface discontinuities. For acceptance, the surface must contain no discontinuities exceeding the following dimensions:

- $1/8$ inch (3 mm) measured in any direction on the surface.
- $3/8$ inch (10 mm)—the sum of the greatest dimensions of all discontinuities exceeding $1/32$ inch (1 mm) but less than or equal to $1/8$ inch (3 mm)
- $1/4$ inch (6 mm)—the maximum corner crack, except when that corner crack resulted from visible slag inclusion or other fusion type discontinuities, then the $1/8$ inch (3 mm) maximum applies
- Specimens with corner cracks exceeding $1/4$ inch (6 mm) with no evidence of slag inclusions or other fusion type discontinuities are disregarded, and a replacement test specimen from the original weldment shall be tested

If the weld has met all of the above requirements, a welder qualification report is completed only after certification by approved agency personnel.

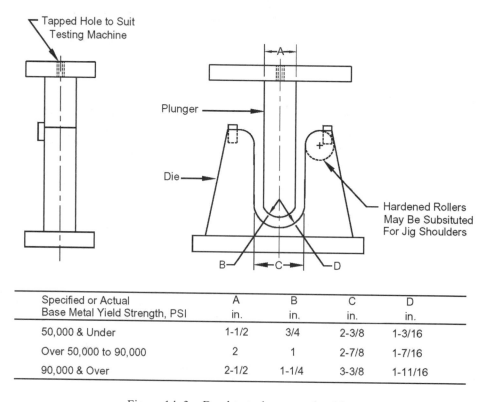

Specified or Actual Base Metal Yield Strength, PSI	A in.	B in.	C in.	D in.
50,000 & Under	1-1/2	3/4	2-3/8	1-3/16
Over 50,000 to 90,000	2	1	2-7/8	1-7/16
90,000 & Over	2-1/2	1-1/4	3-3/8	1-11/16

Figure 14–3   Bend test plunger and guide

What is the required size of a fillet weld break test?

A fillet weld break test requires the weld be continuous the entire length of the joint with a minimum length of 6 inches (150 mm), or if welding pipe, a quarter section shall be loaded in such a way that the root of the weld is in tension. At least one welding start and stop shall be located within the test specimen. The load shall be increased or repeated until it fractures or bends flat upon itself. See Figure 4–4.

How is the fillet weld break test performed?

The fillet weld break test is performed by placing force, as depicted inFigure 14-4, on the back side of the weld this force is usually applied with a hammer or a plunger in a hydraulic press.

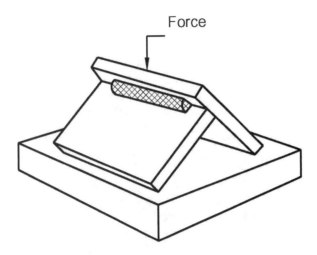

Figure 14–4   Fillet weld break test

What are the acceptable criteria for a fillet weld break test?

The weld must first pass a visual examination prior to the break test. The weld must be reasonably uniform in appearance, free of overlap, cracks, undercut, and have no visible porosity on the weld surface. The broken specimen passes if:

- The specimen bends flat upon itself, or
- The fillet weld, if fractured, has a fracture surface showing complete fusion to the root of the joint with no inclusion or porosity larger than 3/32 inch (2 mm) in greatest dimension, and
- The sum of the greatest dimensions of all inclusions and porosity do not exceed 3/8 inch (10 mm) in the 6-inch (150 mm) long specimen.

What does figure 14-5 describe?

The fillet weld drawing in this figure is showing the size of this fillet weld (5/16" as shown on the welding symbol), the material dimensions, the required stop and start of the fillet weld and how to prepare this weld for breaking and macro-etching.

What is macro-etching?

Macro-etching is the etching of metal surfaces to accentuate the gross structural details and possible discontinuities or defects for observation and evaluation by the unaided eye, or by a magnification not to exceed ten diameters.

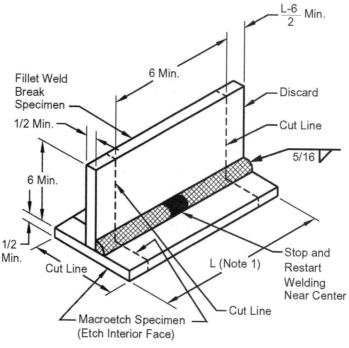

Notes:
1. L=8 min. (welder) 15 min. (welding operator).
2. Either end may be used for the required macroetch specimen. The other end may be discarded.

Figure 14–5   Fillet weld dimensions for a break and or macro-etch test

What testing is required for pipe qualification?

Pipe, after welding, must meet visual requirements similar to plate. Pipe is usually welded with an open root, although some may be welded with welding inserts or internal backing or chill rings. Like plate, depending on pipe wall thickness, pipe is tested by bending four samples either to the side or root and face bending. Location of samples for bend tests is important so the tests are performed at different quadrants of the pipe. The bend tests are performed using the plunger device shown in Figure 14–4. Figure 14-6 shows the locations where test coupons (samples) are taken from pipe welds.

What positions may pipe be welded in?

Pipe, like flat plate, may be welded in the flat, horizontal, vertical and overhead positions. There are two special positions for pipe and those are pipe on a horizontal plane welded either from bottom to top (ASME) pressure vessel welding or top to bottom (API) cross country and distribution petroleum piping. See Chapter 4 Figure 4–13 for pipe welding positions.

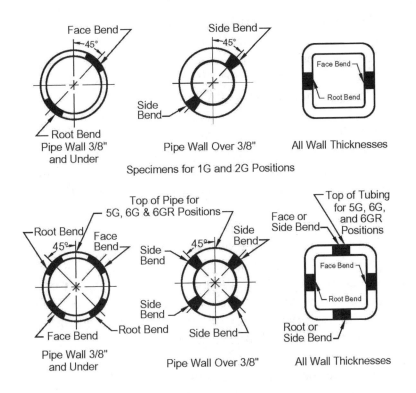

Figure 14–6   Pipe test coupon location

After successfully completing a pipe welding test what is a welder qualified to weld?
After having completed a pipe welding test in positions where the pipe is fixed a welder is considered a master of his trade. Pipe welded in the 6G position qualifies the welder for welding pipe and plate in all positions with only limits on thickness depending on the pipe wall thickness. Wall thickness variables are the same as structural welding of plate; welding pipe with a wall thickness of one inch qualifies the welder for unlimited thickness. Welding with a restriction ring 6GR qualifies the welder for welding not only any position but also any configuration such as production pipe welding of Ts, Ys, and Ks.

Is the welder certified now that he has passed the welder performance qualification?

No. Certification does not occur until an authorized representative of the organization performing the qualification tests certifies (recognizes) the results of this test and issues a welding certificate to the welder. By having a representative of the testing organization certify test results, others outside the welder's company can have confidence in qualification test results. AWS authorized quality control officials are commonly recognized as the testing authority i.e. Certified Welding Inspectors (QC-! CWI).

What is a *certified* welder?

A certified welder is someone who has passed a welder qualification test for a specific weld. Certification is valid for that weld procedure only. The term *certified welder* is widely misused in industry; the term *qualified welder* should be used instead.

If I am a qualified welder, can I weld everything?

Being qualified for one welding procedure does not qualify the welder for all other procedures. When a qualification certificate is issued, it covers only the specific process and welding variables described by the welding procedure specification. The qualification for one procedure may not be transferred to another qualification without further testing.

Who is responsible for the qualification of welders?

It is the employer's responsibility to see that his welders have the proper qualifications and to maintain records of their qualification. However, a welder may take qualification tests on his own to enhance his own employability.

What is the duration of a standard qualification?

The *AWS, D1.1 Structural Code—Steel*, and the *ASME Boiler and Pressure Vessel Code, Section IX (Welding Qualifications)*, have similar rules for the duration of a qualification: A qualification remains in effect indefinitely unless the welder is not engaged in the given process of welding for more than six months, or unless there is some specific reason to question his ability. The *American Petroleum Institute, Standard 1104*, states only that a welder may be required to re-qualify if a question arises about his competence.

Where can I get code books and information about qualification requirements?

Employers who follow a welding code should have codes books on hand for reference. The following organizations publish welding codes and many have informative web sites:

American Association of State Highway and Transportation Officials
444 North Capitol Street, NW, Suite 249
Washington, DC 20001
202 624-5800

Aerospace Industries Association of America
1250 Eye Street, NW, Suite 1200
Washington, DC 20005

American Institute of Steel Construction
One East Wacker Drive
Chicago, IL 60601
312 670-2400
www.aisc.org

American Iron and Steel Institute
1101 17th Street, NW
Washington, DC 20036
202 452-7100
www.steel.org

American National Standards Institute
11 West 42nd Street
13th Floor
New York, NY 10036
212 642-4900
www.ansi.org

American Petroleum Institute
1220 L Street, NW
Washington, DC 20005
202 682-8000
www.api.org

American Society of Mechanical Engineers
3 Park Avenue
New York, NY 10016
800 THE-ASME
www.asme.org

American Water Works Association
6666 West Quincy Avenue
Denver, CO 80235
303 794-7711
www.awwa.org

American Welding Society
550 N. W. Le Jeune Road
Miami, Florida 33126
305 443-9353
800 443-9353
www.aws.org

Association of American Railroads
50 F Street, NW
Washington, DC 20001
202 639-2100
www.aar.org

Society of Automotive Engineers
400 Commonwealth Drive
Warrendale, PA 15096
724 776-4841
www.sae.org

American Railway Engineering Association
50 F Street, NW
Suite 7702
Washington, DC 20001
202 639-2190

WELDER, WELDING OPERATOR OR TACK WELDER QUALIFICATION TEST RECORD

Type of Welder _____

Name _____ Identification No. _____

Welding Procedure Specification No. _____ Rev _____ Date _____

Variables	Record Actual Values Used in Qualification	Qualification Range
Process/Type [Table 4.10, Item (2)]		
Electrode (single or multiple) [Table 4.9, Item (9)]		
Current/Polarity		
Position [Table 4.10, Item (5)]		
Weld Progression [Table 4.10, Item (7)]		
Backing (YES or NO) [Table 4.10, Item (8)]		
Material/Spec. [Table 4.10, Item (1)]	to	
Base Metal		
Thickness: (Plate)		
Groove		
Fillet		
Thickness: (Pipe/tube)		
Groove		
Fillet		
Diameter: (Pipe)		
Groove		
Fillet		
Filler Metal [Table 4.10, Item (3)]		
Spec. No.		
Class		
F-No.		
Gas/Flux Type [Table 4.10, Item (4)]		
Other		

VISUAL INSPECTION (4.8.1)
Acceptable   YES or NO _____

Guided Bend Test Results (4.30.5)

Type	Result	Type	Result

Fillet Test Results (4.30.2.3 and 4.30.4.1)

Appearance _____ Fillet Size _____

Fracture Test Root Penetration _____ Macroetch _____

(Describe the location, nature, and size of any crack or tearing of the specimen.)

Inspected by _____ Test Number _____

Organization _____ Date _____

RADIOGRAPHIC TEST RESULTS (4.30.3.1)

Film Identification Number	Results	Remarks	Film Identification Number	Results	Remarks

Interpreted by _____ Test Number _____

Organization _____ Date _____

We, the undersigned, certify that the statements in this record are correct and that the test welds were prepared, welded, and tested in accordance with the requirements of section 4 of ANSI/AWS D1.1, (_____) Structural Welding Code—Steel.
                                                                                    (year)

Manufacturer or Contractor _____ Authorized By _____

Form E-4   Date _____

Example of the American Welding Society Welder Qualification test record

WELDING PROCEDURE SPECIFICATION (WPS) Yes ☐
PREQUALIFIED _____ **QUALIFIED BY TESTING** _____
or **PROCEDURE QUALIFICATION RECORDS (PQR)** Yes ☐

Identification # _____
Revision _____ Date _____ By _____
Authorized by _____ Date _____
Type---Manual ☐   Semi-Automatic ☐
        Machine ☐   Automatic ☐

Company Name _____
Welding Process(es) _____
Supporting PQR No.(s) _____

JOINT DESIGN USED
Type:
Single ☐   Double Weld ☐
Backing: Yes ☐ No ☐
         Backing Material:
Root Opening _____ Root Face Dimension _____
Groove Angle: _____ Radius (J–U) _____
Back Gouging: Yes ☐ No ☐   Method _____

BASE METALS
Material Spec. _____
Type or Grade _____
Thickness: Groove _____ Fillet _____
Diameter (Pipe) _____

FILLER METALS
AWS Specification _____
AWS Classification _____

SHIELDING
Flux _____ Gas _____
                    Composition _____
Electrode-Flux (Class) _____ Flow Rate _____
_____ Gas Cup Size _____

PREHEAT
Preheat Temp., Min _____
Interpass Temp., Min _____ Max _____

POSITION
Position of Groove: _____ Fillet: _____
Vertical Progression: Up ☐ Down ☐

ELECTRICAL CHARACTERISTICS

Transfer Mode (GMAW)   Short-Circuiting ☐
                        Globular ☐ Spray ☐
Current: AC ☐ DCEP ☐ DCEN ☐ Pulsed ☐
Other _____
Tungsten Electrode (GTAW)
            Size: _____
            Type: _____

TECHNIQUE
Stringer or Weave Bead: _____
Multi-pass or Single Pass (per side) _____
Number of Electrodes _____
Electrode Spacing   Longitudinal _____
                      Lateral _____
                      Angle _____

Contact Tube to Work Distance _____
Peening _____
Interpass Cleaning: _____

POSTWELD HEAT TREATMENT
Temp. _____
Time _____

WELDING PROCEDURE

| Pass or Weld Layer(s) | Process | Filler Metals | | Current | | Volts | Travel Speed | Joint Details |
		Class	Diam.	Type & Polarity	Amps or Wire Feed Speed			

Form E-1 (Front)

Example of the American Welding Society WPS, PQR form (front)

Procedure Qualification Record (PQR) # _____
Test Results

TENSILE TEST

Specimen No.	Width	Thickness	Area	Ultimate tensile load, lb	Ultimate unit stress, psi	Character of failure and location

GUIDED BEND TEST

Specimen No.	Type of bend	Result	Remarks

VISUAL INSPECTION

Appearance_____

Undercut _____

Piping porosity _____

Convexity _____

Test date _____

Witnessed by _____

Other Tests

Radiographic-ultrasonic examination

RT report no.: _____ Result_____

UT report no.: _____ Result_____

FILLET WELD TEST RESULTS

Minimum size multiple pass   Maximum size single pass

Macroetch   Macroetch

1. _____ 3. _____   1. _____ 3. _____

2. _____   2. _____

All-weld-metal tension test

Tensile strength, psi _____

Yield point/strength, psi _____

Elongation in 2 in., % _____

Laboratory test no. _____

Welder's name _____   Clock no. _____   Stamp no._____

Tests conducted by _____   Laboratory

Test number _____

Per _____

We, the undersigned, certify that the statements in this record are correct and that the test welds were prepared, welded, and tested in accordance with the requirements of section 4 or ANSI/AWS D1.1, (_____) Structural Welding Code—Steel.
(year)

Signed _____
Manufacturer or Contractor

By _____

Title _____

Date _____

Form E-1 (Back)

Example of the American Welding Society PQR (back)

Chapter 15

Fabrication & Repair Tips

If a little knowledge is dangerous,
where is the man who has so much as to be out of danger?
Thomas Huxley

Introduction

This chapter contains step-by-step instructions for performing some of the most common welding, brazing, and soldering tasks. There are details on how to make rectangular frames, put legs on a table, and make a three-dimensional solid frame. There are also instructions for repairing cracked truck frames and welding on vehicles. In addition, there is a section on pipe and tubing repairs and tools. Finally, a procedure for soldering copper tubing is given.

Rectangular Frames

What are two ways to make angle iron corners for a rectangular frame?
Mitering and notching. See Figure 15–1. Both methods work, but a beginner might find notching easier, since it is more dimensionally tolerant. Following welding and grinding either fit-up style will result in a good finish.

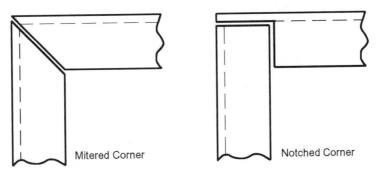

Mitered Corner   Notched Corner

Figure 15–1   A mitered (left) and notched (right) frame corners

Is there another approach to making rectangular frames from angle iron?
Yes, see Figure 15–2. This approach works well when a notching machine is available and lends itself to production work. Getting the correct bend allowance gap is critical, because it provides the extra material needed to go *around* the outside of the corner when the bend is made Figure 15–2 (a). Begin by setting the bend allowance gap to slightly less than the thickness of the angle iron and go from there.

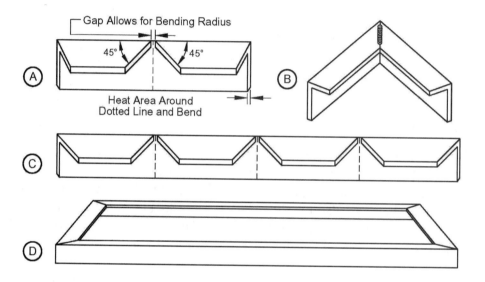

Figure 15–2   Notching and bending to make a single-piece frame: (a) Corner detail before bending, (b) corner detail after bending, welding, and grinding, (c) notched angle iron frame ready for bending, and (d) completed frame

What are the common ways to check for squareness of a frame?
Check for equal diagonals between opposite corners with steel measuring tape. On large frames use a carpenter's square, on smaller ones use a machinist's square. If the sides of the frame are to be plumb and level, a large level can be used. When welding a very large L-shape, where a square is too small and there are no diagonals to measure, use a 3-4-5 triangle: (1) measure off four units (feet, meters or some multiple of same) on one leg, and (2) measure off three units on the other leg, (3) then adjust the hypotenuse, the longest side of the triangle. This procedure makes a perfect right triangle. See Figure 15–3.

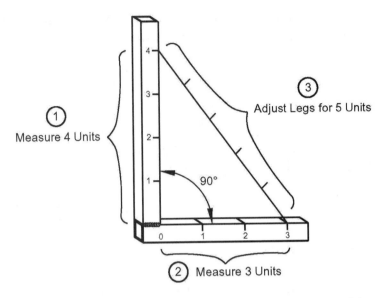

Figure 15–3   How to utilize a 3-4-5 triangle to set members at right angle

How should member length be adjusted when working with larger members?

By reducing the overall length of frame sides to allow for additional joint root spacing otherwise your frame will be oversized.

How can you improve the chances of a welded frame being welded in square?

In *decreasing* order of effectiveness:

- Secure members in a rigid fixture and weld them in it.
- Clamp members to a steel table and then weld them.
- Use a fixture to hold parts for tacking, then weld the tacked parts *outside* the fixture. This fixture can be as simple as a sheet of plywood with wood blocks fixed to it to hold the work in place while the tack welds are made.
- Use Bessy®-type corner clamps. Hint: Begin by tack welding each of the corners together using the clamp each time, then check for square. Bend back into square if needed. If the tack welds are not too large, you'll be able to straighten the frame with moderate force—by hand and without hydraulic jacks. Begin final welds at *opposite* corners. Warning: Making a final weld one corner at a time in a corner clamp will bring poor results— the final two corner pieces are not likely to meet.

- Use magnetic corner tools—these are effective only for light sheet metal as they lack the strength to resist weld-induced distortion forces even with light angle iron, Figure 15–4.

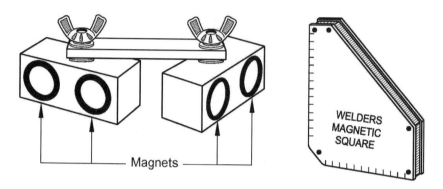

Figure 15–4   Magnetic corner tools

In the field or with large and heavy members, lay the members up square and level on a concrete floor (you may need to shim them to get them flat), tack them together, then weld them. Check for frame to be square and flat after making each tack and bend members back to square and flat *before* making the next tack or weld. Large (and unobtainable) forces may be needed to bring the frame back into square with this method if your tacks are too large. Warning: Welding directly on concrete can cause it to explode violently. Shimming the work off the floor will eliminate this hazard. Distortion control hints: Weld opposite corners first, weld the *same* relative corner or side position in the exact *same* sequence on all four corner joints: First weld all outside faces, then all top corners, finally all bottom faces. Make each weld in the same relative direction: from the outside of the frame to its inside or vice versa. Also, give each weld a moment to cool before making the next.

You have welded a rectangular frame of angle iron (not rectangular tubing) and it does *not* lie flat. Now what?
Follow the steps in Figure 15–5 showing how to bend the horizontal face of the frame to flatten it. Use an open end wrench or fabricate a tool of your own.

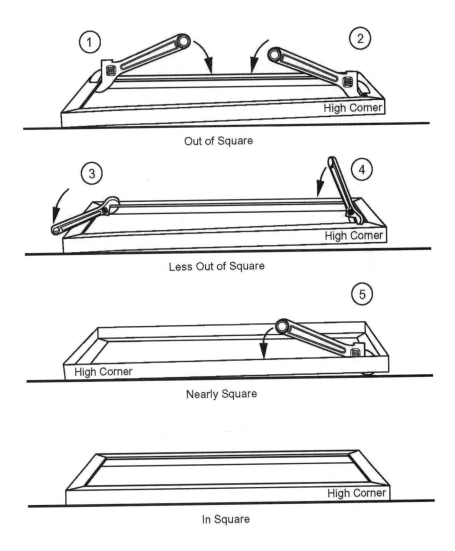

Figure 15–5   Method of adjusting an angle iron frame to lie flat

Mounting Table Legs

You are building a table and have completed the top rectangular frame. How legs (choose angle iron, tubing, or pipe) put on so they are square when welded?

Follow the steps in Figure 15–6.

Put the table frame upside down on a flat surface like the top of the welding table. Use two clamps to lightly secure a leg to both sides of the frame corner.

With a carpenter's square adjust the leg so it is perpendicular to the frame and using a length of steel or wood use two clamps to brace the member to bring it into square. Repeat this squaring/bracing/clamping for the other right angle, Figure 15–6 (b). Fully tighten two clamps holding leg to frame. Re-check for square in both directions and adjust as needed, then weld the leg to the frame, Figure 15–6 (c). Repeat this for each leg.

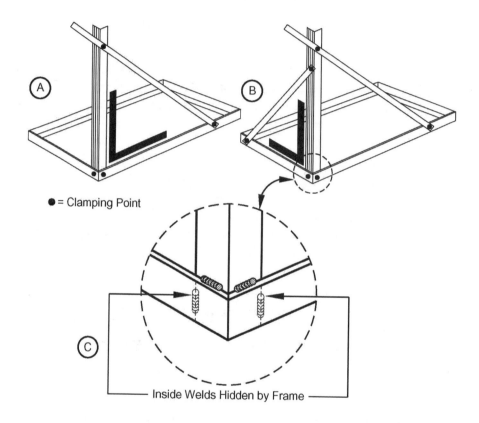

Figure 15–6   Welding table legs on square

Box Frames

You are making a rectangular solid box frame. How should you do this?
Make the upper and lower frames as described previously. Use clamps to secure the four verticals to the lower frame and make them square to the lower frame, Figure 15–6. Place the completed upper frame over the legs. Make

whatever compromises and adjustments are needed to the verticals to make them meet the upper frame. Some tweaking may be necessary. Note that all opposite diagonals, like dotted line X-Y, will be the same length in a *rectangular* box. Clamp the legs to the upper frame. Tack all of the joints, for square/fit, weld all joints. This method will work equally well with angle iron or rectangular tubing. See Figure 15–7.

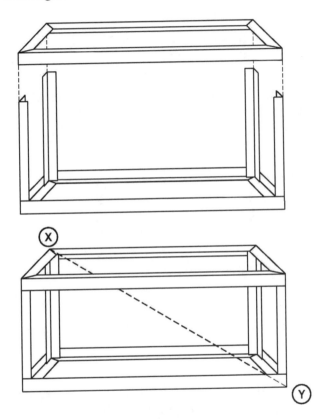

Figure 15–7   Fabrication of a box frame

Cracked Vehicle Frame Repairs

There are prominent labels on the C-channel frames of modern tractors and heavy trucks warning against cutting or welding on them. Why is this warning there?

To save weight thinner, lighter, steel U-channel members with a special heat treatment to provide extra strength were used. Welding and torch cutting on these members destroys the strength of the factory heat treatment. Do not weld, flame cut or drill on these members if they have not failed. If you have to mount something on the frame, use the extra and unused existing holes put in at the factory. However, if C-channels must be repaired, minimize welding on them.

How should a cracked C-channel truck frame be repaired?
Use the following steps:

Clean the repair area. First, steam clean and scrub the entire area surrounding the weld. (This cleaning is particularly important for waste hauling vehicles.) Then use an oxyfuel torch to dry this area and remove remaining mill scale. Finally, wire brush the area down to shiny metal. Compare your failure with those shown in Figure 15–8 to determine which Case your frame failure matches best and then follow the repair steps for that Case.

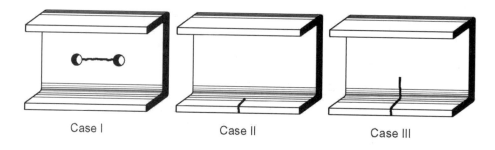

Case I   Case II   Case III

Figure 15–8   Occurrence of typical truck frame cracks

Case I—Horizontal crack along the web between factory-drilled holes.
This is a common case and cracks as long as 10 inches occur. See repair steps in Figure 15–9.

Grind a V-groove to within $1/16$ to $1/8$ inch of the thickness of the web steel along the path of the crack and extending 2 inches beyond the initial crack on each end. Put this V-groove on the inside of the C-channel, Figure 15–9 (b). Use a copper backing plate clamped to the back of the groove to protect the back of the weld from atmospheric contamination. Using SMAW low-hydrogen or iron powder electrodes, fill the V-groove with weld metal and grind it flush. Use enough current for a full penetration weld. See Figure 15–9 (c). Grind the weld flush (on the outside of the channel too if necessary to make it flat) making sure to leave the ground surface as smooth as possible. Any irregularities or scratches are stress raisers. Cut and fit a 1/2 inch thick carbon steel reinforcement plate on the inside the web extending at least 6 inches beyond the ends of the weld repair. Grind this plate on the lower and upper edges so it fits *tightly* against the web and edges of the flanges for its entire distance, Figure 15–9 (d). Using existing factory-drilled C-channel holes if possible, secure the reinforcement plate to the web with bolts matching the diameter of these holes. Holes are usually sized 1/2", 9/16", or 5/8" in diameter. Use a washer under each nut. If no holes are available, drill your own. Stop. The repair is complete. No additional welding is needed. See Figure 15–9 (e).

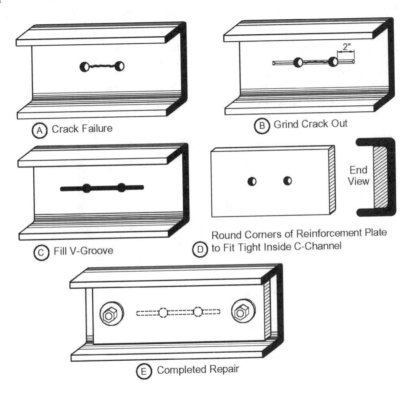

Figure 15–9   Case I: Horizontal crack in truck C-channel between factory-drilled
holes in web. This is a common case and cracks as long as
10 inches (250 mm) can occur

Case II—Crack on bottom flange perpendicular to web only.

Here are the repair steps:

Grind a V-groove half way through the thickness of the flange along the path of the crack. See Figure 15–10 (b). Using SMAW with a low-hydrogen or iron powder electrode, fill the V-groove with weld metal and grind it flush. Use enough current for full penetration of the flange metal. See Figure 15–10 (c). Grind the weld flush (on the inside of the flange too if necessary to make it flat) making sure to leave the ground surface as smooth as possible. Any remaining surface imperfections are stress raisers. Cut a 1/2 × 1 1/2 inch reinforcement bar from mild carbon steel 12 to 15 inches (300 to 380 mm) long. Center it on the crack. Weld this bar on the middle of the flange using a SMAW low-hydrogen or iron powder electrode. The 1 1/2 inch dimension of the bar is vertical. See Figure 15–10 (d). Using a grinder, gouge a bevel groove through the unwelded side of the reinforcement bar to sound weld metal on the other side, Figure 15–10 (d). Place a fillet weld in the gouged groove using SMAW with a low-hydrogen or iron powder electrode to secure the other side of the reinforcement bar. See Figure 15–10 (e). Do *not* make welds perpendicular to the length of the channel at the ends of the reinforcement bar.

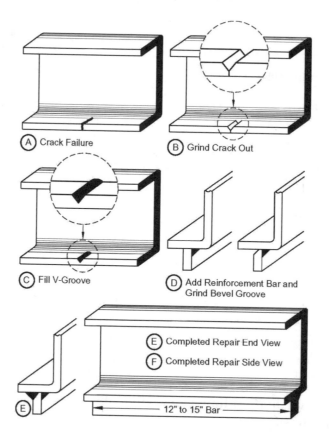

Figure 15–10   Case II: Crack on bottom flange perpendicular to web only

Case III—Crack on bottom flange perpendicular to web and extending up into web.

This is what happens when Case II is left uncorrected. See Figure 15–11.

The repair steps are:
- Grind out crack with a V-groove halfway through the thickness of the flange along the path of the crack.
- Drill a 3/8 to 1/2 inch (10 to 13 mm) diameter hole at the end of the crack in the web to relieve stress.
- Use reinforcement methods of Cases I and II to add both a bottom reinforcement bar and a reinforcement plate inside the channel web.

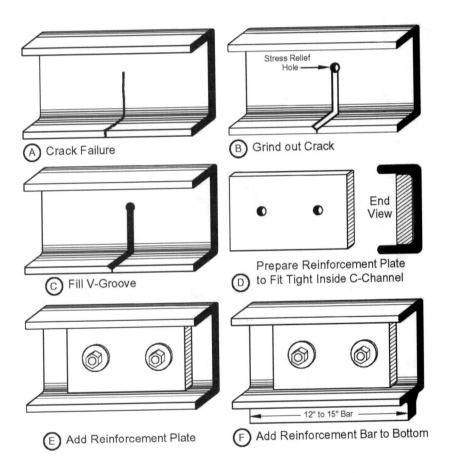

Figure 15–11   Case III: Crack begins on bottom flange and extends
into web of channel

Is there another approach to repairing cracked C-channels?

Another and more traditional C-channel repair approach is to weld reinforcement plates to the web instead of bolting them, or to weld a reinforcement bar along the bottom flange, but this is less commonly done today. If you choose welding instead of bolting for C-channel web repair, be sure to place welds *parallel* to the channel. Do not place any welds perpendicular to the channel. Welds perpendicular to the channel concentrate stresses on just one section of the weld. This is because the end welds on the patch plate prevent beam stress from being distributed evenly along the weld length. They become a new stress raiser and will produce near-term failure. See Figure 15–12.

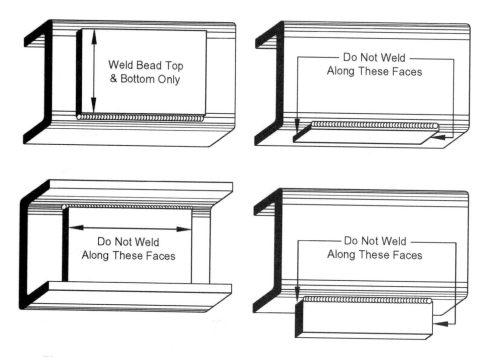

Figure 15–12   Welding reinforcement plates to C-channels: Welds parallel to
C-channels are acceptable; perpendicular welds are not

Caution: Remember that these failures occurred in the first place because the member was stressed beyond its design capacity by excessive loads and fatigue. Dump truck body action stresses, hydraulic cylinder loads, road vibration, and truck overloading all contribute to failure. *Failure is likely to* happen again, usually in the next weakest location, because the member is subject to the same load conditions that caused the initial failure.

You have a relatively new model car—one with the newer high-strength steel—and have tried an oxy-fuel weld repair on a sheet metal body part, but the weld keeps cracking. What is wrong?

This high-strength sheet metal cannot be oxyacetylene welded. It can, however, be successfully GMAW welded. This special steel was used to save weight. Begin the weld bead on the *outside* edge of the crack and work toward the *inside*, which will keep the inherent weakness of the bead-ending crater away from the metal's edge where it would act as a stress raiser and lead to a new failure. On galvanized body parts ER70S-3 electrodes should be used as they contain less silicon than ER70S-6 electrodes which contributes to cracking when mixed with zinc.

What precautions must be observed when mounting a roll bar or safety cage to a unibody vehicle?

While unibody design provides a strong and rigid vehicle frame, because it is made of relatively light sheet metal, it cannot support the forces that roll cage tubing ends can exert on it. A good solution is to use GMAW to weld the tubing to a mounting plate and have the mounting plate distribute the forces over a wider area than the tubing end alone. Use ER705-3 electrode wire. Either an all-around fillet weld or several 1/2 inch (13 mm) diameter plug welds are suitable to hold the mounting plate to the unibody box beam, Figure 15–13 (left). When roll cage tubing supports must be mounted to vehicle flooring, a stiff (1/4 inch or 6.5 mm) mounting plate is required to distribute forces over the thinner sheet metal flooring. This may be welded all around or bolted in a sandwich as shown in Figure 15–13 (right).

Pipe & Tubing

What is the difference between *pipe* and *tubing*?

Pipe usually has a much thicker wall than tubing. These thicker walls permit pipe to accept threads; tubing because it has thinner walls, cannot be threaded. Another difference is that pipe from 1/4 to 12 inches (6 to 300 mm) diameter is specified by its *inside* diameter; pipe 14 inches diameter and larger is specified by its *outside* diameter. Tubing is always specified by its *outside* diameter and wall thickness.

When specifying copper tubing what are the main parameters beside diameter?

The major choices are:

Copper tubing is available in *drawn* or *annealed* condition. Drawn copper tubing cannot be bent (unless it has been annealed) without having its sidewalls collapse. It comes only in straight lengths of 12, 18 or, 20 feet and is ideal for straight runs where appearance is important. Annealed tubing comes in both coils and straight lengths and can be easily bent without tools in smaller diameter sizes without wall collapse. Because it can be formed to bend or fit around obstacles along its path, many fittings (and copper soldering joints) can be eliminated by using a *single* piece of annealed tubing. There are three common wall thicknesses: Type K (heaviest), Type L (standard), and Type M (lightest). All three are used in domestic water service and distribution. Customer budget, preference, and local codes govern this choice.

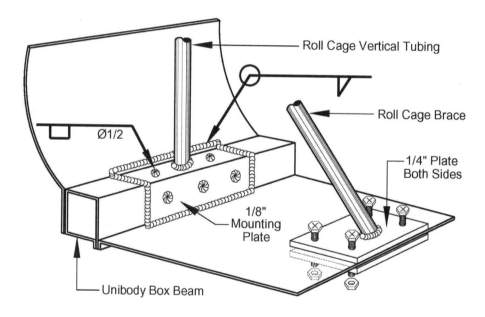

Figure 15–13   Mounting roll bar tubing on a unibody vehicle: Welding support
plate to unibody box beam (right) and bolting mounting plate to
sheet metal when welding is not permitted

You are cutting a large diameter pipe to length and need to mark your cut line. How can you be sure the cut line is square with the axis of the pipe?
Use a *wrap-around*, which is a length of thin, flexible material—vinyl, fiber, or cardboard that is wrapped around the pipe 1 1/2 times as shown in Figure 15–14. Adjust the wrap-around so that the second, overlapping layer lies squarely over the first layer, then hold the wrap-around tight with one hand and

mark the cut line with the other. A typical wrap-around is 1/16 inch thick by 2 1/4 inches wide by 48 inches long (1.6 mm thick, 6 cm wide, by 120 cm long).

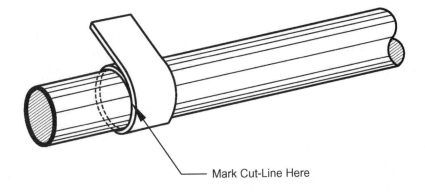

Mark Cut-Line Here

Figure 15–14   Utilizing a wrap-around to mark the cut will help in making a square cut line

You need to join two lengths of pipe and want to align them accurately, but you do not have a commercial pipe welding fixture. How is this done?
Tack two lengths of angle iron to form a double V-base as in Figure 15–15. Many applications will require a longer welding fixture than the one shown. Align the pipes before tack welding, make the root pass, followed subsequent passes.

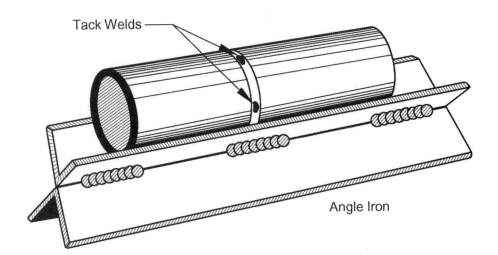

Tack Welds

Angle Iron

Figure 15–15   Using angle iron to align pipe

What device do welders commonly use to make working on pipe easier?
Pipe rollers permit welder to always be welding in the flat position. The welder welds while a helper rotates the pipe to maintain the weld area on top of the pipe so welding is done in the flat position or in field welding the welder may make weld in a fixed or stationary. Using four rollers together keeps the pipes in alignment while they are being welded. On very large pipe or castings, a motor driven positioning devise may take the place of rollers. See Figure 15–16.

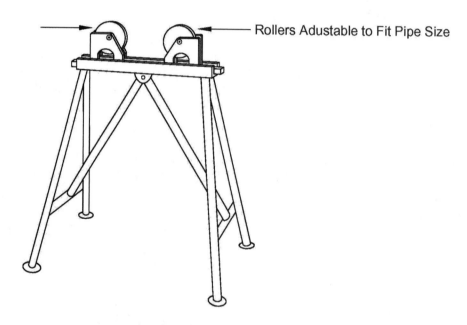

Rollers Adustable to Fit Pipe Size

Figure 15–16   Commercial pipe rollers on stand

What tool can quickly locate and mark any angular point around a pipe?
A Curv-O-Mark® Contour tool shown in Figure 15–17 locates any angle on
the outside of the pipe. Just set the adjustable scale to the angle desired, place
the tool on the pipe to the "level" position of the bubble and strike the built-
in center punch to mark the point. Figure 15–17 shows the contour tool locating the exact top, or zero degree position, the 90 degree position and the 60
degree position.

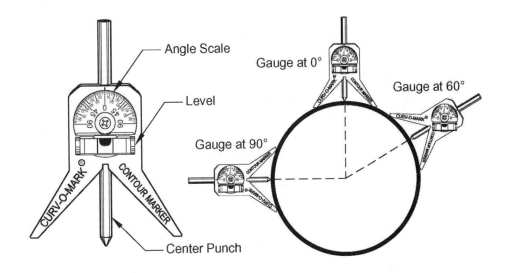

Figure 15–17   Curv-O-Mark® Contour tool for locating angular positions around a pipe

Pipe and Tubing Joints

What is a good way to repair cracked *structural* tubing (carries no fluid)?
- Drill 1/4 inch (6.5 mm) stress relief holes at the ends of the crack.
- Cut or grind away the cracked area between the 1/4 inch holes.
- Cut a patch out of a similar-sized piece of tubing and reshape this patch to
  fit over the cracked tubing area.
- Weld on the patch as shown in Figure 15–18: Use no continuous welds, no
  end welds and no welds closer to the end of the patch piece than 1/4 inch
  (6.5 mm).

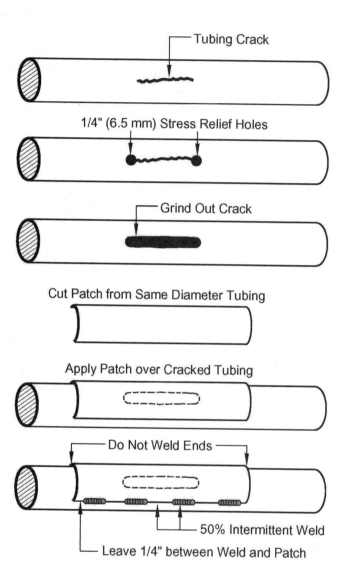

Figure 15–18   Repairing cracked structural tubing

What are four ways to join one pipe inserted into a larger pipe by welding for a *structural* application (no fluid carried)?
See Figure 15–19.

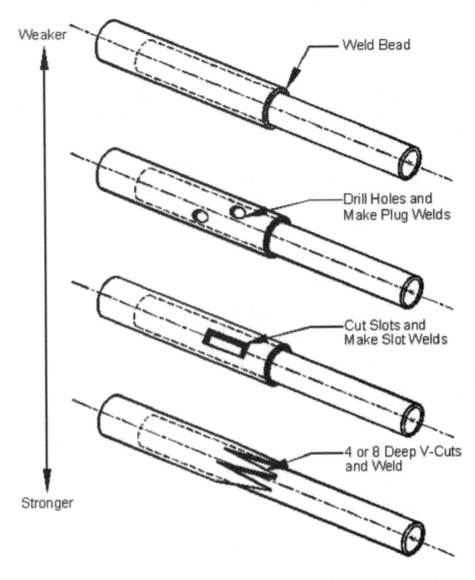

Figure 15–19   Ways to join pipe lengths by welding

You need to join two equal-diameter pipes in a *structural* application which will be subjected to torsion, tension or shear. You want a strong joint and the pipes in coaxial alignment. What to do?

Use the method in Figure 15–20 to insure that the pipes are in alignment, then make a full penetration weld. Such a weld is as strong as the pipe itself. Note that the inner pipe is for proper alignment, not strength.

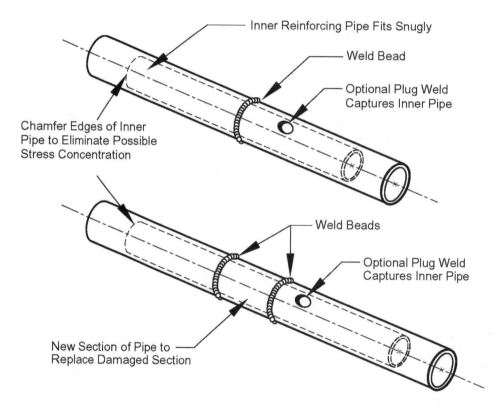

Figure 15–20   Two different methods of splicing structural pipe
which assure concentric alignment

What is a *saddle* and why is it made?

A saddle (or *fishmouth*) is the shaping of the end of one piece of tubing so it meets and fits tightly against another. We do this to make strong welds. The gap between the two tubing pieces should not exceed the diameter of the GMAW/FCAW/GTAW wire used to make the weld. See Figure 15–21.

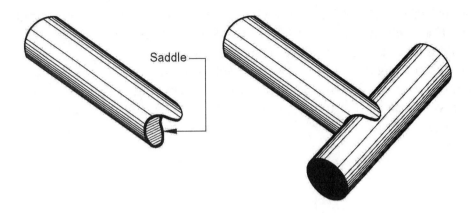

Figure 15–21   Tubing saddle

What are several common ways to make a tubing saddle?

Hack sawing and hand filing. Begin by forming the saddle on a bench or pedestal grinder, making the final adjustments with a file. A hand-operated nibbling tool is a faster way to make a saddle, although some operator skill is needed. Using a saddling hole-saw tool works well, Figure 15–22.

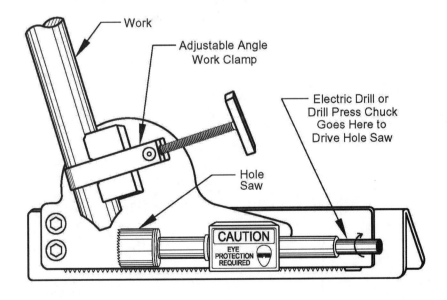

Figure 15–22   Commercial saddle cutting tool

Holding thin-walled tubing, especially stainless steel, in a vise when cutting it with a hacksaw or portable band saw distorts the tubing and the vise jaws damage the smooth sides. How do you avoid this?
First insert a tightly-fitting wooden dowel inside the tubing end to prevent it from collapsing in the vise. Then place soft jaws over the steel jaws of the vise to prevent marring the tubing. The soft jaws are usually made of aluminum, copper, lead, or plastic. See Figures 15–23 and 15–24.

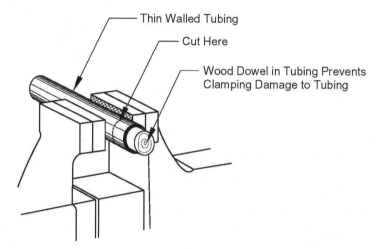

Figure 15–23   Use a dowel when clamping thin-walled tubing
prevents collapsing the tube

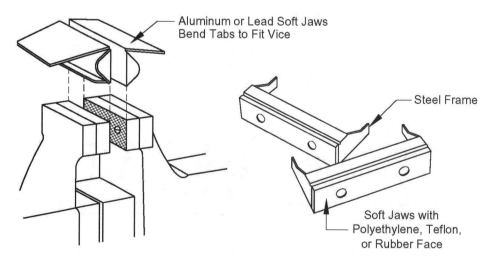

Figure 15–24   Soft jaws in the vise prevent damage to the work

Soldering Copper Tubing

What solders may be used for soldering copper tubing carrying potable water?

By US Federal law only lead-free solders may be used. These solders are commonly 95-5 tin-antimony, but may include a variety of other alloys of about 95% tin combined with copper, nickel, silver, antimony, or bismuth. These lead-free solders can cost 4 to 8 times as much as the 50-50 tin-lead alloy they replace.

What solders may be used for copper carrying *non-drinkable* water or other liquids?

A 50-50 tin-lead solder may be used. It costs less than lead-free solder alloys and has a wider pasty temperature zone, so it handles better.

What torches are suitable for sweating copper tubing?

Air-acetylene, propane, and MAAP® gas torches all work well. The Bernz-O-matic® torches in Figure 15–25 burn either propane or MAAP gas. Changing the gas orifice is all that is required to switch between fuels. MAAP gas is more expensive than propane and has a hotter flame, so can heat the pipe and fitting faster and handle larger diameter tubing. The version with a separate regulator, 48 inch (1.2 m) hose and hand piece is especially convenient. It can put heat into tight quarters and does not stall out when inverted as the single-piece torch does.

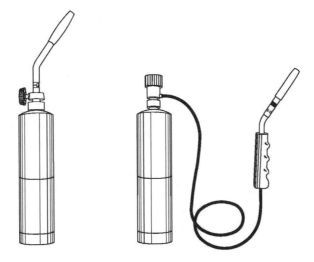

Figure 15–25   Torch-on-cylinder model (left) and hose-based model (right)

What steps are needed to prepare copper tubing for soldering?
- When measuring tubing lengths for cutting, .remember tubing must be long enough to reach to the bottom of its fitting, but not so long as to cause stress in the completed piping.
- Cut the tubing to length, preferably with a disc-type tubing cutter to insure square ends. A hacksaw, an abrasive wheel, or a portable or stationary band saw may also be used.
- Remove burrs from inside the tubing ends with a reamer or a file. Remove outside burrs with a file as they prevent proper seating of the tubing in the fitting cup.
- Using emery cloth, nylon abrasive pads, or male and female stainless steel brushes sized to the tubing as in Figure 15–26, remove the dirt and oxide from the end of the tubing and the inside mating surfaces of the fitting. Get down to fresh, shiny metal. Clean the outside of the tubing from the end of the tubing to about $3/8$ inch (1 cm) beyond where the tubing enters the fitting. The capillary space between the tubing and its fitting is about 0.004" (0.10 mm). Removing too much metal from either the tubing or the fitting will prevent proper flow of solder around the tubing by capillary forces. Keep your fingers off the cleaned copper areas to prevent contamination of the tubing surface with skin oils.

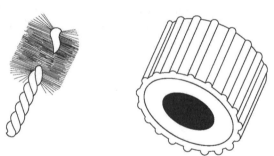

Figure 15–26   Male stainless steel brush for cleaning inside fittings (left) and
female brush for cleaning the outside ends of copper tubing (right)

- Select the flux to match the solder, since a mismatch may cause problems.
- Apply paste flux to the ends of the tubing sections and the fitting's interior mating surfaces using a brush. Keep the flux off your hands.
- Insert the tubing end into the fitting with a twisting motion. Make sure the tubing is properly seated in its fitting. You are now ready to solder the fitting.

How is soldering performed?

- Begin by heating at the bottom, then the sides of the fitting, since the heat will rise and warm the upper portion of the fitting at the same time.
- After several seconds of heating, remove the torch heat and touch the solder to the copper tubing where it meets the joint at the top of the fitting. See Figure 15–27. This test determines whether the joint temperature is hot enough to melt the solder. If more heat is needed, apply the torch again.
- Provided the joint has been properly cleaned and fluxed, and just after the flux begins to smoke, the solder will melt and be drawn into the space between the tubing and the fitting, and will begin to leak out the bottom of the fitting. This will happen rapidly. Run the end of the solder around the joint to insure full solder coverage.
- Remember that the cylindrical or capillary space between the outside of the tubing and the inside of the fitting is about 0.004 inches (0.1 mm), so not much solder will go into the fitting. Now that the fitting is hot, move the torch to the tubing on the other end of the fitting and repeat the soldering process. Because the fitting is already hot, it will not take much time to bring this second joint to temperature.
- On a fitting with an upper and a lower joint as shown in Figure 15–27, the top joint will now be hot, so apply the solder to it.
- To finish the joint, take a clean rag and wipe the excess solder from both ends of the fitting, leaving a meniscus around the tubing where it enters the fitting.
- Allow joints to cool in air without wiggling; do not spray with water or dip fitting in water as cracking may result.
- Finally, wipe the cooled joint with water or alcohol to remove any remaining flux as it is corrosive.

Suggestion: When one of the joints is upside down, solder the inverted joint *first*. If you overheat this joint, the solder will be too thin (runny) to be held in place by capillary attraction and it will run out. Just the right amount of heat will let the solder be drawn up into the joint. Since heat rises, the top joint will then be ready to solder, so apply a little more heat to it and run the solder around the ring where the tubing meets the fitting on the upper joint.

Note 1: Trying to improve a copper solder joint by loading solder on the outside of the fitting where the pipe enters is useless. The strength of the joint is *inside* the fitting and cannot be seen. If the joint was clean, properly fluxed and heated when the solder was applied, it will hold.

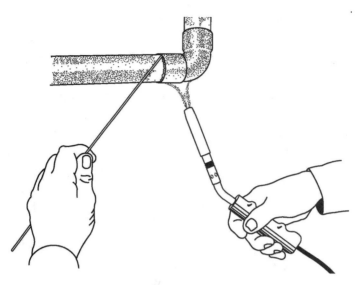

Figure 15–27   Soldering copper tubing

Note 2: Soldering must be done in pipes and fittings that are free of water. Some methods to achieve a dry pipe when the lines have been in service are: Open taps above and below the joint to help pipes drain.

Remove pipe straps on each side of the joint to permit bending the pipes down to let water run out. On vertical pipes containing water, use a short length of 1/4 inch hose to siphon out the water so the water level inside the pipe is at least 8 inches below the joint. Using paper towels and a stick, dowel or long a screwdriver, remove all moisture. Water drops within 8 inches (200 mm) of the soldering can spoil the joint.

Whenever you see steam rising from a fitting after soldering, there is a very good chance that water in the lines has spoiled the joint. Take it apart, fully drain the lines, and begin again. Valves on the lines to be soldered may not shut off completely, or it may be impossible to drain the line enough to prevent all water from entering the soldering area. To keep this water away, stuff fresh white bread into the line(s) to form a plug and push it 8 inches (200 mm) back into the line away from the fitting. Then assemble the copper tubing to the fitting and complete the solder joint. Flush out the bread when the joint is complete.

Note 3: Rule of thumb: The length of the solder needed will be *about* the nominal diameter of the tubing being soldered. For example, 1/2 inch copper tubing will use approximately 1/2 lineal inch of solder, 3/4 inch tubing will require 3/4 lineal inch of solder. Naturally, drip losses, runs inside the pipe, and solder rings on top of the fitting will increase the solder needed. Remember that the only solder that is working is what *cannot* be seen, that is between the outside of the tubing and the inside of the fitting.

You must solder a copper tubing fitting in place in a wall. When the torch is used to apply heat to the fitting the surrounding materials—the 2×4"s and the drywall—catch fire. How do you prevent this?
Use a commercial woven glass heat shield to keep the torch flame off the building materials, Figure 15–28, or use a piece of heavy steel or aluminum to do the same thing. Old timers used pieces of asbestos and these worked even better.

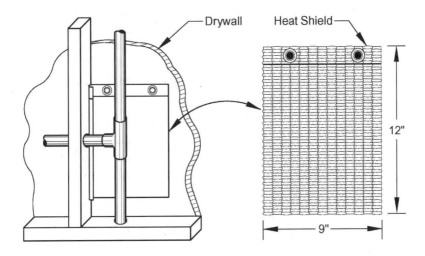

Figure 15–28   A heat shield can be used when working in close quarters

Scenario: An underground copper tubing water supply line has developed a leak as in Figure 15–29 (a). You have located the section of the pipe to replace and cut it out, Figure 15–29 (b). The problem now is how to install a new section of line and get it seated properly when each end of the copper line is firmly embedded in the soil and cannot be moved horizontally. How to proceed?
• Purchase a piece of special diameter patch tubing which is made to just slip *over* the *outside* of the existing tubing; it is sold in short lengths.
• Cut a piece of this patch tubing about 2 inches (50 mm) longer than the gap in the line.
• Prepare the outsides of the ends of the line and the insides of the ends of the patch tube by cleaning then applying flux, Figure 15–29 (c).
• Expose enough of one end of the existing copper tubing so it can be pulled up and the section of patch tubing can be slipped on it, Figure 15–29 (d).

- Center the patch tubing over the break.
- Solder the patch in place and the repair is complete, Figure 15–29 (e).
- This is a permanent repair.

Note: A temporary, emergency repair may be made by cleaning the area of the crack and brazing the joint area with a silver-containing filler metal.

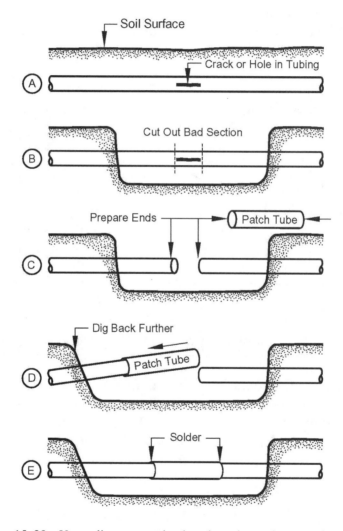

Figure 15–29   Use a slip-over patch tube when pipe ends cannot be separate

Appendix

A — Conversion Factors

	English to Metric	**Metric to English**
Capacity	ft^3 × 0.028316 = m^3	m^3 × 35.315 = ft^3
	in^3 × 0.01639 = liters	liters × 61.02 = in^3
	U.S. gallons × 3.785 = liters	liters × 0.2642 = U.S. gallons
	in^3 × 16.39 = cm^3	cm^3 × 0.06102 = in^3
Length	in × 25.40 = mm	mm × 0.03937 = in
Weight	lb × 0.453592 = kg	kg × 2.20462 = lb
Pressure	psi × 0.06895 = bars	bars × 14.5 = psi
	psi × 6.895 = kPa	kPa × 0.145 = psi
	psi × 0.006895 = MPa	MPa × 145 = psi
	psi × 0.0703 = kg/cm^2	kg/cm^2 × 14.224 = psi
Flow Rate	ft^3/hr × 0.472 = liters/min	liters/min × 2.12 = ft^3/hr

Table Appendix A   Conversion Factors

B — Properties of Elements and Important Compositions

Element	Symbol	Density	Weight (lb/ft^3)	Melting Point	
				°C	°F
Aluminum	Al	2.7	169	660	1221
Antimony	Sb	6.69	418	631	1167
Argon	Ar	1.4*	0.111	−189	−308
Barium	Ba	3.51	219	727	1340
Beryllium	Be	1.85	115	1287	2349
Bismuth	Bi	9.78	610	271	520
Boron	B	2.46	154	2076	3769
Cadmium	Cd	8.65	540	321	610
Carbon	C	2.27	142	3527	6318
Cerium	Ce	6.69	417	795	1463
Cesium	Cs	1.88	117	28	82
Chromium	Cr	7.14	445	1907	3465
Cobalt	Co	8.90	556	1495	2723
Columbium	Cb	7.06	441	1700	3092
Copper	Cu	8.92	556	1084	1984
Gold	Au	19.33	1206	1064	1947
Helium	He	0.14*	0.0111	<−272	<−458
Hydrogen	H	0.070*	0.0053	−259	−434.2
Indium	In	7.31	456	157	315
Iridium	Ir	22.65	1414	2466	4471
Iron	Fe	7.87	491	1538	2800
Lead	Pb	11.34	707	327	621
Lithium	Li	0.534	33	180	356
Magnesium	Mg	1.74	109	650	1202
Manganese	Mn	7.47	466	1246	2275
Mercury	Hg	13.55	846	−38.8	−37.9
Molybdenum	Mo	10.28	642	2623	4753
Nickel	Ni	8.80	549	1455	2651

* Density compared with air, not water.   (www.webelements.com)

Table Appendix B–1   Properties of some elements important in metallurgy

Element	Symbol	Density	Weight (lb/ft³)	Melting Point	
				°C	°F
Nitrogen	N	0.97*	0.063	−210	−346
Osmium	Os	22.61	1411	3033	5491
Oxygen	O	1.10*	0.0866	−218	−360
Palladium	Pd	12.02	750	1555	2831
Phosphorous	P	1.82	114	44	111.6
Platinum	Pt	21.09	1317	1768	3215
Potassium	K	0.856	53	63.4	146
Selenium	Se	4.82	310	221	430
Silicon	Si	2.33	145	1414	2577
Silver	Ag	10.49	655	962	1763
Sodium	Na	0.968	60	97.7	207.9
Sulfur	S	1.96	122	115.2	239
Tantalum	Ta	16.65	1039	3017	5463
Thorium	Th	11.73	732	1842	3348
Tin	Sn	7.31	456	232	450
Titanium	Ti	4.51	282	1668	3034
Tungsten	W	19.25	1202	3422	6192
Uranium	U	19.05	1189	1132	2070
Vanadium	V	6.11	381	1910	3470
Zinc	Z	7.14	446	419	786
Zirconium	Zr	6.51	406	1853	3371

* Density compared with air, not water.   (www.webelements.com)

Table Appendix B–1 (continued)   Properties of some elements
important in metallurgy

	Density Specific Gravity	Weight (lb/ft^3)	Melting Point	
			°C	°F
Bronze (90% Cu-10% Sn)	8.78	548	850-1000	1562-1832
Brass (90% Cu-10% Zn)	8.60	540	1020-1030	1868-1886
Brass (70% Cu-30% Zn)	8.44	527	900-940	1652-1724
Bronze (90% Cu-9% Al)	7.69	480	1040	1905
Bronze, Phosphor (90% Cu-10% Sn)	8.78	551	1000	1830
Bronze, Silicon (96% Cu-3% Si)	8.72	542	1025	1880
Iron, Cast	7.50	450	1260	2300
Iron, Wrought	7.80	485	1510	2750
Steel, high-carbon	7.85	490	1374	2500
Steel, low-alloy	7.85	490	1430	2600
Steel, low-carbon	7.84	490	1483	2700
Steel, medium-carbon	7.84	490	1430	2600

Table Appendix B–2   Properties of important metal compositions

C — AWS Letter Designations
for Welding & Allied Processes

Group	Welding Process	AWS Letter Designation
Arc Welding	Carbon arc	CAW
	Electrogas welding	EGW
	Flux cored arc welding	FCAW
	Gas metal arc welding	GMAW
	- Pulsed arc	GMAW-P
	- Short-circuiting arc	GMAW-S
	Gas tungsten arc	GTAW
	- Pulsed arc	GTAW-P
	Plasma arc welding	PAW
	Shielded metal arc welding	SMAW
	Stud arc welding	SW
	Submerged arc welding	SAW
Brazing	Block brazing	BB
	Carbon arc brazing	CAB
	Diffusion brazing	DB
	Dip brazing	DB
	Flow brazing	FLB
	Furnace brazing	FB
	Induction brazing	IB
	Infrared brazing	IRB
	Resistance brazing	RB
	Torch brazing	TB
Other welding processes	Electron beam welding	EBW
	Electroslag welding	ESW
	Flow welding	FLOW
	Induction welding	IW
	Laser beam welding	LBW
	Percussion welding	PEW
	Thermit welding	TW
Oxyfuel gas welding	Oxyfuel welding	OFW
	Air acetylene welding	AAW
	Oxyacetylene welding	OAW
	Oxyhydrogen welding	OHW
	Pressure gas welding	PGW
Resistance welding	Flash welding	FW
	Projection welding	PW
	Resistance seam welding	RSEW
	Resistance spot welding	RSW

Table Appendix C   Welding and brazing processes and their AWS letter designations

Soldering	Dip soldering	DS
	Furnace soldering	FS
	Induction soldering	IS
	Infrared soldering	IRS
	Iron soldering	INS
	Resistance soldering	RS
	Torch soldering	TS
	Wave soldering	WS
Solid state welding	Cold welding	CW
	Diffusion welding	DFW
	Explosion welding	EXW
	Forge welding	FOW
	Friction welding	FRW
	Hot pressure welding	HPW
	Roll welding	ROW
	Ultrasonic welding	USW
Thermal cutting – Arc	Air carbon arc cutting	AAC
	Carbon arc cutting	CAC
	Gas metal arc cutting	GMAC
	Gas tungsten arc cutting	GTAC
	Plasma arc cutting	PAC
	Shielded metal arc cutting	SMAC
Thermal cutting	Electron beam cutting	EBC
	Laser beam cutting	LBC
Thermal cutting – Oxygen	Metal powder cutting	POC
	Oxyfuel gas cutting	OFC
	- Oxyacetylene cutting	OFC-A
	- Oxyhydrogen cutting	OFC-H
	- Oxynatural gas cutting	OCF-N
	- Oxypropane cutting	OFC-P
	Oxygen arc cutting	AOC
	Oxygen lance cutting	LOC
Thermal spraying	Arc spraying	ASP
	Flame spraying	FLSP
	Plasma spraying	PSP

Table Appendix C (Continued)   Welding and brazing processes and their
AWS letter designations

Glossary
Nomenclatura

A

acceptable weld: A weld that meets the applicable requirements.
soldadura aceptable: Una soldadura que satisface las exigencias aplicables.

acetone: A colorless, flammable, volatile liquid used as a paint remover and as a solvent for oils and other organic compounds. Used in acetylene cylinders to saturate the monolithic filler material to stabilize the acetylene.
acetona: Líquido incoloro, volátil, que se usa para remover pintura, y como disolvente de aceites y de otras substancias orgánicas. Para estabilizar el acetileno, se usa en las bombonas para que sature el material de aportación.

acetylene feather: The intense white, feathery-edged portion adjacent to the cone of a carburizing oxyacetylene flame.
pluma de acetileno: En una llama carburante de un soplete oxiacetilénico, la zona muy blanca, en forma de pluma, adyacente al cono de la llama.

actual throat: The shortest distance between the weld root and the face of a fillet weld.
garganta actual: En un cordón de soldadura, la distancia más corta entre la raíz de la soldadura y la superficie del cordón.

adhesion: A state of being stuck together. The joining together of parts that are normally separate.
adhesión: La condición de estar pegado. La unión de partes normalmente separadas.

AISI: The American Iron and Steel Institute.
AISI: Organización en los EE.UU. llamada Instituto Americano del Hierro y del Acero.

aluminum: One of the chemical elements, a silvery, lightweight, easily worked metal that resists corrosion.
aluminio: Elemento químico, de color plateado, ligero de peso, fácil de trabajar mecanicamente, y muy resistente a la corrosión.

ampere: A unit of electrical current measuring the rate of flow of electrons through a circuit. One ampere is equivalent to the current produced by one volt applied across a resistance of one ohm.

amperio: *La unidad de corriente eléctrica que mide el flujo de electrones en un circuito eléctrico. Un amperio corresponde a la corriente contínua producida por una tensión de un voltio, a través de una resistencia de un ohmio.*

annealing: A process of heating then cooling metal to acquire desired qualities such as ductility.
templaje: *El proceso de calentar y luego enfriar un metal para que adquiera ciertas propiedades, como la ductilidad, por ejemplo.*

anode: The positive terminal of an electrical source.
ánodo: *El polo positivo de una fuente de electricidad.*

arc cutting: A group of thermal cutting processes that severs or removes metal by melting with the heat of an arc between an electrode and workpiece.
corte con arco: *Un grupo de procesos térmicos para cortar metales. El metal se separa o se pierde fundiéndolo con el calor de un arco entre un electrodo y la pieza que se está cortando.*

alloy: A substance with metallic properties, composed of two or more chemical elements of which at least one is a metal.
aleación: *Una substancia con propiedades metálicas, que está compuesta de dos o más elementos químicos, de los cuales al menos uno de ellos es un metal.*

alloying element: Elements added in a large enough percentage to change the characteristics of the metal. Such elements may be chromium, manganese, nickel, tungsten, or vanadium; these elements are added to produce specific physical properties such as hardness, toughness, ductility, strength, resistance to corrosion, or resistance to wear.
elemento de aleación: *Elemento que se añade a un metal en proporciones suficiente-mente altas para cambiar las características del metal. Algunos de estos elementos son cromo, manganeso, tungsteno o vanadio, etc. Estos elementos producen ciertas propiedades físicas como la dureza, la tenacidad, la ductilidad, la resistencia mecánica, y la resistencia a la corrosión y al desgaste.*

alloy steel: A plain carbon steel to which another element, other than iron and carbon, has been added in a percentage large enough to alter its characteristics.
acero de aleación: *Un simple acero al carbono al que se le añade un elemento – que no sea hierro o carbono – en cantidad suficiente para alterar sus propiedades.*

alternating current (AC): An electric current that reverses its direction periodically.
corriente alterna (C.A.): *Una corriente eléctrica que cambia de dirección periodicamente.*

alternative fuels: Propane, methylacetylene propadiene (MPS), natural gas, or fuel gases, other than acetylene, used for welding or cutting.
combustibles optativos: *Propano, metilacetileno propadieno, gas natural, o gases de combustión – menos el acetileno – que pueden usarse para soldar o cortar metales.*

arc blow: The deflection of an arc from its normal path because of magnetic forces.
soplo del arco: *La deviación de su trayectoria sufrida por un arco eléctrico debida a fuerzas mágneticas.*

arc force: The axial force developed by an arc plasma.
fuerza del arco: *La fuerza axial producida por el plasma de un arco.*

arc gap: A nonstandard term used for the arc length.
abertura del arco: *Una expresión, no aprobada, la cual se usa para designar longitud de arco.*

arc gouging: Thermal gouging that uses an arc cutting process variation to form a bevel or groove.
estriar con arco: *Escarbar un metal termicamente para formar un chaflán o una ranura, usando una variación del proceso de cortar metales con un arco.*

arc plasma: A gas that has been heated by an arc to at least a partially ionized condition, enabling it to conduct electric current.
plasma de arco: *Un volumen de gas que ha sido calentado por un arco a una temperatura tal que el gas está suficientemente ionizado para conducir una corriente eléctrica.*

arc spraying: A thermal spraying process using an arc between two consumable electrodes of surfacing materials as a heat source and a compressed gas to atomize and propel the surfacing material to the substrate.
rocío de arco: *El proceso de rociar en caliente usando un arco entre dos electrodos consumibles de materiales de alisamiento que actúan como fuentes de calor, y un gas comprimido que pulveriza ese material mandándolo al substrato.*

arc strike: A discontinuity resulting from an arc, consisting of any localized re-melted metal, heat-affected metal, or change in the surface profile of any metal object.
golpe de arco: *Discontinuidad producida por un arco, que consiste de porciones de metal refundido, metal afectado por el calor, o cambio en la rugosidad de la superficie de una pieza metálica.*

arc time: The time during which an arc is maintained in making an arc weld.
duración del arco: *El tiempo durante el cual un arco es mantenido para hacer una soldadura por arco.*

arc welding: Arc welding is a group of welding processes in which fusion is produced by heating with an electric arc or arcs with or without the application of pressure and with or without the use of filler metal.
soldadura con arco: *Grupo de procedimientos para soldar, en los cuales fusión de los metales ocurre por medio de arco(s) eléctrico(s) con o sin la aplicación de presión y con o sin el uso de material de aportación.*

arc voltage: The voltage across the arc.
tensión de arco: *Diferencia de potencial a través del arco.*

as welded: The condition of weld metal, welded joint, and weldments after welding, but prior to any subsequent thermal or mechanical treatment.
como soldado: *La codición del metal de soldadura, la juntura soldada, y las soldaduras después de soldado pero antes de ser sometido a tratamientos térmicos o mecánicos subsiguientes.*

ASTM: The American Society for Testing and Materials.
ASTM: Organización en los EE.UU. llamada Sociedad Americana para Ensayos y Materiales.

austenite: One of the basic steel microstructures wherein carbon is dissolved in iron. Austenite forms at elevated temperatures.
austenita: Una de las structuras básicas del acero en las que el carbono está disuelto en el hierro. Austenita se forma a temperaturas elevadas.

autogenous weld: A fusion weld made without using a filler material.
soldadura autógena: Soldadura a fusión hecha sin usar material de aportación.

AWS: The American Welding Society.
AWS: Organización en los EE.UU. llamada Sociedad Americana para Soldadura.

axis of a weld: A line through the length of a weld, perpendicular to and at the geometric center of its cross-section.
eje de una soldadura: Una línea a lo largo de una soldadura, perpendicular a su sección, y que pasa por su centro geométrico.

B

back bead: A weld bead resulting from a back weld pass. Back beads are made after the primary weld is completed.
retroreborde: Un cordón de soldadura , el resultado de un pase invertido. Estos cordones son siempre aplicados después de haberse hecho la soldadura principal.

back fire: The momentary recession of the flame into the welding tip, or cutting tip followed by immediate reappearance or complete extinction of the flame, accompanied by a loud popping report.
retroquema: La desaparición momentánea de la llama del soplete dentro de la boquilla de soldar o de cortar. Esta desaparición, a veces, causa la extinción de la llama, y otras veces la llama reaparece inmediatamente, pero en ambos casos, está acompañada de una explosión muy sonora.

backgouging: The removal of weld metal and base metal from the weld root side of a welded joint to facilitate complete fusion and complete joint penetration upon subsequent welding from that side.
Escarbando por detrás: La remoción de porciones de metal de base y metal de soldadura del lado de la raíz, para promover una fusión completa y una penetración completa de la juntura al hacer un pase subsiguiente, de ese lado.

backhand welding: A welding technique in which the welding torch or gun is directed opposite to the progress of welding.
soldando con el dorso: Una técnica de soldar en la que se hace que el soplete o la pistola esté orientado en dirección opuesta a aquella en la que progresa la soldadura.

backing: Material or device placed against the back side of a joint to support and retain molten weld-metal. The material may be partially fused or remain unfused during welding and may be either metal or nonmetal (metal strip, asbestos, carbon, copper, inert gas, ceramics).
soporte: Un aparato o dispositivo que se monta en la parte trasera de la juntura para sopor-

tar y retener el metal de aportación fundido. El aparato puede a veces fundirse parcial-
mente, o puede permanecer separado durante la soldadura; y puede ser hecho de metal o
de otra substancia (fibras metálicas, asbesto, carbón, cobre, un gas inerte, o cerámica).

backing bead: A weld bead resulting from a backing pass. Backing beads are completed
before welding the primary weld.
cordón de respaldo: El cordón que resulta de un pase de soldadura de respaldo. Estos cor-
dones se completan antes de empezar la soldadura principal.

backing pass: A weld pass made to provide a backing for the primary weld.
pase de respaldo: El pase de soldadura hecho en una soldadura de respaldo.

backing ring: Backing in the form of a ring, generally used in the welding of pipe.
anillo de soporte: Soporte, en forma de anillo, usado generalmente en la soldadura de tubos.

backing strip: Non-standard term used to describe a backing on the root side of the weld in
the form of a strip.
tira de soporte: Un término, no aprobado, que se usa para describir un soporte, en el lado
de la raíz, que tiene la forma de una tira.

back-step sequence: A longitudinal sequence in which weld passes are made in the direction
opposite weld progression, usually used to control distortion.
secuencia "pase atrás": Una secuencia longitudinal en la cual algunos pases se hacen en la
dirección opuesta a la dirección en la que la soldadura avanza. Esta técnica se usa prin-
cipalmente para controlar la deformación de la pieza.

back weld: A weld made at the back of a single groove weld.
soldadura trasera: Una soldadura hecha en la parte trasera de una soldadura con una ranu-
ra nada más.

bainite: A steel microstructure that is harder than pearlite, cementite, or ferrite, and more
ductile than martensite.
bainita: Una estructura del acero que es más dura que la perlita, la cementita, o la ferrita, y
más dúctil que la martensita.

base material: The material that is welded, brazed, soldered, or cut.
material de base: El material que va a ser soldado con arco,, soldado fuerte o débil, o cortado.

base metal: The metal or alloy that is welded, brazed, soldered, or cut.
metal de base: El metal o la aleación que va a ser soldado con arco, soldado fuerte o débil,
o cortado.

bead weld: A term used for surfacing welds.
soldadura a cordón: Otro nombre para las soldaduras de alisamiento.

bevel: An edge preparation, the angular edge shape.
chaflán: La preparación, de forma angular, del borde de una pieza.

bevel angle: The angle between the bevel of a joint member and a plane perpendicular to the
surface of the member.

ángulo del chaflán: *El ángulo formado por el plano que contiene el chaflán de una de las piezas de una juntura y un plano perpendicular a la superficie de la pieza.*

body-centered cubic (BCC): One of the common types of unit cells described as a cube with an atom at each of the eight corners and a single atom at the center of the cell. This arrangement is typical of the ferritic form of iron. Among the common BCC metals are iron, carbon steel, chromium, molybdenum, and tungsten.

CCC: Red cúbica centrada en el cubo: *Una de las más comunes disposiciones atómicas. La unidad celular consiste de un átomo en el centro de un cubo en cuyas ocho esquinas se hallan sendos átomos. Esta disposición es típica del hierro en su fase ferrítica. Hierro, acero al carbono, cromo, molíbdeno, y tungsteno son algunos de los metales más comunes con esa disposición.*

boxing: The continuation of a fillet weld around a corner of a member as an extension of the principle weld.

encajonamiento: *La continuación de un cordón de soldadura en torno de una esquina de una pieza, como extensión de la soldadura principal.*

braze: A weld produced by heating an assembly to the brazing temperature using filler metals having a liquidus above 840°F (450°C) and below the solidus of the base metal. The filler metal is distributed between the closely fitted faying surfaces of the joint by capillary action.

soldadura fuerte: *Una soldadura calentando el ensamblaje a la temperatura del metal de aportación, el cual debe tener una temperatura de fusión de más de 840°F (450°C) pero menor que el solidus del metal de las piezas. Ese metal se distribuye en el espacio delimitado por las caras estrechamente yuxtapuestas por medio de la acción de la capilaridad.*

braze metal: The filler metal used to make the joint of a brazed socket having a liquidus above 850°F (450°C).

metal de aportación: *El metal que se usa para que se difunda por la juntura por capilaridad, y que tiene un liquidus de más de 840°F (450°C).*

braze welding: A welding process that uses a filler metal with a liquidus above 840°F (450°C) and below solidus of the base metal. The base metal is not melted. Unlike brazing, in braze welding the filler metal is not distributed in the joint by capillary action.

soldadura blanda o débil: *Un tipo de soldadura que usa un metal de aportación con un liquidus de más de 840°F (450°C) pero menor que el solidus del material de base. El metal de base no se funde. Contrariamente a la soldadura fuerte, en la solda-dura blanda, el metal de aportación no se difunde en la juntura por capilaridad.*

brazing: A group of welding processes that produces coalescence of materials by heating them to the brazing temperature in the presence of a filler metal having a liquidus above 850°F (450°C) and below the solidus of the base metal. The filler metal is distributed between the closely fitted faying surfaces of the joint by capillary action.

soldadura: *Un grupo de procesos de union que producen la coalescencia de materiales, aumentando sus temperaturas , en la presencia de un metal de aportación cuyo liquidus es más de 840°F (450°C) y menos que el solidus del metal de base. El metal de aportación se difunde entre las superficies estrechamente yuxtapuestas por medio de la acción de la capilaridad.*

brazing filler metal: The metal or alloy used as a filler metal in brazing, which has liquidus above 850°F (450°C) and below the solidus of the base metal.
metal de aportación para soldaduras fuertes: *El metal o la aleación que se usa como metal de aportación en soldaduras fuertes, el cual tiene un liquidus por encima de 840°F (450°C) pero por debajo del solidus del metal a soldar.*

Brinell hardness test: A common testing method using a ball penetrator. The diameter of the indentation is converted to units of Brinell hardness number (BHN).
ensayo de dureza tipo Brinell: *Un método muy común que emplea como penetrador una bola de acero. El diámetro de la deformación causada por la bola se convierte en unidades de dureza Brinell.*

buckling: Bending or warping caused by the heat of welding.
comba: *Flexión o arqueamiento causado por el calor producido al soldar.*

buttering: A surfacing variation that deposits surfacing metal on one or more surfaces to provide metallurgically compatible weld metal for the subsequent completion of the weld.
emplastar: *Una variación de alisamiento que deposita el metal de alisamiento en una o más superficies para crear compatibilidad metalúrgica en el metal de soldadura en las subsiguientes operaciones hasta completar la soldadura.*

butting member: A joint member that is prevented by the other member from movement in one direction perpendicular to its thickness dimension.
miembro de tope: *Un miembro de la juntura al que le está prohibido, por otro miembro, moverse en una dirección perpendicular a la dirección de su espesor.*

butt joint: A joint between two members aligned approximately in the same plane.
junta a tope: *Una juntura de dos miembros alineados aproximadamente en un mismo plano.*

C

capacitor: A device consisting of two or more conducting plates separated from one another by an insulating material and used for storing an electrical charge.
condensador: *Un dispositivo que contiene dos o mas platos conductores separados uno del otro por un material aislante, y usado para almacenar una carga eléctrica.*

capillary action: The force by which liquid in contact with a solid is distributed between closely fitted faying surfaces of the joint to be brazed or soldered.
acción capilar: *La fuerza por la que un líquido en contacto con un sólido se distribuye entre las superficies estrechamente yuxtapuestas de una juntura que va a ser soldada.*

carbon: A nonmetallic chemical element that occurs in many inorganic and all organic compounds. Carbon is found in diamond and graphite, and is a constituent of coal, petroleum, asphalt, limestone, and other carbonates. In combination, it occurs as carbon dioxide and as a constituent of all living things. Adjustment of the amount of carbon in iron produces steel.
carbono: *Un elemento químico no metálico que se encuentra en muchos compuestos inorgánicos y en todos los compuestos orgánicos. En su estado natural se encuentra como diamante y como grafito, y es un componente del carbón, petróleo, asfalto, piedra caliza, y otros carbonatos. En combinación, se halla en el dióxido de carbono, y como*

constituyente en todas las substancias o cosas animadas. La añadidura de carbono al hierro produce el acero.

carbon steel: A steel containing various percentages of carbon. Low-carbon steel contains a maximum of 0.15% carbon; mild steel contains 0.15% to 0.35% carbon; medium-carbon steel contains 0.35% to 0.60% carbon; high-carbon steel contains from 0.60% to 1.0% carbon.

acero al carbono: Acero que contiene varios porcentajes de carbono. Acero con poco carbono tiene un máximo de 0,15% de carbono; acero dulce tiene de 0,15% a 0,35% de carbono; acero a medio carbono tiene de 0,35% a 0,60% de carbono; y el acero a alto carbono tiene entre 0,60% y 1,0% de carbono.

carburizing flame: A reducing oxygen-fuel gas flame in which there is an excess of fuel gas, resulting in a carbon-rich zone extending around and beyond the inner cone of the flame.

llama carburante: Una llama reductora de oxígeno en la que hay un exceso de combus-tible, lo cual resulta en haber una zona alrededor del cono de la llama rica en carbono.

cast iron: A family of alloys, containing more than 2% carbon and between 1% and 3% silicon. Cast irons are not malleable when solid, and most have low ductility and poor resistance to impact loading. There are four basic types of cast iron gray, white, ductile, and malleable.

hierro fundido: Una familia de aleaciones que tienen más del 2% de carbono y entre el 1% y el 3% de silicio. Los hierros fundidos, en estado sólido, no son maleables, y la mayoría no son dúctiles y tienen poca resistencia a las cargas de impacto. Hay cuatro clases de hierro fundido, a saber: gris, blanco o especular, dúctil, y maleable.

cathode: The negative terminal of a power supply; the electrode when using direct current electrode negative (DCEN).

cátodo: El polo negativo de una fuente de eléctricidad. Cuando se usa el método de electrodo negativo a corriente contínua, el electrodo es el cátodo.

caulking: Plastic deformation of weld and adjacent base metal surfaces by mechanical means to seal or obscure discontinuities.

recalque: Deformación plástica de una soldadura y las superficies contiguas del metal de base, por medios mecánicos, para sellar u ocultar discontinuidades.

cementite: A very hard form of low-temperature steel that contains more than 0.8% carbon. Cementite occurs in steel that has not been previously heat treated or in steel that has been cooled slowly after being transformed into austenite.

cementita: Una estructura muy dura del acero a baja temperatura que tiene más de 0,8% de carbono. La cementita se encuentra en aceros que no han sido previamente tratados termicamente; o en aceros que han sido enfriados lentamente después de haber sido transformados en austenita.

chain intermittent weld: An intermittent weld on both sides of a joint where the weld increments on one side are approximately opposite those on the other side.

soldadura intermitente a cadena: Una soldadura intermitente en ambos lados de una juntura, en la que los puntos de soldadura en un lado están aproximadamente opuestos a los puntos en el otro lado.

Charpy V-notch test: An impact test used to determine the notch toughness of materials.
ensayo Charpy de muesca en V: Un ensayo con carga de impacto en la muesca para determinar la tenacidad del material .

chill plate: A piece of metal placed behind material being welded to correct overheating.
plancha enfriadora: Pieza de metal que se pone por detrás del material que está siendo soldado, para corregir el sobrecalentamiento.

chill ring: A non-standard term for a backing ring.
anillo enfriador: Expresión, no aprobada, para anillo de soporte

chromium: A lustrous, hard, brittle, steel-gray metallic element used to harden steel alloys, in production of stainless steel, and as a corrosion resistant plating.
cromo: Elemento metálico, brillante, duro, frágil, del color del acero, que se usa para endurecer aceros de aleación, en la producción de acero inoxidable, y para hacer enchapados resistentes a la corrosión.

cladding: A surfacing variation that deposits or applies surfacing material usually to improve corrosion or heat resistance.
revestimiento: Una variación del proceso de alisamiento en la que se depositan ciertos materiales, generalmente para aumentar la resistencia a la corrosión y al calor.

coalescence: The growing together or growth into one body of the materials being welded.
coalescencia: En el caso de materiales que se están soldando, la acción de crecer juntos para formar un solo cuerpo.

coefficient of thermal expansion: The increase in length per unit length for each degree a metal is heated.
coeficiente de expansión térmica: El aumento de la longitud por cada unidad de longitud, y por cada grado de aumento de la temperatura de un metal

cohesion: Cohesion is the result of a perfect fusion and penetration when the molecules of the parent material and the added filler materials thoroughly integrate as in a weld.
adhesión: Es el resultado de una fusión y una penetración perfectas; cuando las moléculas del material soldado y las de los materiales de aportación se mezclan resueltamente, como en soldaduras.

cold crack: A crack that develops after solidification is complete.
grieta fría: La grieta que aparece después que la solidificación ha terminado.

cold work: Cold working refers to forming, bending, or hammering a metal well below the melting point. Cold working of metals causes hardening, making them stronger but less ductile.
trabajo en frío: El acto de formar, encorvar, o martillear los metales, lo cual causa un endurecimiento que los hace más fuertes pero menos dúctiles.

cold soldered joint: A joint with incomplete coalesence caused by insufficient application of heat to the base metal during soldering.
soldadura fría: Una juntura con coalescencia incompleta debido al no haber calentado la pieza suficientemente durante la soldadura.

complete fusion: Fusion over the entire fusion faces and between all adjoining weld beads.

fusión completa: *Se dice cuando la fusión ocurre sobre todas las superficies y entre los cordones contiguous.*

complete joint penetration: A root condition in a groove weld in which weld metal extends through the joint thickness.

penetración completa de la juntura: *Una condición en la raíz de la juntura de una soldadura en ranura, en la cual el material de soldadura se extiende sobre todo el espesor de la juntura.*

composite: A material consisting of two or more discrete materials with each material retaining its physical identity.

compuesto: *Un material hecho de dos o más materiales discretos que retienen sus identidades físicas.*

composite electrode: A generic term for multi-component filler metal electrodes in various physical forms such as stranded wires, tubes, or covered wire.

electrodo compuesto: *Un término genérico para electrodos que contienen metales de aportación en diversas formas, v. gr., alambre retorcido, tubos, e hilos cubiertos.*

concavity: The maximum distance from the face of a concave fillet weld perpendicular to a line joining the weld toes. A concave fillet weld will have a face that is contoured below a straight line between the two toes of a fillet weld.

concavidad: *La distancia máxima desde la faz de un cordón cóncavo, perpendicular-mente, hasta la línea que toca las dos orillas de la soldadura. La superficie del cordón cóncavo está por debajo del plano que toca las dos orillas del cordón.*

conductor: A device, usually a wire, used to connect or join one circuit or terminal to another.

conductor: *Un dispositivo, generalmente un alambre, que sirve para conectar el terminal de un circuito al terminal de otro circuito.*

cone: The conical part of an oxygen-fuel gas flame adjacent to the tip orifice.

cono: *La parte cónica de una llama de oxígeno y gas combustible, adyacente al orificio de la boquilla.*

constant-current (CC) power source: An arc welding power source with a volt-ampere relationship yielding a small welding current change from a large arc voltage change.

fuente de corriente contínua (CC): *Un suministro de energía para soldadura con arco con una relación voltio/amperio que produce un cambio muy pequeño de corriente para un cambio grande en la tensión del arco.*

constant-voltage (CV) power source: An arc welding power source with a volt-ampere relationship yielding a large welding current change from a small arc voltage change.

fuente de tensión contínua (VC): *Un suministro de energía para soldadura con arco con una relación voltio/amperio que produce un cambio grande de corriente para un cambio muy pequeño en la tensión del arco.*

constricted arc: A plasma arc column that is shaped by the constricting orifice in the nozzle of the plasma arc torch or plasma spraying gun.

arco restringido: Un arco de columna de plasma que surge en la forma producida por la forma de la boquilla del soplete o de la pistola del rociador de plasma.

consumable electrode: An electrode that provides filler metal, therefore is consumed in the arc welding process.

electrodo consumible: Un electrodo que es el material de aportación a la misma vez; consecuentemente, el electrodo se consume en el proceso de la soldadura con arco.

consumable insert: Filler metal that is placed at the joint root before welding, and is intended to be completely fused into the joint root to become part of the completed weld.

embutido consumible: Metal de aportación que se coloca a lo largo de la raíz de la juntura antes de empezar a soldar, para que se convierta en parte de la soldadura.

contact resistance: Resistance to the flow of electric current between two work-pieces or an electrode and the work-piece.

resistencia de contacto: La oposición al flujo de una corriente eléctrica entre dos piezas o entre un electrodo y la pieza.

contact tube: A device that transfers current to a continuous electrode.

tubo de contacto: Un aparato que transfiere corriente a un electrodo contínuo.

contact tube setback: The distance from the contact tube to the end of the gas nozzle. This term is used in gas metal arc and gas shielded flux cored arc welding.

enganche para el tubo de contacto: La distancia desde el tubo de contacto hasta el fin del tubo del gas en la boquilla. Este término se usa en soldaduras con arco de metal y gas, y en soldaduras con arco con material fundente blindado y centrado en el cañón del soplete.

convexity: The maximum distance from the face of a convex fillet weld perpendicular to a line joining the toes.

convexidad: La distancia máxima desde la faz de un cordón convexo, perpendicular-mente, hasta la línea que toca las dos orillas de la soldadura.

corner joint: A joint between two members located approximately at right angles to each other in the form of an L.

juntura de esquina: La unión de dos miembros situados aproximadamente a 90 grados uno del otro, formando una L.

corrosive flux: A flux with a residue that chemically attacks the base metal. It may be composed of inorganic salts and acids, organic salts and acids, or activated rosin.

fundente corrosivo: Un fundente con un residuo que ataca quimicamente el metal de base. Puede estar compuesto de sales y ácidos inorgánicos, o de sales y ácidos orgá-nicos, o de resina activada.

cosmetic pass: A weld pass made primarily to enhance appearance.

pasada cosmética: Una pasada de soldadura para mejorar la apariencia.

covered electrode: A composite filler metal electrode consisting of a core of a bare electrode or metal cored electrode to which a covering sufficient to provide a slag layer on the weld metal has been applied. The covering may contain materials providing such functions as shielding from atmosphere, deoxidation, and arc stabilization, and can serve as a source of metallic additions to the weld.

electrodo cubierto: *Un electrodo de metal de aportación compuesto que consiste de un electrodo desnudo, o un electrodo cubierto con metal, central, y el cual se recubre para que produzca una capa de escoria en el metal de soldar. Esta capa puede contener materiales para aislar la soldadura de la atmósfera, reducción, y estabilización del arco; y puede servir de fuente de añadiduras metálicas a la soldadura.*

cover plate: A removable pane of colorless glass, plastics coated glass, or plastics that covers the filter plate and protects it from weld spatter, pitting, or scratching.

cubierta: *Un panel removible de vidrio claro, vidrio con una capa plástica, o de plástico, que cubre el filtro y lo protege contra salpicadas, picaduras de óxido, o rasguños.*

crack: A fracture-type discontinuity characterized by a sharp tip and high ratio of length and width to opening displacement.

grieta: *Una discontinuidad del tipo de fractura caracterizada por tener los extremos bien afilados, y por altas razones de longitud-a-desplazamiento y de espesor-a-desplaza-miento de la apertura.*

cracking a valve: Rapidly opening and closing a valve to clear the orifice of unwanted foreign material.

"abre-y-cierra" una llave: *El abrir y cerrar una válvula rapidamente para limpiar el orificio de suciedades.*

crater: A depression in the weld face at the termination of a weld bead.

cráter: *Una depresión en la cara o parte visible de una soldadura al final del cordón de soldadura.*

crater crack: Radial cracks formed in a weld crater as the weld pool solidifies and shrinks.

grietas de cráter: *Grietas en disposición radial que aparecen en un cráter a medida que la balsa de soldadura se solidifica y se encoge.*

crystalline structure: The orderly arrangement of atoms in a solid in a specific geometric pattern. Sometimes called a lattice structure.

estructura cristalina: *La disposición métodica de los átomos de un sólido en un modelo geométrico. También se le dice estructura entrelazada.*

cutting attachment: A device for converting an oxygen-fuel gas welding torch into an oxygen-fuel cutting torch.

acesorio para cortar: *Un dispositivo para cambiar un soplete de gas y oxígeno para soldar a un soplete de gas y oxígeno para cortar.*

cutting head: The part of a cutting attachment to which the cutting torch or tip may be attached.

cabezal de cortar: *La parte del accesorio de cortar donde se puede montar el soplete o la boquilla.*

cutting tip: An attachment to an oxygen cutting torch from which the gases exit.

boquilla de cortar: *Un accesorio que se monta en un soplete de cortar con oxígeno, y por donde salen los gases.*

cycle: The duration of alternating current represented by the current increase from an initial value to a maximum in one direction then to a maximum in the reverse direction and its return to the original initial value.

ciclo: *La duración de una corriente alterna representada por un aumento de un valor inicial hasta llegar a un valor máximo en una dirección, y después, un regreso, en la dirección contraria, al valor inicial.*

cylinder manifold: A multiple header for interconnection of gas sources with distribution points.

conector de bombonas: *Una válvula de distribución para conectar diversos suministros de gas con sus puntos de uso.*

D

defect: A discontinuity or discontinuities that by nature or accumulated effect render a part or product unable to meet minimum applicable acceptance standards or specifications. The term designates rejection.

defecto: *Una o varias discontinuidades que por naturaleza, o por efecto cumulativo, hacen que una pieza o conjunto sea incapaz de satisfacer el mínimo estándar de aceptación o el mínimo límite de las especificaciones para las cuales fué diseñado. El término implica rechazamiento.*

deposited metal: Filler metal that has been added during welding, brazing, or soldering.

metal depositado: *El metal de aportación que ha sido añadido durante el proceso de soldadura.*

deposition rate: The weight of filler material deposited in a unit of time.

velocidad de depósito: *El peso de material depositado en la unidad de tiempo.*

depth of bevel: The perpendicular distance from the base metal surface to the root edge or the beginning of the root face.

profundidad del chaflán: *La distancia perpendicular entre la superficie del material de base y la orilla de la raíz, o el comienzo de la cara de la raíz.*

depth of fusion: The distance that fusion extends into the base metal or previous bead from the surface melted during welding.

profundidad de la fusión: *La distancia que la fusión se extiende desde la superficie fundida hasta el metal de base o hasta un cordón de soldadura aplicado previamente.*

diode: Diodes are check valves for electricity. They will pass current in only one direction, from plus to minus, and are used to convert AC to DC.

diodo: *Un dispositivo eléctrico que tiene función análoga a una válvula unidireccional: deja pasar la corriente en una dirección nada más, de positivo a negativo; y por eso se usa en la conversión de C.A. a C.C.*

dip brazing (DB): A brazing process that uses heat from a molten salt or metal bath. When a molten salt is used, the bath may act as a flux. When a molten metal is used, the bath provides the filler metal.

soldadura blanda por inmersión: *Un proceso que usa el calor de un baño de sales o metal.*

Cuando se usan sales, el baño actúa como fundente; cuando se usa metal fundido, el baño provee el metal de aportación.

direct current electrode negative (DCEN): The arrangement of direct current arc welding cables in which the electrode is the negative pole and the workpiece is the positive pole of the welding arc.

electrodo negativo en corriente contínua (ENCC): *El sistema de los cables del arco a corriente contínua en el que el electrodo es el polo negativo y la pieza a soldar es el polo positivo, del arco de soldadura.*

direct current electrode positive (DCEP): The arrangement of direct current arc welding cables in which the electrode is the positive pole and the workpiece is the negative pole of the welding arc.

electrodo positivo en corriente contínua (EPCC): *El sistema de los cables del arco a corriente contínua en el que el electrodo es el polo positivo y la pieza a soldar es el polo negativo, del arco de soldadura.*

discontinuity: An interruption of the typical structure of a material, such as a lack of homogeneity in its mechanical, metallurgical, or physical characteristics. A discontinuity is not necessarily a defect.

discontinuidad: *Una interrupción en la estructura típica de un material, como por ejemplo, carencia de homogeneidad en sus propiedades mecánicas o metalúrgicas, o en sus características físicas.*

distortion: Non-uniform expansion and contraction of metal caused by heating and cooling during the welding process.

deformación: *Expansiones y contracciones no uniformes producidas por el calenta-miento y enfríamiento del metal durante el proceso de soldadura.*

downhill: Welding in a downward progression.

hacia abajo: *La soldadura cuando es ejecutada de arriba hacia abajo.*

drag: During thermal cutting, the offset distance between the actual and straight line exit points of the gas stream or cutting beam measured on the exit surface of the base metal.

arrastre: *En el proceso de corte térmico, el desalineamiento que ocurre entre la línea que sigue la corriente de gas o el haz de corte en actualidad y la línea recta teórica, medido en la superficie del metal ya cortado, o en salida.*

drag angle: The travel angle when the electrode is pointing in a direction opposite to the progression of welding. This angle can also be used to partially define the positions of guns, torches, and rods.

ángulo de arrastre: *El ángulo que existe cuando el electrodo está orientado en dirección opuesta a la de la progresión de la soldadura. Este ángulo se puede usar para definir, al menos parcialmente, las posiciones de las pistolas, sopletes, varillas, y el haz de corte.*

ductility: The tendency to stretch or deform appreciably before fracturing.

ductilidad: *La tendencia a estirarse o deformarse mucho, antes de llegar a la fractura.*

duty cycle: The percentage of time during an arbitrary test period that a power source or its accessories can be operated at rated output without overheating. Most welding machines are rated in intervals of ten minutes meaning that a duty cycle of 50% means the machine can be operated at a given amperage setting for five continuous minutes without damage to the equipment. 60% would give six minutes; 70% would give seven minutes.

ciclo de servicio: El porcentaje del tiempo, en un período de prueba arbitrario, en el que el suministro de energía o sus accesorios pueden ser operados al máximo valor permi-tido sin que le ocurra un sobrecalentamiento. La mayoría de las máquinas soldadoras tienen una capacidad recomendada de intervalos de diez minutos, o sea que un ciclo de servicio de 50% quiere decir que la máquina puede operar a su corriente recomendada por cinco minutos sin que la máquina sufra ningún daño; 60% daría seis minutos; 70%, siete minutos, etc.

E

edge joint: A joint between the edges of two or more parallel or nearly parallel members.

juntura de orilla: Una juntura entre las orillas de dos o más miembros paralelos o casi paralelos.

edge preparation: The preparation of the edges of the joint members, by cutting, cleaning, plating, or other means.

preparación de la orilla: La preparación de las orillas de los miembros de la juntura, cortando, puliendo, enchapando, etc. las orillas.

effective throat: The minimum distance, minus any convexity, between the weld root and the face of a fillet weld.

garganta efectiva: La distancia mínima (sin contar la convexidad) entre la raíz de la soldadura y la cara del cordón de soldadura.

electrode: A component of the electrical welding circuit that terminates at the arc, molten conductive slag, or base metal.

electrodo: Un componente del circuito eléctrico de una soldadura, que termina en el arco, la escoria fundida conductora, o el metal de base.

electrode angle: The angle of the electrode in relationship to the surface of the material being welded; the electrode's perpendicular angle to the metals' surface leaning toward the direction of travel.

ángulo del electrodo: El ángulo del electrodo con respecto a la superficie del material que se está soldando; el ángulo del electrodo perpendicular a la superficie del metal, inclinado hacia la dirección en que progresa la soldadura.

electrode classification: A means of identifying electrodes by their usability, flux coverings, and chemical make up. The American Welding Society has published a series of specifications for consumables used in welding processes.

clasificación de los electrodos: Un medio de identificar los electrodos, donde pueden usarse, fundentes, y constitución química. La American Welding Society (AWS) ha publicado una lista con especificaciones de los materiales consumibles a usar en el proceso de soldaduras.

electrode extension: In gas metal arc welding, flux cored arc welding, electrogas welding, and submerged arc welding, it is the length of electrode extending beyond the end of the contact tube. In gas tungsten arc welding and plasma arc welding, it is the length of the tungsten extending beyond the end of the collet.

extensión del electrodo: En el caso de soldadura con arco de gas y metal, soldadura con arco con fundente, soldadura con electro-gas, soldadura con arco sumergida, se refiere a la longitud del electrodo más allá del fin del tubo de contacto. Finalmente, si se habla de soldadura con arco de gas y tungsteno, o con arco de plasma, es la longitud del tungsteno más allá de la boquilla.

electrode holder: A device used for mechanically holding and conducting current to an electrode during welding or cutting.

boquilla de electrodo: Un dispositivo que soporta el electrodo mecanicamente, y le trae la corriente durante la soldadura o el corte.

electrode lead: The electrical conductor between the source of arc welding current and the electrode holder.

cables de electrodo: El alambre eléctrico que conecta la fuente que suministra la corriente para la sodadura con arco y la boquilla que sujeta el electrodo .

elongation: The amount of permanent extension in the vicinity of a fracture in a tension test; usually expressed in a percentage of original gauge length.

elongación: En un ensayo de tensión, la cantidad de extensión permanente en la vecindad de la fractura; generalmente expresada como porcentaje de la dimension original del calibre.

eutectic composition: The composition of an alloy system that has two descending liquidus curves; the lowest possible melting point for that mixture of metals.

composición del eutectoide: La composición de una aleación tiene dos curvas liquidus que representan las temperaturas de fusión más bajas para cada combinación de metales.

F

face bend test: A test in which the weld face is on the convex surface of a specified bend radius.

prueba de flexión: Un ensayo en el que la cara del cordón está en la superficie convexa de un cierto valor de radio de flexión especificado.

face-centered cubic (FCC): One of the common types of unit cells in which atoms are located on each corner and the center of each face of a cube. Among the common FCC metals are aluminum, copper, nickel, and austenitic stainless steel. This arrangement is typical of the austenitic form of iron.

empaquetamiento cúbico más compacto (ECC): Uno de los tipos más comunes de las estructuras cristalinas en la que los átomos se hallan en las ocho esquinas y en el centro de cada una de las seis caras de un cubo. Entre los metales más comunes se encuentra el aluminio, el cobre, el níquel, el acero inoxidable austenítico, así como en la forma austenítica del hierro.

face reinforcement: Weld reinforcement on the side of the joint from which welding was done.

reenforzamiento de la cara: Soldadura en el lado de la juntura por donde se había hecho la soldadura inicial para reforzar el mismo.

fatigue strength: Ability of a material to withstand repeated loading.

resistencia a la fatiga: La habilidad de ciertos materiales de resistir una carga repetitiva.

faying surface: The mating surface of a member that is in contact with or in close proximity to another member to which it is to be joined.

superficie de empalme: La superficie compañera de un miembro que está en contacto o en proximidad con otro miembro al cual se va a unir por soldadura.

ferrite: A form of low-temperature steel that contains a very small percentage of carbon. Ferrite occurs in steel that has not been previously heat treated or in steel that has been cooled slowly after being transformed to austenite.

ferrita: La estructura del acero de baja temperatura y con poco carbono. Ella se encuentra en aceros que no han sido templados, o en aceros que han sido enfriados muy lentamente después de haberlos cambiado a austenita.

filler material: The material, metal, or alloy to be added in making a welded, brazed, or soldered joint.

material de aportación: El material, metal, o aleación que se añade durante el proceso de soldar una juntura.

filler metal: The metal also known as brazing filler metal, consumable insert, diffusion aid, filler material, solder, welding electrode, welding filler metal, welding rod, and welding wire.

metal de aportación: El metal conocido también por los nombres: metal de aportación para soldaduras blandas, embutido consumible, promotor de difusión, material de aportación, soldadura, electrodo soldador, varilla soldadora, y alambre soldador.

fillet weld: A weld of approximately triangular cross-section joining two surfaces approximately at right angles to each other in a lap joint, T-joint, or corner joint.

soldadura angular: Un cordón de sección aproximadamente triangular, que une dos superficies colocadas a 90 grados uno del otro en una juntura a T, a empalme o de esquina.

fillet weld break test: A test in which the specimen is loaded so that the weld root is in tension.

ensayo de ruptura del cordón de soldadura angular: Una prueba en la que la carga produce una tensión en la raíz.

fillet weld leg: The distance from the joint root to the toe of the fillet weld.

pierna de la soldadura angular: La distancia entre la raíz de la juntura y la orilla del cordón con la pieza.

fillet weld size: For equal-leg fillet welds, the leg lengths of the largest isosceles right triangle that can be inscribed within the fillet weld cross-section. For unequal leg fillet welds, the leg lengths of the right triangle can be inscribed within the fillet weld cross-section.

tamaño de la soldadura angular: Si las dos piernas son iguales (eso es, crean una sección en forma de triángulo isósceles), el tamaño se define como la longitud de la pierna del triángulo isósceles más grande que quepa dentro del cordón. Si las piernas son desiguales, el tamaño es la longitud de la pierna más larga del triángulo recto que se pueda inscribir en la sección.

filter plate: An optical material that protects the eyes against excessive ultraviolet, infrared, and visible radiation. Also called filter glass or filter lens.

plato filtro: Material óptico para proteger la vista a la exposición a demasiada radiación ultravioleta, infraroja, y visible. Tambiéen se le llama vidrio filtro o lente filtro.

filter plate shade: Refers to the lens darkness number, which indicates the darkness of the lens.
tinte del plato filtro: Un número que mide el poder filtrante o de absorción de ciertas radiaciones de la lente.

5F: A welding test position designation for a circumferential fillet weld applied to a joint in pipe, with its axis approximately horizontal, in which the weld is made in the horizontal, vertical, and overhead welding positions. The pipe remains fixed until the welding of the joint is complete.
5F: Una sigla que designa la posición de prueba de la soldadura en el caso de una soldadura angular circunferencial de tubos, en la que el tubo está siempre con su eje aproximadamente horizontal y la soldadura se hace en posiciones horizontal, vertical, y por encima. El tubo queda inmobilizado hasta haberse completado la soldadura.

5G: A welding test position designation for a circumferential groove weld applied to a joint in a pipe with its axis horizontal, in which the weld is made in the flat, vertical, and overhead welding positions. The pipe remains fixed until the welding of the joint is complete.
5G: Una sigla que designa la posición de prueba de la soldadura en el caso de una soldadura de ranura circunferencial de tubos, en la que el tubo está siempre con su eje horizontal y la soldadura se hace en posiciones horizontal, vertical, y por encima. El tubo queda inmobilizado hasta haberse completado la soldadura.

fixture: A device designed to hold and maintain parts in proper relation to each other.
plantilla: Un aparato que sirve para sujetar varias partes en sus posiciones relativas y fijas entre ellas.

flame propagation rate: The speed at which flame travels through a mixture of gases.
velocidad de propagación de la llama: La velocidad con la que viaja una llama por una mezcla de gases.

flare-V-groove weld: A weld in a groove formed by two members with curved surfaces.
soldadura de ranura en V ensanchada: Soldadura hecha en una ranura formada por dos miembros con superficies curvas.

flashback: A recession of the flame into or back of the mixing chamber of the oxygen fuel gas torch or flame spraying gun.
contraquema: La reversión de la llama hacia la cámara donde se mezcla el gas combustiblle con el oxígeno en sopletes y en pistolas rociadoras de llama.

flashback arrester: A device to limit damage from a flashback by preventing propagation of the flame from beyond the location of the arrester.
arrestador de contraquema: Un aparato que sirve para reducir el daño causado por la retroquema, evitando que la llama viaje hacia atrás.

flat welding position: The welding position used to weld from the upper side of the joint at a point where the weld axis is approximately horizontal, and the weld face lies in an approximately horizontal plane.
posición de soldar plana: La posición que se usa para soldar desde la parte alta de la juntu-

ra, en un punto donde el eje de la soldadura es aproximadamente horizontal y la cara del cordón se halla en un plano aproximadamente horizontal.

flaw: An undesirable blemish or discontinuity in a weld such as a crack or porosity.
defecto: Una imperfección indeseable o discontinuidad como una grieta o porosidad.

flux: A material used to hinder or prevent the formation of oxides and other undesirable substances in molten metal and on solid metal surfaces, and to dissolve or otherwise facilitate the removal of such substances.
fundente: Un material que sirve para retrasar o prevenir la formación de óxidos y otras substancias indeseables en el metal fundido y en las superficies de la juntura; y para disolver o ayudar en la remoción de esas substancias.

flux cored electrode: A composite tubular filler metal electrode consisting of a metal sheath and a core of various powdered materials producing an extensive slag cover on the face of a weld bead. External shielding may be required.
electrodo con núcleo de fundente: Un electrodo con metal de aportación, tubular compuesto, que consiste de una envoltura cilíndrica de metal y, en el centro, varios materiales pulverizados que producen una capa de escoria que protege la cara del cordón de soldadura. Es posible que se requiera protección externa.

flux cutting (FOC): An oxygen cutting process that uses heat from an oxyfuel gas flame with a flux in the flame to aid cutting.
corte con fundente: Un proceso de corte con oxígeno que usa el calor generado por una llama de oxígeno + combustible, y con fundente que ayuda en el corte.

forehand welding: A welding technique in which the welding torch or gun is directed toward the progress of welding.
soldando con la palma: Una técnica de soldar en la cual se hace que la boquilla del soplete esté orientada en la misma dirección que aquella en la que progresa la soldadura.

fuel gas: A gas such as acetylene, natural gas, hydrogen, propane, stabilized methylacetylene propadiene, and other fuels normally used with oxygen in one of the oxyfuel processes and for heating.
gas combustible: Un gas como acetileno, gas natural, hidrógeno, propano, metil-acetileno propadieno, y otros, usado normalmente con oxígeno en los procesos llamados en inglés "oxyfuel," y para calentar.

furnace brazing (FB): A brazing process in which the work-pieces are placed in a furnace and heated to the brazing temperature.
soldadura fuerte con horno: Un proceso de soldadura fuerte en el que las piezas son calentadas hasta que lleguen a la temperatura predeterminada.

furnace soldering (FS): A soldering process in which the work-pieces are placed in a furnace and heated to the soldering temperature.
soldadura blanda con horno: Un proceso de soldadura blanda en el que las piezas son calentadas hasta que lleguen a la temperatura predeterminada.

fusible plug: A metal alloy plug that closes the discharge channel of a gas cylinder and is designed to melt at a predetermined temperature permitting the escape of gas.

tapón fusible: *Un tapón hecho de metal de aleación que tapa la salida del gas de una bombona, y no lo deja salir hasta que haya llegado a cierta temperatura.*

fusion: The joining of base material, with or without filler material, by melting them together.
fusión: *La unión de materiales de base, con o sin material de aportación. Esta unión se hace fundiendo las piezas juntas.*

fusion face: A surface of the base metal that will be melted during welding.
cara de fusión: *Una superficie del metal de base, que será fundida en la soldadura.*

fusion welding: Any welding process that uses fusion of the base metal to make the weld.
soldadura a fusión: *Cualquier proceso en el que la soldadura se obtiene fundiendo el metal de base.*

fusion zone: The area of base metal as determined on the cross-section of a weld.
zona de fusión: *El área del metal de base determinada analizando la sección transversal de la soldadura.*

G

gas cylinder: A portable container used for transportation and storage of compressed gas.
bombona (o cilindro) de gas: *Un recipiente portátil para la transportación y el almacenamiento de gas comprimido.*

gas nozzle: A device at the exit end of the torch or gun that directs shielding gas.
boquilla de gas: *Un dispositivo montado a la salida del soplete o pistola que dirije el gas protector.*

gas regulator: A device for controlling the delivery of gas at some substantially constant pressure.
regulador de gas: *Un dispositivo para controlar que la salida del gas ocurra a una presión más o menos constante.*

globular transfer: In arc welding, the transfer of molten metal in large drops from a consumable electrode across the arc.
transferencia globular: *En soldaduras con arco, el traspaso de metal fundido en grandes gotas, del electrodo consumible através del arco.*

GMAW: The welding process Gas Metal Arc Welding; non-standard terms for this process are MIG (metal inert gas), MAG (metal active gas), wire feed, hard wire welding.
GMAW: *Sigla americana para denotar Soldadura con Arco usando Metal y Gas. Otros nombres (no aprobados) son: MIG (Metal y Gas Inerte), MAG (Metal y Gas Activo), alimentado con alambre, y soldadura con alambre duro.*

goggles: Protective glasses equipped with filter plates set in a frame that fits snugly against the face and used primarily with oxygen fuel gas welding processes.
anteojos de seguridad: *Anteojos equipados con lentes filtrantes montados en una montura que se pega muy bien en la cara, para proteger los ojos. Se usan mayormente cuando se está soldando con oxígeno/combustible.*

groove angle: The total included angle of the groove between workpieces.

ángulo de ranura: *El ángulo formado por la ranura que aparece al juntar dos piezas para hacer una soldadura.*

groove face: The surface of a joint member included in the side of the groove from root to toe.
cara de la ranura: *La superficie de un miembro de una juntura incluído en el lado de la ranura, desde la raíz hasta la orilla.*

groove radius: The radius used to form the shape of a J- or U-groove weld.
radio de la ranura: *El radio usado para hacer una soldadura de ranura en la forma de una J o de una U.*

groove weld: A weld made in a groove between the workpieces. See welding symbols.
soldadura de ranura: *La soldadura formada en la ranura creada entre dos piezas. Véanse los símbolos de soldadura en el texto.*

groove weld size: The joint penetration of a groove weld. Also groove throat or effective throat.
tamaño de la soldadura de ranura: *En una soldadura de ranura, la penetración de la juntura. También se le llama garganta de la soldadura, o garganta efectiva.*

ground connection: An electrical connection of the welding machine frame to the earth for safety.
Conexión a tierra: *Una conexión eléctrica de la máquina soldadora a la tierra, para seguridad.*

GTAW: The Gas Tungsten Arc Welding process; non-standard terms are Heliarc™, and TIG (tungsten inert gas).
GTAW: *Sigla americana que denota soldadura con arco usando tungsteno y un gas inerte. Otros nombres (no aprobados) son: heliarc y TIG (tungsteno y gas inerte).*

H

hardfacing: A surfacing variation in which hard material is deposited to reduce wear.
revestimiento duro: *Una variación de revestimiento en la que material es depositado en la superficie para reducir el desgaste.*

heat-affected zone (HAZ): The portion of the base metal whose mechanical properties or microstructure have been altered by the heat of welding, brazing, soldering, or thermal cutting.
zona afectada por el calor: *Aquella porción del metal de base cuyas propiedades mecánicas o su micro-estructura han sido alteradas por el calor creado al soldar o al cortar termicamente.*

hexagonal close packed (HCP): A unit cell in which two hexagons (six-sided shapes) form the top and bottom of the prism. An atom is located at the center and at each point of the hexagon. Three atoms, one at each point of a triangle, are located between the top and bottom hexagons. Among the common HCP metals are zinc, cadmium, and magnesium.
red hexagonal centrada en las caras: *Una celda en la que dos hexágonos forman las bases de un prisma. Las doce esquinas llevan sendos átomos, un átomo ocupa el centro del prisma, y tres átomos más ocupan las tres esquinas de un triángulo localizado en medio*

de las dos bases hexágonales. Entre los materiales con este tipo de estructura átomica se hallan el cinc, el cadmio y el magnesio.

high carbon steel: See carbon steel.

acero a alto crbono: Véase debajo de la definición en inglés de "carbon steel."

horizontal welding position: In a fillet weld, the welding position in which the weld is on the upper side of an approximately horizontal surface and against an approximately vertical surface. In a groove weld, the welding position in which the weld face lies in an approximately vertical plane and the weld axis at the point of welding is approximately horizontal.

posición de soldadura horizontal: En una soldadura angular, la posición de soldar en la que el cordón se aplica en el lado de arriba de una superficie aproximadamente horizontal, y contra otra superficie aproximadamente vertical. En la sodadura de una ranura, la posición en la que la cara de la soldadura está contenida en un plano aproximadamente vertical, y el eje de la soldadura al punto de la soldadura es aproximadamente horizontal.

hydrogen: The lightest chemical element, colorless, odorless, and tasteless. It is found in combination with other elements in most organic compounds and many inorganic compounds. Hydrogen combines readily with oxygen in the presence of heat, and forms water.

hidrógeno: El elemento más ligero; incoloro e insípedo. Se halla naturalmente en combinación con otros elementos en la mayoría de compuestos orgánicos y en muchas substancias inorgánicas. En la presencia de calor, tiene gran afinidad con el oxígeno con quien se combina para formar agua.

I

impact strength: The ability of a material to resist shock, dependent on both strength and ductility of the material.

resistencia al impacto: La habilidad de un material a resistir una carga de choque y que depende de su resistencia y su ductilidad.

inclusion: Entrapped foreign solid material, such as slag, flux, tungsten, or oxide.

inclusión: Partícula extraña de metal sólida, así como escoria, fundente, tungsteno, u óxidos, que se halla atrapada en la soldadura.

incomplete fusion: A weld discontinuity in which fusion did not occur between weld metal and fusion faces or adjoining weld beads.

fusión incompleta: Una discontinuidad en la que la fusión entre el material fundido y las caras de fusión o los cordones de soldadura más cercanas.

incomplete joint penetration: A joint root condition in a groove weld in which weld metal does not extend through the joint thickness.

penetración incompleta de la juntura: En una soldadura de ranura, una condición de la raíz de la soldadura en la que el metal de soldadura no se extiende a todo lo ancho de la juntura.

inert gas: A gas that normally does not combine chemically with other elements or compounds.

gas inerte: Un gas que normalmente no se combina quimicamente con ningún otro material.

infrared radiation: Electromagnetic energy with wave lengths 770 to 12,000 nanometers.
radiación infrarroja: Ondas electro-magnéticas cuyas longitudes de onda van de 770 a 12.000 nanometros.

injector torch: An injector-type torch is used to increase the effective use of fuel gases supplied at pressures of 2 psi (14 kPa), or lower. The oxygen is supplied at pressures ranging from 10 to 40 psi (70 to 275 kPa), the pressure increasing to match the tip size. The relatively high velocity of the oxygen flow is used to aspirate or draw in more fuel gas than would normally flow at the low supply pressures of the fuel gases.
soplete con inyector: Un tipo de soplete que se usa para aumentar el uso eficiente de gases combustibles suministrados a presiones de 2 psi (14 kPa) o menos. El oxígeno es suministrado a una presión que varía de 10 a 40 psi (70 a 275 kPa), subiendo para aparearse con el tamaño de la boquilla. La velocidad del gas, siendo más o menos alta, absorbe más gas combustible que lo que normalmente resultaría con la baja velocidad de los gases combustibles.

intermittent weld: A weld in which the continuity is broken by recurring unwelded spaces.
soldadura intermitente: Una soldadura en la cual la continuidad está interrumpida por espacios que no están soldados.

interpass temperature: In a multipass weld, the temperature of the weld area between weld passes.
temperatura entre pases: En una soldadura con muchos pases, se refiere a la temperatura de la zona de soldadura entre pases.

inverter power supply: A welding power supply with solid-state electrical components that change the incoming 60 Hz power to a higher frequency. Changing the frequency results in greatly reducing the size and weight of the transformer. Inverter power supplies can be used with all arc welding processes.
alimentador eléctrico a inversión: Un suministrador de enrgía para soldar, con componentes electrónicos de estado sólido que convierten la entrada de corriente de 60 Hertz a una frequencia muy alta, para poder usar transformadores más pequeños y más ligeros. Este tipo de alimentador puede usarse en todos los procesos para soldar.

iron carbide: A binary compound of carbon and iron; it becomes the strengthening constituent in steel.
carburo de hierro: Compuesto binario de carbono que contiene más elementos electropositivos, y que en combinación con el hierro, es el constituyente que da la resistencia al acero.

iron soldering: A soldering process in which the heat required is obtained from a soldering iron.
soldadura con hierro de soldar: Un proceso de soldadura en el que el calor para soldar se obtiene de un hierro de soldar.

iron carbon phase diagram: A graphical means of identifying different structures of steel and percentages of carbon occurring in steel at various temperatures.
diagrama de fases de la aleación hiero/carbono: Un método gráfico de identificar las differentes estructuras cristalinas de la aleación en función del porcentaje de carbono añadido, y a varias temperaturas.

isothermal transformation diagram: A graph that identifies different austenitic transformation products that occur over a period of cooling time at isothermal conditions. Also referred to as I-T diagram and T-T-T diagram meaning time-temperature-transformation diagram.

diagrama de transformaciones isotérmicas: *Un gráfico que muestra los diferentes productos austeníticos que ocurren durante el período de enfriamiento en condiciones isotérmicas. También es llamado diagrama IT; y diagrama T-T-T, sigla que quiere decir tiempo y temperatura de transformación.*

Izod test: A test performed on a specimen of metallic material to evaluate resistance to failure at a discontinuity and evaluate the resistance of a comparatively brittle material during extension of a crack. The test is performed using a small bar of round or square cross-section held as a cantilevered beam in a gripping anvil of a pendulum machine. The specimen is broken by a single overload impact of the swinging pendulum, and the energy absorbed in breaking the specimen is recorded by a stop pointer moved by the pendulum.

ensayo de Izod: *Una prueba que se hace con una muestra de material metálico para evaluar la resistencia del material a fallar en una discontinuidad; y para evaluar la resistencia de un material muy frágil a fallar durante la propagación de una grieta. La prueba se lleva a cabo montando una pequeña barra del material, con sección redonda o cuadrada, montada a cantilever, es decir, con un extremo empotrado y el otro libre, en el yunque de una máquina con péndulo. La muestra se rompe con una sola carga de impacto con el péndulo, y la energía absorbida por la muestra se registra en una escala graduada sobre la cual se desliza un indicador por fuerza del péndulo.*

J

J-groove weld: A type of groove weld where one side of the joint forms a J.

soldadura en ranura en J: *Un tipo de soldadura en el que un lado de la juntura tiene la forma de una jota mayúscula.*

joint: The junction of members or the edges of members that are to be joined or have been joined.

juntura: *La unión de dos miembros, o las orillas de los miembros que van a ser soldados.*

joint clearance: The distance between the faying surfaces of a joint in brazing or soldering.

juego en la juntura: *La distancia entre las superficies a unirse en soldaduras fuertes o blandas.*

joint design: The shape, dimensions, and configuration of the joint.

diseño de una juntura: *La forma, dimensiones y configuración de una juntura.*

joint efficiency: The ratio of strength of a joint to the strength of the base metal expressed in percent.

eficacia de una juntura: *la razón de la resistencia mecánica de una juntura a la resistencia del metal de base.*

joint filler: A metal plate inserted between the splice member and thinner joint member to accommodate joint members of dissimilar thickness in a spliced butt joint.

relleno de juntura: *En una juntura de tope a empalme, el plato que se introduce entre el empalme y el miembro más delgado, para acomodarlo con el miembro mas grueso.*

joint geometry: The shape and dimensions of a joint in cross-section prior to welding.
geometría de la juntura: La forma y las dimensiones de la sección de una juntura antes de ser soldada.

joint penetration: The distance the weld metal extends from the weld face into a joint, exclusive of weld reinforcement.
penetración de la juntura: la distancia que el material de soldadura se ha extendido de la cara del cordón hasta el interior de la juntura, sin incluir ninguna soldadura de refuerzo.

joint root: That portion of a joint to be welded where the members approach closest to each other. In cross-section, the joint root may be either a point, a line, or an area.
raíz de la juntura: Aquella porción de la juntura a soldarse, donde los miembros están más cerca uno al otro. Viéndola en sección, la raíz de una juntura puede ser un punto, o una línea, o un plano.

joint spacer: A metal part, such as a strip, bar, or ring, inserted in the joint root to serve as a backing and to maintain the root opening during welding.
separador de juntura: Una pieza metálica, v. gr. tira delgada de relleno, o una barra, o un anillo, la cual causa que la raíz esté abierta durante la soldadura.

joint type: A weld joint classification based on five basic joint configurations such as a butt joint, corner joint, edge joint, lap joint, and T-joint.
tipo de juntura: Clasificación de junturas basada en cinco configuraciones, a saber: juntura de tope, juntura de esquina, juntura de orilla, juntura de empalme, y juntura en forma de T.

K

kerf: The width of a cut produced during a cutting process.
entalladura: El espesor del corte producido durante el corte.

keyhole welding: A technique in which a concentrated heat source penetrates partially or completely through a workpiece, forming a hole (or keyhole) at the leading edge of the weld pool. As the heat source progresses, the molten metal fills in behind the hole to form the weld bead.
soldadura a bocallave: Una ejecución en la que un aparato generador de calor se introduce, parcial or totalmente, através de la pieza creando un hueco (de ahí el nombre de bocallave) en la orilla de la balsa de soldadura. A medida que la introducción del calentador progresa, el metal líquido se cuela por el hueco, formando un cordón.

killed steel: A molten steel that has been held in a ladle, furnace, or crucible until no more gas is evolved and the metal is perfectly quiet.
acero calmado: Un acero, en estado líquido, que se ha dejado en un caldero, o en el horno, o en un crisol hasta que los gases internos se hayan disipado, y el metal se quede perfectamente tranquilo.

L

lamellar tear: A subsurface terrace and step-like crack in the base metal with a basic orientation parallel to the wrought surface. Such items are caused by tensile stresses in the through-thickness direction of the base metals when they have been weakened by the presence of small dispersed, planar shaped, nonmetallic inclusions parallel to the metal surface.

desgarramiento lamelar: La formación de terrazas por debajo de la superficie exterior y de agrietado en forma de escaleras con una orientación más o menos paralela a la superficie labrada, causada por esfuerzos tensiles en la dirección del espesor de los metales de base, los cuales han sido debilitados por la presencia de inclusiones no metálicas, de forma plana, paralelas a la superficie externa del metal de base.

lamination: A type of discontinuity with separation or weakness generally aligned parallel to the worked surface of a metal.

laminación: Un tipo de discontinuidad con separaciones o debilidad normalmente alineadas en un plano paralelo a la superficie externa que esta siendo labrada.

lap joint: A joint between two overlapping members in parallel planes.

juntura a empalme: Una juntura formada por dos miembros asolapados en planos paralelos.

laser: A device that produces a concentrated, coherent light beam by stimulated electronic or molecular transitions to lower energy levels. Laser is an acronym for Light Amplification by Stimulated Emission of Radiation.

laser (pronúnciese leiser): Un dispositivo que produce un haz de luz coherente estimulando la transición de electrones o moléculas de una órbita a alto nivel de energía a otra de nivel más bajo. El nombre es un acrónimo de luz amplificada por estimulación de la emisión de radiación.

laser beam cutting (LBC): A thermal cutting process that severs metal by locally melting or vaporizing with the heat from a laser beam.

corte con rayo laser: Un proceso térmico que separa material fundiendo o vaporizando el metal localmente con el calor de un rayo laser.

laser beam welding: A welding process that produces coalescence with the heat from a laser beam impinging on the joint. The process is used without a shielding gas and without the application of pressure.

soldadura con rayo laser: Un proceso de soldadura que produce unión de la juntura aplicando el calor generado por un rayo laser apuntado hacia la juntura. El proceso funciona sin necesidad de gas aislante o de presión entre los miembros.

lens shade: See filter plate shade.

sombra de lente: Véase debajo de filter plate shade arriba.

linear discontinuity: A discontinuity with a length that is substantially greater than its width.

discontinuidad linear: Una discontinuidad cuya longitud es mucho más grande que su espesor.

liquidus: The lowest temperature at which a metal or an alloy is completely liquid.

liquidus: La mínima temperatura a la que un metal o una aleación se mantiene líquida.

longitudinal crack: A crack with its major axis orientation approximately parallel to the weld axis.

grieta longitudinal: Una grieta cuyo eje mayor está orientado aproximadamente paralelo al eje del cordón de soldadura.

M

macroetch test: A test in which a specimen is prepared with a fine finish, etched, and examined under low magnification.

ensayo de macro-grabado: *Una prueba en la que una muestra se prepara con un acabado muy fino, es grabada al aguafuerte, y después es examinada con baja amplificación.*

manganese: A gray-white nonmagnetic metallic element resembling iron, except it is harder and more brittle. Manganese can be alloyed with iron, copper, and nickel, for commercial alloys. In steel it increases hardness, strength, wear resistance, and other properties. Manganese is also added to magnesium-aluminum alloys to improve corrosion resistance.

manganeso: *Un metaloide, gris-blancuzco, se asemeja al hierro excepto que es más duro y frágil que él. Se liga con el hierro, el cobre, y el níquel para productos comerciales. Añadido al acero, aumenta su dureza, su resistencia mecánica y al desgaste, y otras propiedades. También se le añade a la aleación magnesio-aluminio para aumentar sus propiedades anti-corrosivas.*

manifold: See cylinder manifold.

connector: *Véase debajo de cylinder manifold arriba.*

manual welding: Welding with the torch, gun, or electrode holder held and manipulated by hand. Accessory equipment, such as part motion devices and manually controlled filler material feeders may be used.

soldadura manual: *Soldadura hecha teniendo el soplete, o la boquilla del electrodo en la mano. Accesorios como dispositivos para mover la pieza, alimentadores de material de aportación operados manualmente son permitidos.*

MAPP® gas: A trade name for a fuel gas methacetylene-propadiene.

gas MAPP®: *Marca registrada para el gas combustible metilacetileno-propadieno*

martensite: A very hard, brittle microstructure of steel produced when steel is rapidly quenched after being transformed into austenite.

martensita: *Estructura granular del acero, muy dura y muy frágil, que aparece cuando el acero es templado muy rapidamente después de haber sido transformado en austenita.*

mechanized welding: Welding with equipment that requires manual adjustment of the equipment controls in response to visual observation of the welding with the torch, gun, or electrode holder held by a mechanical device.

soldadura mecanizada: *Método de soldar que requiere el ajuste manual de los controles de parte del operador respondiendo a indicaciones visuales que él tenga de la soldadura, con la pistola, el soplete, o la boquilla del electrodo soportada por medios mecánicos.*

medium steel: Refer to carbon steel.

acero mediano en carbono: *Refiérase a debajo de carbon steel arriba.*

melt-through: Visible root reinforcement produced in a joint welded from one side.

fundido completo: *Refuerzo visible en la raíz, producido por la soldadura de una juntura en un lado nada más.*

metal: A class of chemical elements that are good conductors of heat and electricity, usually malleable, ductile, lustrous, and more dense than other elemental substances.

metal: *Una clase de elementos químicos que son buenos conductores de calor y de electricidad. Normalmente son maleables, dúctiles, brillantes, y más densos que cualquier otro elemento.*

metal-cored electrode: A composite tubular filler metal electrode consisting of a metal sheath and a core of various powdered materials.
electrodo con núcleo de metal: *Un electrodo tubular compuesto, con metal de aportación, que consiste de una vaina metálica y un núcleo de varios materiales pulverizados.*

metal electrode: A filler or non-filler metal electrode used in arc welding or cutting, which consists of a metal wire or rod that has been manufactured by any method and that is either bare or covered with a suitable covering or coating.
electrodo de metal: *Un electrodo de metal — con o sin metal de aportación – usado en corte y soldadura con arco, que consiste de un alambre o varilla metálica, construído por cualquier método, y que está desnudo o cubierto apropiadamente.*

metallic bond: The principal atomic bond that holds metals together.
afinidad metálica: *El enlace principal que mantiene a los metales enteros.*

metallurgy: The science explaining the properties, behavior, and internal structure of metals.
metalurgia: *La ciencia que trata de las propiedades, el comportamiento, y la estructura interna de los metales.*

methylacetylene propadiene: A family of alternative fuel gases that are mixtures of two or more gases (propane, butane, butadiene, methylacetylene, and propadiene). Methylacetylene propadiene is used for oxyfuel cutting, heating, brazing, and soldering.
metilacetileno propadieno: *Una familia de gases combustibles que son mezclas de dos o más gases (propano, butano, butadieno, metilacetileno, y propadieno).*

microstructure: A term use to describe the structure of metals. Visual examination of etched metal surfaces and fractures reveal some configurations in etched patterns that relate to structure, but magnification of minute details yields considerably more information. Microstructures are examined with low-power magnifying glass, optical microscope, or electron microscope.
micro-estructura: *Término usado para describir la estructura interna de los metales. Examinaciones visuales de superficies grabadas con aguafuerte y de fracturas revelan ciertas configuraciones en plantillas grabadas que son relacionadas con la estructura; pero más información se obtiene magnificando detalles microscópicos. Esas micro-estructuras pueden ser analizadas con lupas de bajo poder, microscopios ópticos, y microscopios electrónicos.*

microetch test: A test in which the specimen is prepared with a polished finish, etched, and examined under high magnification.
ensayo micro-grabado: *Una prueba en la que la pieza se prepara con un acabado pulido, es grabada al aguafuerte, y luego examinada con alta magnificación.*

mild steel: Refer to carbon steel.
acero dulce: *Refiérase a debajo de carbon steel arriba.*

mixing chamber: That part of a welding or cutting torch in which a fuel gas and oxygen are mixed.

región de mezcla: La parte del soplete de cortar o soldar en donde el gas combustible y el oxígeno se mezclan.

modulus of elasticity: The ratio of stress to strain in material; also referred to as Young's modulus.

módulo de elasticidad: La razón del esfuerzo a la deformación elástica concomitante, en un metal. También se le llama módulo de Young.

molybdenum: A hard, silver-white metal, a significant alloying element in producing engineering steels, corrosion resistant steels, tool steels, and cast irons. Small amounts alloyed in steel promote uniform hardness and strength.

molibdeno: Un metal, duro, plateado blanco, un elemento de aleación muy importante en la producción de aceros para ingeniería, aceros inoxidables, aceros para utensilios, y hierro de fundición. Pequeñas cantidades añadidas al acero producen dureza y resistencia mecánica uniforme.

multipass welding: A weld requiring more than one pass to ensure complete and satisfactory joining of the metal pieces.

soldadura de varios pases: Una soldadura que requiere más de un pase para asegurar la unión satisfactoria de las piezas metálicas.

multiple welding position: An orientation for a non-rotated circumferential joint requiring welding in more than one welding position, as in welding a pipe or tube in a fixed position. (5F, 5G position is pipe on a horizontal plane and not moved or turned during welding; 6F, 6G position is pipe at 45° off the horizontal plane and not moved or turned during welding).

soldadura en varias posiciones: Una orientación para una juntura circunferencial que no puede ser rotada, y que requiere soldadura en más de una posición, como, por ejemplo, soldar un tubo que está montado en una posición fija. (La posición 5F,5G es para soldar un tubo fijo en posición horizontal; la posición 6F, 6G es para soldar un tubo fijo orientado a 45 grados con respecto a la horizontal).

N

NEMA: The National Electrical Manufacturers' Association.

NEMA: Organización en los EE.UU. llamada Asociación Nacional de Manufactura-dores (de aparatos) Eléctricos.

neutral flame: An oxyfuel gas flame that has characteristics neither oxidizing nor reducing.

llama neutra: Una llama de oxígeno + gas combustible que no tiene tendencias ni a oxidar ni a reducir.

nitrogen: A gaseous element that occurs freely in nature and constitutes about 78% of earth's atmosphere. Colorless, odorless, and relatively inert, although it combines directly with magnesium, lithium, and calcium when heated with them. Produced either by liquefaction and fractional distillation of air, or by heating a water solution of ammonium nitrate.

nitrógeno: Elemento gaseoso muy común en la naturaleza terrestre: representa el 78% de la atmósfera que circunda el planeta. Es incoloro, inodoro y relativamente inerte. Sin embargo, se combina facilmente con magnesio, el litio, y calcio en la presencia de calor.

Se obtiene por licuefacción, seguida por destilación fraccional, del aire; o calentando una solución ácuea de nitrato de amonio.

non-consumable electrode: An electrode that does not provide filler metal, as used in the GTAW process.

electrode no consumible: Un electrodo que no provee el metal de aportación. Un ejemplo es el electrodo para corte con arco de carbono.

non-corrosive flux: A soldering flux that in either its original or residual form does not chemically attack the base metal. It usually is composed of rosin-based materials.

fundente no corrosivo: Un fundente que no combina con el metal de base, en su forma original o su forma residual. Normalmente deriva de materiales resinosos.

non-destructive examination: The act of determining the suitability of some material or component for its intended purpose using techniques that do not affect its serviceability.

prueba no destructiva: El acto de determinar si un material o un componente es adecuado para lo que fué diseñado, sin que por ello pierda su capacidad original.

normalizing: The process of heating a metal above a critical temperature and allowing it to cool slowly under room temperature conditions to obtain a softer and less distorted material.

normalizar: El proceso de calentar un metal por encima de su temperatura crítica, y dejar que se enfríe lentamente a la temperatura del laboratorio, para obtener un material que es más blando y menos deformado.

O

ohm: A unit of electrical resistance. An ohm is equal to resistance of a circuit in which a potential difference of one volt produces a current of one ampere.

ohmio: La unidad de resistencia eléctrica. Un ohmio es la resistencia de un circuito en el que una diferencia de potencial de un voltio, genera una corriente de un amperio

open-circuit voltage: The voltage between the output terminals of the power source when no current is flowing to the torch or gun.

tensión con circuito abierto: El voltaje en los terminales de salida del alimentador de potencia cuando no hay ninguna corriente pasando por la pistola o el soplete.

open root joint: An unwelded joint without backing or consumable insert.

juntura con raíz abierta: Una juntura no soldada, sin soporte ni embutido de consumibles.

overlap: The protrusion of weld metal beyond the weld toe or weld root.

asolapado: El metal de la soldadura que sobresale más allá de la orilla con el metal de base, ode la raíz de la soldadura.

oxidizing flame: An oxyfuel flame in which there is an excess of oxygen, resulting in an oxygen-rich zone extending around and beyond the cone.

llama oxidante: Una llama de oxígeno/gas combustible en la que hay exceso de oxígeno, lo cual resulta en haber una zona alrededor y cerca del cono rica en oxígeno.

oxygen: A colorless, odorless, tasteless, gaseous chemical element, the most abundant of all elements. Oxygen occurs free in the atmosphere, forming 1/5 of its volume, and in combination in water, sandstone, limestone, etc.; it is very active being able to combine with nearly all other elements and is essential to life.

oxígeno: *Elemento químico, incoloro, inodoro, insípido, gaseoso; el elemento más abundante en la Tierra. Se encuentra, libre, en la atmósfera, de la que forma la quinta parte de su volúmen. También se halla, en combinación, en el agua, piedra de arena, piedra de cal, etc. Es muy activo, y tiende a combinarse con practicamente todos los demás elementos. Por último, es indispensable para la vida.*

oxygen lance: A length of pipe used to convey oxygen to the point of cutting in oxygen lance cutting.

lanceta de oxígeno: *En el corte con lanceta de oxígeno, tubo que transporta el oxígeno al punto de corte.*

oxyhydrogen cutting (OFC-H): An alternative fuel gas cutting process that uses hydrogen as the fuel source.

corte con oxi-hidrógeno: *Una alternativa del corte con gas combustible, en la que el gas es hidrógeno.*

oxyhydrogen welding (OHW): An alternative fuel gas welding process that uses hydrogen as the fuel source.

soldadura con oxi-hidrógeno: *Una alternativa de la soldadura con gas combustible, en la que el gas es hidrógeno.*

oxynatural gas cutting (OFC-N): An alternative fuel gas cutting process that uses natural gas as the fuel source.

corte con oxi-gas natural: *Una alternativa del corte con gas combustible, en la que el gas es gas natural.*

oxypropane cutting (OFC-P): An alternative fuel gas cutting process that uses propane gas as the fuel source.

corte con oxi-propano: *Una alternativa del corte con gas+combustible, en la que el gas es propano.*

P

parent metal: A non-standard term referring to the base metal.
metal padre: *Un término, no aprobado, para referirse al metal de base.*

partial joint penetration weld: A joint root condition in a groove weld in which incomplete joint penetration exists.

soldadura con penetración parcial de la juntura: *En una soldadura de ranura, la condición en la raíz de la juntura en la que existe una penetración parcial de la soldadura.*

pass: A single progression of welding along a joint, resulting in a weld bead or layer.
pase: *Una sola progresión de soldadura a lo largo de una juntura, lo que resulta en una capa o cordón.*

pearlite: A mixture of ferrite and cementite that contains approximately 0.8% carbon. Pearlite occurs in low-temperature steel that has not been previously heat treated or in steel that has been cooled slowly after being transformed into austenite.

perlita: *Una combinación de ferrita y cementita que tiene aproximadamente un 0,8% de carbono. La perlita aparece en aceros de baja temperatura que no han sido calentados previamente, o en aceros que han sido enfríados lentamente después de haber sido transformados en austenita.*

peening: The mechanical working of metals using impact blows.

martilleado: *El trabajo mecánico hecho en una pieza metálica por medio de una serie seguida de martillazos.*

penetration: A non-standard term used in describing depth of fusion, joint penetration, or root penetration.

penetración: *Un término, no aprobado, para describir la profundidad de una fusión, penetración de una juntura, o penetración de la raíz.*

phase diagram: A graph that identifies alloy phases occurring at various temperatures and percentages of alloying elements. Also referred to as an equilibrium diagram.

diagrama de fases: *Un gráfico que muestra, en un metal de aleación, las fases por las que pasa en función de su temperatura y del porcentaje de los elementos de aleación. Es también llamado diagrama de equilibrios.*

phase transitions: When metals or metal alloys go from solid to liquid or the reverse, this is a phase transition. Iron phase transitions are: at room temperature to 1,670°F (910°C) iron is body-center cubic, 1670°F (910°C) to 2535°F (1388°C) iron is face-center cubic, and 2535°F (1390°C) the melting point of iron to 2800°F (1538°C) iron is again body-center cubic. These changes are also called allotropic transformations.

transiciones de fase: *Una transición de fase ocurre, por ejemplo, cuando un metal de aleación pasa de su fase sólida a su fase líquida, o vice versa. Las trnsiciones de phase del hierro son así: De la temperatura ambiental hasta 1.670°F (910°C) el cristal de hierro es de red cúbica centrada en el cubo; de 1.670°F (910°C) a 2.535°F (1.390°C) cambia a red cúbica centrada en las caras; y, por último, de 2.535°F (1.390°C) (que es el punto de fusión del hierro) hasta 2.800°F (1.538°C) regresa a la red cúbica centrada en el cubo. Estas transiciones se llaman también transformaciones alotrópicas.*

phosphorous: A highly reactive, toxic, non-metallic element used in steel, glass, and pyrotechnics. It is almost always found in combination with other elements such as minerals or metal ores. Found in steel and cast iron as an impurity. In steel it is reduced to 0.05% or less otherwise phosphorous causes embrittlement and loss of toughness, however small amounts in low-carbon steel produce a slight increase in strength and corrosion resistance.

fósforo: *Elemento químico, metaloide, muy activo, tóxico, usado en el acero, el vidrio, y en la pirotecnia. Casi siempre se encuentra en combinación con otros elementos, como minerales o menas. En el acero y el hierro de fundición aparece como una impuridad. De hecho, en acero hay que reducirlo a 0,05% o menos, o causa fragilidad y pérdida de tenacidad. Sin embargo, añadidura de cantidades mínimas al acero a bajo carbono produce un pequeño aumento de la resistencia mecánica, y la resistencia a la corrosión.*

pilot arc: A low-current arc between the electrode and the constricting nozzle of the plasma arc torch to ionize the gas and facilitate the start of the welding arc.

arco piloto: *Un arco a baja corriente entre el electrodo y la boquilla constrictiva en el*

soplete con arco de plasma para ionizar el gas, y así facilitar el inicio del arco de sol-
dadura.

plasma arc cutting (PAC): An arc cutting process that uses a constricted arc and removes the
molten metal with a high-velocity jet of ionized gas issuing from the constricting orifice.
corte con arco de plasma: *Un proceso de corte con arco que usa un arco constringido y*
remueve el metal fundido con un chorro a alta velocidad de gas ionizado que sale por el
orificio constringido.

plasma arc welding (PAW): An arc welding process that uses a constricted arc between a
non-consumable electrode and the weld pool or between the electrode and the constricting nozzle. Shielding is obtained from the ionized gas issuing from the torch, and may be
supplemented by an auxiliary source of shielding gas.
soldadura con arco de plasma: *Un proceso de soldadura con arco que usa un arco restringi-*
do entre un electrodo no consumible y la balsa de soldadura, o entre el electrodo y la
boquilla constringente. El gas ionizado que emana del soplete aisla, y puede ser suple-
mentado con una fuente auxiliar de gas aislante.

plug weld: A weld made in a circular hole in one member of a joint fusing that member to
another member.
soldadura de tapón: *Una soldadura en un orificio circular de un miembro de una juntura*
fundiendo ese miembro con el otro.

polarity: The condition of being positive or negative with respect to some reference point or
object. In welding the terminals of the power supply are designated negative and positive.
Which terminal is hooked to the electrode determines polarity.
polaridad: *La condición de ser positivo o negativo con respecto a un punto u objeto de refer-*
encia. En las máquinas para soldadura, los terminales del alimentador de energía están
marcados positivo y negativo. El terminal al que se conecta el electrodo determina la
polaridad.

porosity: A discontinuity formed by gas entrapment during solidification or in a thermal
spray deposit.
porosidad: *Una discontinuidad causada por gases que quedan entrapados durante la solidi-*
ficación, o por depósitos de un rociador térmico.

positive pressure torch: The positive pressure torch requires that gases be delivered at pressures above 2 psi (14 kPa). In the case of acetylene, the pressure should be between 2 and
15 psi (14 to 103 kPa). Oxygen is generally supplied at approximately the same pressure
for welding.
antorcha a presión: *Un soplete el cual se debe alimentar con gases a presiones de 2 psi (14*
kPa) o mayor. En el caso del acetileno, la presión debe ser entre psi (14 kPa) y 15 psi
(103 kPa). El oxígeno también se suministra a estas presiones para soldaduras.

post-flow time: The time interval from current shut off to either shielding gas or cooling
water shut off.
retardo en apagarse: *El intervalo entre el momento en que se apaga la corriente y el*
momento en que el gas aislante o el agua refrigerante se cierran.

post-heating: The application of heat to an assembly after welding, brazing, soldering, thermal spraying, or thermal cutting.

post calentamiento: La aplicación de calor a un conjunto después de haberlo soldado, rociado, o cortado.

power factor: The ratio of true power (watts) to the apparent power (volts times amperes). The power factor is equal to the cosine of the angle of lag between the alternating current and voltage wave.

factor de potencia: La razón de la verdadera potencia (medida en vatios) y la potencia aparente (medida en volt-amperios). Su valor es igual al valor del coseno del ángulo de fase entre la onda de la corriente alterna y la del voltaje alterno.

power source: An apparatus for supplying current and voltage suitable for welding, thermal cutting, or thermal spraying.

fuente de energía: Un aparato para el suministro de la corriente y tensión apropiadas para soldar, cortar termicamente o rociar termicamente.

precipitate: To cause to become insoluble, with heat or a chemical reagent, and separate out from a solution.

precipitar: Causar a una substancia en solución que se vuelva insoluble usando calor o un reactivo químico, y que se deposite en forma sólida.

precipitation hardening: A multiphase heat treatment process that strengthens alloys by causing phases to precipitate at various temperatures and cooling rates.

endurecimiento por precipitación: Tratamiento térmico de varias fases que endurecen las aleaciones haciendo que las fases se precipiten a diversas temperaturas y diversas velocidades de enfriamiento.

pre-flow time: The time interval between start of shielding gas flow and arc starting.

retardo en prenderse:l inervalo entre el momento en que el gas aislante comienza a fluir y el comienzo del arco.

preform: Brazing or soldering filler metal fabricated in a shape or form for a specific application.

preforma: La forma que se le da a un metal de aportación en su fabricación para su uso en un caso particular.

preheat: The heat applied to the base metal or substrate to attain and maintain preheat temperature.

pre-calentamiento: El calor aplicado a un metal de base o substrato para obtener y mantener una temperatura de calentamiento predeterminada.

pressure regulator: A device designed to maintain a nearly constant supply pressure. Regulators may be attached to pressurized cylinders, gas generators, or pipe lines to reduce pressure as desired to operate equipment.

regulador de presión: Un dispositivo para mantener el suministro de una presión casi constante. Los reguladores pueden ser conectados a bombonas de gas comprimido, generadores a gas, o líneas de tuberías para reducir la presión al valor necesario para operar la maquinaria.

pre-qualified welding procedure specification: A welding procedure specification that complies with the stipulated conditions of a particular welding code or specification and is therefore acceptable for use under that code or specification without a requirement for qualification testing.
especificaciones para un proceso de soldadura pre-calificado: especificaciones que satisfacen las condiciones estipuladas por un código de soldadura; y es, consecuente-mente, aceptable para su uso dentro de ese código sin requerir que se haga una prueba de calificación

primary windings: The windings connected to and receiving power from an electrical circuit.
bobina primaria: el arrollamiento de alambre de un transformador que se conecta al suministro de enrgía para recibir su potencia de ese.

procedure qualification: The demonstration that welds made by a specific procedure can meet prescribed standards.
calificación de un procedimiento: La prueba que indica que soldaduras hechas con un procedimiento específico puede pasar los estándars impuestos.

process: A grouping of basic operational elements used in welding, thermal cutting, brazing, or thermal spraying.
proceso: Un grupo de elementos operativos básicos usados en soldaduras, cortes con calor, soldadura fuerte, o rociador térmico.

protective atmosphere: A gas or vacuum envelope surrounding the workpieces, used to prevent or reduce the formation of oxides and other detrimental surface substances, and to facilitate their removal.
atmósfera protectora: Una zona de gas o de vacío que envuelve las piezas a soldarse, y se usa para evitar o reducir la formación de óxidos y otras substancias de detrimento a sus superficies; y para facilitar su remoción.

pulsed-power welding: An arc welding process variation in which the power is cyclically programmed to pulse so that effective but short duration values of power can be utilized. Such short duration values are significantly different from the average value of power. Equivalent terms are pulsed-voltage or pulsed-current welding.
soldadura a pulsos: Una variación del proceso de soldadura con arco en la cual la energía es suministrada en pulsos a intervalos de corta duración pero muy potentes. Estos pulsos son mucho más altos que la potencia media. Otros términos son: soldadura a voltage pulsado o soldadura a corriente pulsada.

purging: The removing of any unwanted gas or vapor from a container, chamber, hose, torch, or furnace.
purgar: La remoción de gases o vapores indeseables de un recipiente, cámara, mangas, sopletes u hornos.

push angle: The travel angle when the electrode is pointing in the direction of the weld progression. This angle can also be used to partially define the positions of welding guns.
ángulo de empuje: El ángulo de curso cuando el electrodo está inclinado en la misma dirección del progreso de la soldadura. Este ángulo se puede usar para definir, en parte, la posición de los sopletes.

Q

qualification: A specific set of procedures designed to test a welder's ability; followed by a welder qualification test. After passing a particular qualification test a welder is then qualified to weld to the variables of that qualification.

calificación: Un grupo específico de procedimientos diseñado para comprobar la habilidad de un soldador, seguido por un exámen de calificación. Al haber pasado una cierta prueba, el soldador está entonces calificado para hacer soldaduras definidas en esa calificación.

quenching: The sudden cooling of heated metal by immersion in water, oil, or other liquid. The purpose of quenching is to produce desired strength properties in hardenable steel.

temple: En un metal muy caliente, el enfriamiento rápido obtenido por inmersión del metal en un baño de agua, aceite, u otro líquido. El propósito es el de obtener ciertas características en las propiedades de los aceros endurecibles.

R

reactor: A device used in arc welding circuits to minimize or smooth irregularities in the flow of the welding current; also called an inductor.

reactancia: Un dispositivo eléctrico que se usa en soldaduras con arco para eliminar o minimizar fluctuaciones de la corriente del arco. También se le llama inductancia.

reducing flame: An oxyfuel flame with an excess of fuel gas.

llama reductora: una llama de oxígeno/gas combustible con exceso de combustible.

residual stress: Stress present in a joint member or material that is free of external forces or thermal gradients.

esfuerzo residual: Esfuerzo que se halla en un miembro de una juntura o en materiales que no sufren la acción de fuerzas externas o a variaciones de temperatura.

resistance brazing (RB): A brazing process using heat from the resistance to electric current flow in a circuit of which the workpieces are a part.

soldadura fuerte con resistencia: Una soldadura fuerte que usa el calor generado por la resistencia al flujo de corriente en un circuito del que las piezas a soldar forman parte.

resistance soldering (RS): A soldering process using heat from the resistance to electric current flow in a circuit of which the workpieces are a part.

soldadura blanda con resistencia: Una soldadura que usa el calor generado por la resistencia al flujo de corriente en un circuito del que las piezas a soldar forman parte.

resistance welding (RW): A group of welding processes that with the application of pressure produces coalescence of the faying surfaces with the heat obtained from resistance of the workpieces to the flow of the welding current in a circuit of which the workpieces are a part.

soldadura a presión con resistencia: Un grupo de procesos de soldadura que produce coalescencia de las superficies yuxtapuestas por medio del calor generado por la resistencia al flujo de la corriente de soldadura en el circuito del cual las piezas forman parte, y por medio de la aplicación de presión a la misma vez.

resistor: A device with measurable, controllable, or known electrical resistance used in electronic circuits or in arc welding circuits to regulate the arc amperes.

resistor: *Un componente eléctrico con una resistencia elétrica que puede ser medida, contro-lada, y conocida. Se usa en circuitos eléctricos; y en circuitos de soldaduras con arco, sirve para controlar el amperaje del arco.*

Rockwell hardness test: The most common hardness testing method. This procedure uses a minor load to prevent surface irregularities from affecting results. There are nine different Rockwell hardness tests corresponding to combinations of three penetrators and three loads.
ensayo de dureza tipo Rockwell: *El método más comun para medir la dureza de metales. Consiste en la aplicación de una pequeña carga para evitar que irreularidades de la superficie afecten el resultado. Hay nueve escalas que corresponden al uso de tres pene-tradores diferentes con tres cargas diferentes.*

root bead: A weld bead that extends into or includes part or all of the joint root.
cordón de raíz: *Un cordón de soldadura que se extiende dentro de la raíz de la juntura incluyendo parte de, o toda ella.*

root bend test: A test in which the weld root is on the convex surface of a specified bend radius.
prueba de flexión de la raíz: *Una prueba en la que el cordón de la raíz está de la parte con-vexa de una superficie con un cierto radio de curvatura.*

root face: That portion of the groove face within the joint root.
ccara de la raíz: *La porción de la cara de la ranura que se encuentra dentro de la raíz de la juntura.*

root opening: A separation at the joint root between the workpieces.
apertura de la raíz: *El espacio, en la raíz de la juntura, entre los miembros de la juntura.*

root penetration: The distance the weld metal extends into the joint root.
penetración de la raíz: *La distancia que el metal de soldadura se extiende dentro de la raíz de la juntura.*

root reinforcement: Weld reinforcement opposite the side from which welding was done.
refuerzo de la raíz: *Cordón de refuerzo en el lado opuesto al de la soldadura original.*

runoff weld tab: Additional material that extends beyond the end of the joint, on which the weld is terminated.
pestaña del fin the soldadura: *Exceso de material que se extiende más allá del fin de la jun-tura, y donde la soldadura termina.*

S

SAE: The Society of Automotive Engineers.
SAE: *Organización en los EE.UU., llamada Sociedad de Ingenieros Automotivos.*

safety disc: A disc in the back side of a high pressure cylinder valve designed to rupture and release gas to the atmosphere preventing cylinder rupture if the cylinder is mishandled.
disco de seguridad: *En una bombona de gas comprimido, un disco en la parte trasera de la válvula de la bombona hecho de modo que se rompa y deje escapar el gas más bien que dejar que la bombona explote por mal tratamiento.*

seal weld: Any weld designed primarily to provide a specific degree of tightness against leakage.
cordón de sello: Un cordón diseñado para producir cierto grado de impermeabilidad.

seam weld: A continuous weld made between or upon overlapping members, in which coalescence may start and occur on the faying surfaces, or may have proceeded from the outer surface of one member. The continuous weld may consist of a single weld bead or a series of overlapping spot welds.
soldadura de costura: Una soldadura contínua hecha entre, o por encima de miembros de una juntura asolapados, y en las que coalescencia puede empezar a ocurrir en las superficies de la juntura, o puede que venga de la superficie externa de uno de los miembros. La soldadura contínua puede consistir de un cordón simple, o de una serie de soldaduras por puntos.

semiautomatic: Pertaining to the manual control of a process with equipment that controls one or more of the process conditions automatically.
semi-automático: Referiéndose al control manual de un proceso con aparatos que controlan automaticamente una o más de las variables del proceso

shear: To tear or wrench by shearing stress; to cut through using a cold cutting tool when shearing metal.
romper con fuerza cortante: Rasgar o deslocar por medio de esfuerzo cortante. Cuando se trata de metales, cortar a través de una pieza usando una herramienta para cortar en frío.

shear strength: The characteristic of a material to resist shear forces.
resistencia a la cizalladura: La característica de un material de resistir cargas en la forma de fuerzas cortantes.

shielding gas: Protective gas used to prevent or reduce atmospheric contamination.
gas aislante: Gas protector, que elimina, o reduce, la contaminación atmosférica.

short-circuiting transfer: Metal transfer in which molten metal from a consumable electrode is deposited during repeated short circuits.
transferencia con corto-circuitos: El caso en que metal fundido de un electrodo consumible es depositado durante corto-circuitos repetidos.

side bend test: A test in which the side of a transverse section of the weld is on the convex surface of a specified bend radius.
prueba de flexión lateral: Una prueba en la que el lado de una sección transversal de la soldadura está de la parte convexa de una superficie con un cierto radio de curvatura.

silicon: A non-metallic element resembling graphite in appearance, used extensively in alloys. It is the second most common element on earth. Silicon is usually found in the oxide (silicate) form. Silicon contributes to the strength of low-alloy steels and increases hardenability along with performing the valuable function of a deoxidizer, eliminating trapped gas.
silicio: Elemento no metálico, muy parecido al grafito, muy usado como elemento de aleación. Se encuentra como óxido (silicato). El silicio contribuye a mejorar la resistencia de aceros de baja aleación, y aumenta su cementación; y a la misma vez desempeña

la importante función de desoxigenante, eliminando los gases que están atrapados en el metal.

silicon rectifier: A silicon semiconductor device that acts like a check valve for electricity and is used to change alternating current to direct current.

***rectificador de silicio**: Un dispositivo hecho del elemento semiconductor silicio, que funciona como una válvula unidireccional para la electricidad, por lo que se usa para convertir una corriente alterna en corriente contínua.*

single-phase: A generator or circuit in which only one alternating current voltage is produced.

***monofásico**: Un generador o circuito en el que se produce una sola corriente o voltaje alterno.*

6F: A welding test position designation for a circumferential fillet weld applied to a joint in pipe, with its axis approximately 45° from horizontal, in which the weld is made in flat, vertical, and overhead welding positions. The pipe remains fixed until welding is completed.

***6F**: Una sigla que designa la posición de la soldadura en el caso de una soldadura angular circunferencial de tubos, en la que el tubo está siempre con su eje aproxima-damente a 45°con la horizontal; y la soldadura se hace en posiciones horizontal, vertical,y por encima. El tubo queda inmobilizado hasta haberse completado la soldadura.*

6G: A welding test position designation for a circumferential groove weld applied to a joint in pipe, with its axis approximately 45° from horizontal, in which the weld is made in the flat, vertical, and overhead welding positions. The pipe remains fixed until welding is completed.

***6G**: Una sigla que designa la posición de la soldadura en el caso de una soldadura de ranura circunferencial de tubos, en la que el tubo está siempre con su eje aproximada-mente a 45° con la horizontal y la soldadura se hace en posiciones horizontal, vertical,y por encima. El tubo queda inmobilizado hasta haberse completado la soldadura.*

6GR: A welding test position designation for a circumferential groove weld applied to a joint in pipe, with its axis approximately 45° from horizontal, in which the weld is made in the flat, vertical, and overhead positions. A restriction ring is added adjacent to the joint to restrict access to the weld. The pipe remains fixed until welding is completed.

***6GR**: Una sigla que designa la posición de la soldadura en el caso de una soldadura de ranura circunferencial de tubos, en la que el tubo está siempre con su eje a 45° con la horizontal y la soldadura se hace en posiciones horizontal, vertical,y por encima. Un aro de restricción se añade para limitar el acceso a la soldadura. El tubo queda inmobilizado hasta haberse completado la soldadura.*

slag: A nonmetallic product resulting from the mutual dissolution of flux and non-metallic impurities in some welding and brazing processes.

***escoria**: El producto, no metálico, de la disolución del fundente y de impurezas no metá-licas que ocurre en algunos tipos de soldadura.*

slope: A term used to describe the shape of the static volt-ampere curve of a constant-voltage welding machine. Slope is caused by impedance and is usually introduced by adding substantial amounts of inductance to the welding power circuit. As more inductance is added to a welding circuit, there is a steeper slope to the volt-ampere curve. The added inductance limits the available short-circuit current and slows the rate of response of the welding machine to changing arc conditions.

inclinación: *Término que describe la forma de la curva estática de volt-amperios en una máquina soldadora a voltaje constante. La inclinación es causada por la impedancia que resulta del aumento de inductancia en el circuito que suministra la energía. A más inductancia, más tiende la curva a ser vertical. Añadiendo inductancia limita la corriente en corto-circuitos y retarda la reacción de la máquina a cambios en la condición del arco.*

slot weld: A weld made in an elongated hole in one member of a joint fusing that member to another member. The hole may be open at one end.

soldadura de ranura: *Soldadura hecha en un orificio alargado en un miembro de la juntura, fundiéndolo sobre otro miembro. El orificio puede estar abierto en un extremo.*

slugging: The unauthorized addition of metal, such as a length of rod, to a joint before welding or between passes, often resulting in a weld with incomplete fusion.

aporreando: *La adición, no aprobada, de metal, como por ejemplo un pedazo de varilla, en una juntura antes de soldarla, o entre pases, y que resulta en una soldadura con fusión incompleta.*

solder: The metal or alloy used as a filler metal in soldering, which has a liquidus not exceeding 840°F (450°C) and below the solidus of the base metal.

soldadura: *El metal o aleación que se usa como material de aportación en soldaduras blandas, cuyo liquidus no sobrepasa los 840°F (450°C) y es menor que el liquidus del metal de base.*

soldering: A group of welding processes that produce coalescence of materials by heating them to the soldering temperature and by using a filler metal having a liquidus not exceeding 840°F (450°C) and below the solidus of the base metals. The filler metal is distributed between closely fitted faying surfaces of the joint by capillary action.

soldar: *Un grupo de procesos de soldadura que produce coaleescencia de materiales calentándolos hasta llegar a la temperatura de soldar y usando un metal de aportación cuyo liquidus no sobrepasa los 840°F (450°C) y es menor que el liquidus de los materiales de base. El metal de aportación se distribuye por todas las superficies de la juntura en contacto íntimo por efecto de capilaridad.*

soldering iron: A soldering tool having an internally or externally heated metal bit usually made of copper.

hierro de soldar: *Un utensilio que tiene una punta metálica calentada internamente o externamente , que muchas veces es hecha de cobre.*

solder interface: The interface between solder metal and the base metal in a soldered joint.

superficies colindantes de una soldadura: *el espacio entre el metal de soldadura y el metal de base. Se le llama in-ter-féis en inglés.*

solder metal: That portion of a soldered joint that has melted during soldering.

metal de soldar: *Aquella porción de la juntura soldada que se fundió durante el proceso de soldarla.*

solidus: The highest temperature at which a metal or an alloy is completely solid.

sólidus: En un metal o en una aleación, la máxima temperature a la que el material está todavía completamente sólido.

solutionizing: The process of dispersing one or more liquid, gaseous, or solid substances in another, usually a liquid, so as to form a homogeneous mixture.
"solucionizando": El proceso de dispersar una o más substancias líquidas, gaseosas, o sólidas en otra substancia, generalmente otro líquido, con el propósito de formar una mezcla más homogenea.

spacer strip: A metal strip or bar prepared for a groove weld and inserted in the joint root to serve as a backing and to maintain the root opening during welding. It can also bridge an exceptionally wide root opening due to poor fit.
tira de separación: Una tira o barra de metal preparada para montarse en la raíz de la juntura de una soldadura de ranura. Sirve de soporte y también mantiene la raíz abierta durante la soldadura. Puede ser montada en casos de raíces muy amplias debido a una juntura mal montada con mucho juego.

spatter: The metal particles expelled during fusion welding that do not form a part of the weld.
salpicadura: Partículas metálicas lanzadas durante la soldadura a fusión, y que no forman parte de la soldadura.

spheroidizing: A stress relieving method of long-term heating of high-carbon steel at or near the lower transformation temperature, followed by slow cooling to room temperature.
esferoidizar: Un método de aliviar los esfuerzos residuales en acero a alto carbono. Esto se hace calentando el acero hasta que llegue a una temperatura por debajo de la menor temperatura de transformación del acero, seguido por un enfriamiento lento hasta llegar a la temperatura ambiente.

spliced joint: A joint in which an additional workpiece spans the joint and is welded to each member.
juntura empalmada: Juntura en la que se monta otra pieza que cubre la juntura y sus bordes se sueldan a cada miembro.

spool: A filler metal package consisting of a continuous length of welding wire in coil form wound on a cylinder (called a barrel) which is flanged at both ends. The flange contains a spindle hole of smaller diameter than the inside diameter of the barrel.
bobina: Un paquete de metal de aportación en forma de alambre enrollado sobre un cilindro (llamado barril) con rebordes en los lados. El reborde tiene un orificio cuyo diámetro es menor que el del barril.

spot weld: A weld made between or upon overlapping members in which coalescence may start and occur on the faying surfaces or may proceed from the outer surface of one member. The weld cross-section is approximately circular.
soldadura por puntos: Una soldadura hecha a una juntura asolapada en la que coalescencia puede empezar en las superficies de la juntura, o en la superficie externa de uno de los miembros. La sección de la soldadura es aproximadamente circular.

spray transfer: Metal transfer in which molten metal from a consumable electrode is propelled axially across the arc in small droplets.

traslado con rocío: Trnsferencia de metal en el que el metal fundido proviene del electrodo consumible, y es impelido axialmente a través del arco en la forma de gotas pequeñas.

stack cutting: Thermal cutting of stacked metal plates arranged so that all the plates are severed by a single cut.
corte de pila: Corte térmico de una pila de planchas metálicas hecha de tal forma que todas las planchas sean cortadas en un solo pase.

staggered intermittent weld: An intermittent weld on both sides of a joint with the weld increments on one side alternating with respect to those on the other side.
soldadura escalonada intermitente: Una soldadura intermitente de ambos lados de una juntura y con los puntos de soldadura de un lado intercalados entre los puntos del otro lado.

standoff distance: The distance between a welding nozzle and the workpiece.
distancia de compensación: La distancia de la boquilla del soplete a la pieza.

steel: A material composed primarily of iron, less than 2% carbon, and (in an alloy steel) small percentages of other alloying elements.
acero: Material compuesto principalmente de hierro, menos de 2% de carbono, y (en los aceros de aleación) porcentajes pequeños de otros elementos de aleación.

step-down transformer: A transformer that reduces the incoming voltage.
transformador reductor: Un transformador que reduce la (alta) tension, o corriente de entrada en un valor bajo de las mismas.

step-up transformer: A transformer that increases the incoming voltage.
transformador elevador: Un transformador que eleva la (baja) tension, o corriente de entrada en un valor alto de las mismas.

stickout: In GTAW, the length of the tungsten electrode extending beyond the end of the gas nozzle. In GMAW and FCAW, the length of the unmelted electrode extending beyond the end of the contact tube.
extensión: Usando GTAW, la distancia que el electrodo de tungsteno se extiende después de pasar la boquilla de gas. Usando GMAW y FCAW, la longitud del electrodo que se extiende después de pasar el tubo de contacto.

strain: Distortion or deformation of a metal structure due to stress.
deformación: Distorsión o deformación de la estructura metálica debida a los esfuerzos.

stress: A force causing or tending to cause deformation in metal. A stress causes strain.
esfuerzo: Condición interna de una substancia elástica causada por cargas externas, y que va acompañada por una codición de deformaciones en la substancia.

stringer bead: A type of weld bead made without appreciable weaving motion.
cordón de zanca: Un tipo de cordón de soldadura depositado sin zigzagueo notable.

stub: The short length of filler metal electrode, welding rod, or brazing rod that remains after its use for welding or brazing.
colilla: El pedazito de electrodo con metal de aportación o varilla de soldadura, que queda después de haber sido usados para hacer la soldadura.

stud welding (SW): A general term for joining a metal stud or similar part to a workpiece. Welding may be accomplished by arc, resistance, friction, or other process with or without external gas shielding.

soldadura de perno: *Un término general para describir la unión de un perno de metal a una pieza. La soldadura puede hacerse con arco, resistencia, fricción, u otro proceso, con o sin aislamiento de gas externo.*

submerged arc welding (SAW): An arc welding process that uses an arc or arcs between a bare metal and the weld pool. Molten metal is shielded by a blanket of granular flux on the workpieces. The process is used without pressure and with filler metal from the electrode and sometimes from a supplemental source (welding rod, flux, or metal granules).

soldadura con arco sumergido: *Proceso de soldadura con arco en el que se usa uno o más arcos entre el/los elctrodo(s) de metal, desnudos, y la balsa de soldadura. El metal fundido está cubierto por una capa de granos de fundente en las piezas. Se puede usar o sin presión, y con metal de aportación del electrodo, y algunas veces del exrno (v.gr., varilla de soldar, fundente, o gránulos de metal.*

substrate: Any material to which a thermal spray deposit is applied.

substrato: *Todo material al que se le ha aplicado un depósito con rocío térmico.*

sulfur: A pale yellow, odorless, brittle, nonmetallic element found underground either in the solid state or as a molten sulfur.

sulfuro: *Elemento no metálico, inodoro, de color amarillo. Se encuentra bajo tierra, en estado sólido, o en estado líquido, como sulfuro fundido.*

surface preparation: The operation necessary to produce a desired or specified surface condition.

preparación de superficie: *La operación necesaria para producir una condición de superficie deseada o especificada.*

surfacing: The application by welding, brazing, or thermal spraying of a layer of material to a surface to obtain desired properties or dimensions, as opposed to making a joint.

alisamiento: *La aplicación, por medio de soldaduras, o rocío térmico, de un estrato de material a una superficie para obtener ciertas dimensiones; en vez de soldar una juntura.*

surfacing material: The material that is applied to a base metal or substrate during surfacing.

material de alisamiento: *El material que se aplica a un metal de base o a un substrato durante el alisamiento.*

surfacing weld: A weld applied to a surface, as opposed to making a joint, to obtain desired properties or dimensions.

cordón de alisamiento: *Un cordón aplicado a una superficie – en vez de para hacer una juntura – para obtener ciertas propiedades o dimensiones.*

sweat soldering: A soldering process variation in which workpieces that have been pre-coated with solder are reheated and assembled into a joint without the use of additional solder (also called sweating).

sudar: *Un proceso de soldadura en el que piezas que han sido previamente cubiertas con soldadura, son recalentadas y montadas como una juntura sin tener que usar soldadura adicional.*

T

tack weld: A weld made to hold the parts of a weldment in proper alignment until the final welds are made.

soldadura provisional a puntos: Una soldadura que es hecha solamente para tener las partes de una soldadura en alineamiento hasta que la soldadura final sea hecha.

tensile strength: The resistance to breaking exhibited by a material when subjected to a pulling stress. Measured in lb/in^2 or kPa.

resistencia a la tensión: La resistencia de un material a romperse cuando fuerzas tensiles están actuando sobre él. Sus unidades son: lb/in^2 o KiloPascal.

tension test: A test in which a specimen is loaded in tension until failure occurs.

prueba de tensión: Una prueba en la que una muestra se somete a una carga de tensión hasta que la muestra se rompe.

theoretical throat: The distance from the beginning of the joint root perpendicular to the hypotenuse of the largest right triangle that can be inscribed within the cross-section of a fillet weld. This dimension is based on the assumption that the root opening is equal to zero.

garganta teórica: La distancia perpendicular del inicio de la raíz de la juntura a la hipotenusa del triángulo recto más grande que se pueda inscribir completamente en la sección de un cordón de soldadura.

thermal conductivity: The ability of a material to transmit heat.

conductividad de calor: La habilidad de un material a transmitir energía calórica.

thermal cutting (TC): A group of cutting processes that severs or removes metal by localized melting, burning, or vaporizing of the workpieces.

corte térmico: Un grupo de procesos para cortar que separa o remueve metal fundiendo el metal localmente, quemando, o evaporando las piezas.

thermal expansion: The expansion of materials caused by heat input.

expansión térmica: La expansión de elementos debida a la absorción de calor.

thermal spraying (THSP): A group of processes in which finely divided metallic or non-metallic surfacing materials are deposited in a molten or semi-molten condition on a substrate to form a thermal spray deposit. The surfacing material may be in the form of powder, rod, cord, or wire.

rocío térmico: Un grupo de procesos en los que se depositan materiales no metálicos finamente divididos, en estado de completa fundición of casi fundidos, sobre un substrato para así formar un depósito de rocío térmico. Estos materiales vienen ya preparados en forma de polvo, varilla, cuerda, o alambre.

thermal stress relieving: A process of relieving stresses by uniform heating of a structure or a portion of a structure, followed by uniform cooling.

reducción de esfuerzos térmicos: El proceso para aliviar los esfuerzos en una estructura o parte de ella, calentándola uniformemente seguido de un enfriamiento uniforme.

three-phase power: A generator or circuit delivering three voltages that are $1/3$ of a cycle apart in reaching maximum value. Three-phase current is usually used for circuits of 220 volts or more.

energía trifásica: Un generador, o circuito que suministra tres tensiones que están desfasadas 1/3 de ciclo (o sea que cada una tiene su valor máximo 1/3 de ciclo despues de la precedente, o 1/3 de ciclo antes de la subsiguiente). Generalmente, corriente trifásica se usa cuando la tensión es de 200 voltios o más.

time temperature transformation (TTT): See isothermal transformation diagram definition.

tiempo-temperatura de transformación: Véase debajo de isothermal transformation diagram arriba

tinning: A non-standard term for pre-coating.

"estañar": Término, no aprobado, que quiere decir revestimiento preliminar.

3F: A welding test position designation for a linear fillet weld applied to a joint in which the weld is made in the vertical welding position.

3F: Una sigla que designa la posición de prueba de la soldadura en el caso de una soldadura angular aplicada a una juntura en la cual el cordón se hace en la posición de soldadura vertical.

3G: A welding test position designation for a linear groove weld applied to a joint in which the weld is made in the vertical welding position.

3G: Una sigla que designa la posición de prueba de la soldadura en el caso de una soldadura de ranura linear aplicada a una juntura en la cual el cordón se hace en la posición de soldadura vertical.

T-joint: A joint between two members located approximately at right angles to each other in the form of a T.

juntura en T: Una juntura en la que los miembros se colocan formando la figura de una T aproximadamente .

torch brazing (TB): A brazing process that uses heat from a fuel gas flame.

soldadura fuerte con soplete: Un proceso de soldadura fuerte en la que el calor viene de la llama de un gas combustible.

torch oscillation: Moving a torch in a back and forth motion.

oscilación del soplete: Movimiento del soplete con una oscilación adelante-atrás.

torch soldering (TS): A soldering process that uses heat from a fuel gas flame.

soldadura blanda con soplete: Un proceso de soldadura blanda en la que el calor viene de la llama de un gas combustible.

torsion: The stress produced in a body, such as a rod or wire, by turning or twisting one end while the other is held firm or twisting in the opposite direction.

torsión: El esfuerzo producido en un cuerpo, v. gr., una varilla, o un alambre, teniendo un extremo fijo y torciendo el otro.

transferred arc: A plasma arc established between the electrode of the plasma arc torch and the workpiece.

arco transferido: Un arco de plasma establecido entre el electrodo del arco de plasma y la pieza.

transverse crack: A crack with its major axis oriented approximately perpendicular to the weld axis.

grieta transversal: Grieta cuyo eje mayor es aproximadamente perpendicular al eje de la soldadura.

travel angle: The angle less than 90° between the electrode axis and a line perpendicular to the weld axis, in a plane determined by the electrode axis and the weld axis. This angle can also be used to partially define the positions of welding guns, torches, rods, and beams.

ángulo de viaje: El ángulo (menos de 90°) entre el eje del electrodo y una línea perpendicular al eje de la soldadura, en un plano determinado por las dos líneas. Este plano puede usarse para definir, parcialmente, la posición de pistolas de soldar, sopletes, varas y vigas.

travel angle pipe: The angle of less than 90° between the electrode axis and a line perpendicular to the weld axis at its point of intersection with the extension of the electrode axis, in a plane determined by the electrode axis and a line tangent to the pipe surface at the same point. This angle can also be used to partially define the positions of welding guns, torches, rods, and beams.

tubo con ángulo de viaje: El ángulo (menos de 90°) entre el eje del electrodo y una línea perpendicular al eje de la soldadura, en su punto de intersección con la extensión del eje del electrodo, en un plano determinado por el eje del electrodo y una línea tangente a la superficie del tubo. Este plano puede usarse para definir, parcialmente, la posición de pistolas de soldar, sopletes, varas y vigas.

ultimate tensile strength: The final measurement of material placed in tension at the point of breaking.

máxima resistencia a la tracción: La última lectura de datos de un material que, puesto en tensión, ha llegado al punto de rotura.

tungsten electrode: A non-filler metal electrode used in arc welding, arc cutting, and plasma spraying, made principally of tungsten.

electrodo de tungsteno: Electrodo metálico, sin material de aportación, usado en solda-dura con arco, corte con arco, rociado de plasma, y hecho principalmente de tungsteno.

2F pipe: A welding test position designation for a circumferential fillet weld applied to a joint in pipe, with its axis approximately vertical, in which the weld is made in the horizontal welding position.

tubo 2F: Una sigla que designa la posición de prueba de la soldadura en el caso de una soldadura circunferencial de ranura linear aplicada a un tubo con su eje aproximada-mente vertical, en la cual el cordón es hecho en la posición de soldadura horizontal.

2F plate: A welding test position designation for a linear fill weld applied to a joint in which the weld is made in the horizontal welding position.

plato 2F: Una sigla que designa la posición de prueba de la soldadura en el caso de una soldadura angular, cuando es aplicada en la posición de soldadura horizontal.

2FR: A welding test position designation for a circumferential fillet weld applied to a joint in pipe, with its axis approximately horizontal, in which the weld is made in the horizontal welding position by rotating the pipe about its axis.

2FR: *Una sigla que designa la posición de prueba de la soldadura en el caso de una soldadura angular circunferencial aplicada a una juntura de tubo, con su eje aproximadamente horizontal, y en la que la soldadura se hace en la posición de soldadura horizontal rotando el tubo en su eje.*

2G pipe: A welding test position designation for a circumferential groove weld applied to a joint in a pipe, with its axis approximately vertical, in which the weld is made in the horizontal welding position.

tubo 2G: *Una sigla que designa la posición de prueba de la soldadura en el caso de una soldadura circunferencial de ranura linear aplicada a una juntura de tubo con su eje aproximadamente vertical, en la cual el cordón es hecho en la posición de soldadura horizontal.*

2G plate: A welding test position designation for a linear groove weld applied to a joint in which the weld is made in the horizontal welding position.

plato 2G: *Una sigla que designa la posición de prueba de la soldadura en el caso de una soldadura de ranura linear aplicada a una juntura en la cual el cordón es hecho en la posición de soldadura horizontal.*

U

U-groove weld: A type of groove weld.
soldadura de ranura en U: *Un tipo de soldaduras de ranura.*

under-bead crack: A crack in the heat-affected zone generally not extending to the surface of the base metal.
grieta debajo del cordón: *Una grieta en la zona afectada por el calor, y que general-mente no llega hasta la superficie del material de base.*

undercut: A groove melted into the weld face or root surface and extending below the adjacent surface of the base metal.
rebaja: *Una ranura fundida dentro de la cara del cordón o la superficie de la raíz, y que se extiende por debajo de la superficie adyacente del metal de base.*

Underfill: A condition in which the weld face or root surface extends below the adjacent surface of the base metal.
bajo-relleno: *Una condición en la que la cara del cordón o de la superficie de la raíz se extiende por debajo de la superficie del metal de base.*

uphill: Welding with an upward progression.
hacia arriba: *Soldadura vertical que procede de abajo hacia arriba.*

V

vertical welding position: The welding position in which the weld axis, at the point of welding, is approximately vertical, and the weld face lies in an approximately vertical plane.
posición de soldadura vertical: *La posición en la que el eje de soldadura, en este punto de la soldadura, es aproximadamente vertical, y la cara del cordón resta en un plano aproximadamente vertical.*

vertical up: A nonstandard term for uphill welding.
verticalmente: Un término, no aprobado, equivalente a soldadura hacia arriba.

V-groove weld: A type of groove weld.
soldadura de ranura en V: Un tipo de soldadura de ranura.

volt: A unit of electrical force or potential.
voltio: Unidad de tensión eléctrica o diferencia de potencial entre dos puntos de un circuito eléctrico.

W

waster plate: A piece of metal used to initiate thermal cutting.
plato de ayuda: Pedazo de plancha que sirve para iniciar el corte térmico.

watt: A unit of electric power equal to voltage multiplied by amperage. One horsepower is equal to 746 watts.
vatio: Unidad de potencia eléctrica. Se obtiene multiplicando la tensión por la corriente. Un "caballo inglés" (hp) es igual a 746 vatios.

wave soldering (WS): An automatic soldering process where workpieces are passed through a wave of molten solder.
soldadura de onda: Proceso de soldar automático en el que las piezas son pasadas por una onda de soldadura fundida.

weave bead: A type of weld bead made with transverse oscillation.
cordón zigzag: Un tipo de soldadura en la cual se hace oscilar a la boquilla en dirección transversal a la dirección de progreso de la soldadura.

weld: A localized coalescence of metal or nonmetals produced either by heating the materials to the welding temperature, with or without the application of pressure, or by the application of pressure alone, with or without the use of filler material.
soldadura: La coalescencia local de metales o metaloides, el resultado de haber calentado los materiales a la temperatura de soldadura, y con o sin la aplicación de presión; o del haber aplicado presión solamente, y con o sin el uso de materiales de aportación.

weldability: The capacity of material to be welded under imposed fabrication conditions into a specific suitably designed structure and to perform satisfactorily in the intended service.
soldabilidad: La capacidad de un material para ser soldado bajo las condiciones de fabricación estipuladas, para crear una estructura adecuada y que pueda desempeñar el servicio para el que fué diseñado satisfactoriamente.

weld axis: A line through the length of the weld, perpendicular to and at the geometric center of its cross-section.
eje de soldadura: Una línea que percorre la longitud de la soldadura atravesando per-pendicularmente el centro geométrico de cada sección transversal de la soldadura.

weld bead: A weld resulting from a pass.

botón de soldadura: *El cordón de soldadura que resulta después de haber hecho un pase por la juntura.*

weld crack: A crack located in the weld metal or heat-affected zone.
grieta de soldadura: *Una grieta en el metal de soldadura o en la zona afectada por el calor.*

welder certification: Written verification that a welder has produced welds meeting a prescribed standard of welder performance.
certificado de soldador: *Verificación, por escrito, que atesta que la persona ahí nombrada ha hecho soldaduras que satisfacen los estándares prescritos de ejecución.*

welder performance qualification: The demonstration of a welder's ability to produce welds meeting prescribed standards.
calificación de ejecución como soldador: *La demostración de la habilidad para hacer soldaduras que satisfagan ciertos estándares prescritos.*

weld face: The exposed surface of a weld on the side from which welding was done.
cara del cordón: *La superficie externa del cordón de soldadura del lado por el que se hizo la soldadura.*

weld groove: A channel in the surface of a workpiece or an opening between two joint members that provides space to contain a weld.
ranura para soldadura: *Un pequeño canal o muesca en una pieza, o la apertura entre dos miembros de una juntura a soldar, que provee el espacio para hacer el cordón.*

welding: A joining process that produces coalescence of materials by heating them to the welding temperature with or without the application of pressure, or by the application of pressure alone with or without the use of filler metal.
soldadura: *Un proceso de unión de partes que resulta en la coalescencia de materiales calentándolos a la temperatura de soldadura, con o sin la aplicación de presión; o aplicando presión solamente; y con o sin el uso de material de aportación.*

welding arc: A controlled electrical discharge between the electrode and the workpiece that is formed and sustained by the establishment of a gaseous, conductive medium called an arc plasma.
arco para soldar: *Una descarga eléctrica, regulada, entre el electrodo y la pieza a soldar. Es generada y mantenida creando un medio gaseoso, conductor, que se llama arco de plasma.*

welding electrode: A component of the welding circuit through which current is conducted and that terminates at the arc, molten conductive slag, or base metal.
electrodo para soldar: *Un componente del circuito eléctrico para soldar, por el cual pasa una corriente que termina en el arco, o la escoria fundida, o el metal de base.*

welding filler metal: The metal or alloy to be added in making a weld joint that alloys with the base metal to form weld metal in a fusion welded joint.
metal de aportación para soldar: *El metal o liga de metales que se aporta cuando se va a hacer la soldadura de una juntura. Este material se liga con el metal de base para formar un metal de soldadura en lo que se llama una juntura soldada por fusión.*

welding helmet: A device equipped with a filter plate designed to be worn on the head to protect eyes, face, and neck from arc radiation, radiated heat, spatter, or other harmful matter expelled during some welding and cutting processes.

casco de protección: *Una parte del vestuario del soldador que se pone en la cabeza para proteger los ojos y el cuello contra radiación del arco, calor, chispas, u otro material que pueda brotar durante una soldadura o un corte con arco. Está equipado con un filtro de vidrio para los ojos.*

welding leads: The workpiece lead (cables) and electrode lead (cables) of an arc welding circuit.

alambres de conexión: *El par de cables que conectan los terminales del suministrador de energía con la pieza a soldar y con el electrodo en un circuito para soldar con arco.*

welding operator: One who operates adaptive control, automatic, mechanized, or robotic welding equipment.

operario soldador: *Persona que hace funcionar equipo que requiere ajuste, equipo automático, equipo mecanizado, o equipo de robots, todos aplicados a la soldadura.*

welding positions: The relationship between the weld pool, joint, joint members, and welding heat source during welding.

posiciones para soldar: *La relación espacial que existe entre la balsa de soldadura, la juntura, los miembros de la juntura, y el aparato suministrador de calor, durante la soldadura.*

welding power source: An apparatus for supplying current and voltage suitable for welding.

suministrador de energía para soldar: *Aparato que produce la tensión y la corriente adecuadas para soldar.*

welding procedure: The detailed methods and practices involved in the production of a weldment.

procedimiento para soldar: *Los métodos y la práctica, en detalle, que se emplean en la producción de una soldadura.*

welding procedure qualification record (WPQR): A record of welding variables used to produce an acceptable test weldment and the results of tests conducted on the weldment of a qualified welding procedure specification.

catálogo de procedimientos para soldadura y sus calificaciones: *Una lista de las variables que se han usado para obtener una soldadura aceptable; y los resultados de las pruebas hechas en la soldadura para calificarla como una especificación de un procedimiento para soldadura.*

welding procedure specification (WPS): A document providing the required welding variables for a specific application to assure repeatability by properly trained welders and welding operators.

especificación de procedimiento para soldar: *Un documento que contiene la selección de las variables necesarias para una aplicación específica, y para asegurar su repetición siempre y cuando esta sea hecha por soldadores y operarios de soldadura.*

welding rectifier: A device, usually a semiconductor diode, in a welding power source for converting alternating current to direct current.

rectificador para soldaduras: Un dispositivo – generalmente un diodo semiconductor – en un suministrador de energía para soldaduras, que convierte corriente alterna en corriente contínua.

welding rod: A form of welding filler metal, normally packaged in straight lengths, that does not conduct the welding current.
varilla de soldar: Una forma de empaque del metal de aportación, normalmente en varas de cierta longitud,y que no conduce la corriente de soldadura.

welding schedule: A written statement, usually in tabular form, specifying values of parameters and welding sequence for performing a welding operation.
plan de soldaduras: Una planilla que enumera los valores de los parámetros y la secuencia de las soldaduras requeridas para hacer un trabajo específico.

welding sequence: The order of making welds in a weldment.
secuencia de soldaduras: El orden en que se deben hacer las soldaduras en cada caso.

welding symbol: A graphical representation of a weld.
símbolo de soldadura: Un icón que representa, graficamente, un tipo de soldadura.

welding tip: That part of an oxyfuel gas welding torch from which the gases issue.
boquilla: La parte final del soplete para soldar con arco por donde salen los gases.

welding transformer: A transformer used for supplying current for welding.
transformador para soldar: Un transformador diseñado para suministrar la corriente necesaria para soldar.

welding wire: A form of welding filler metal, normally packaged as coils or spools, that may or may not conduct electrical current depending upon the welding process with which it is used.
alambre para soldar: Una forma de metal de aportación, normalmente empacado en abobinado o en carretes que pueden ser – o no – conductores, según el tipo de soldadura en la que se usa.

weld interface: The interface between weld metal and base metal in a fusion weld, between base metals in a solid-state weld without filler metal, or between filler metal and base metal in a solid-state weld with filler metal.
áreas colindantes: El espacio entre el metal de la soldadura y el metal de base, en una soldadura a fusión; o entre metales de base en soldaduras de estado sólido pero sin metal de aportación; o entre metal de aportación y el metal de base en soldaduras de estado sólido con metal de aportación.

weld interval: The total of heat and cool times and upslope time used in making one multiple-impulse weld (resistance welding).
intervalo de soldadura: La suma de los tiempos de calentamiento, enfriamiento, y el tiempo de "rampa arriba" usados en la soldadura a resistencia para hacer una solda-dura de impulso múltiple.

weld joint mismatch: Misalignment of the joint members.
desalineamiento de juntura: Desalineamiento de los miembros de una juntura para soldar.

weldment: An assembly whose component parts are joined by welding.
soldadura: Un ensamblaje mantenido por la soldadura de sus miembros

weld metal: The portion of a fusion weld that has been completely melted during welding.
metal de soldadura: En soldaduras a fusión, la porción que se ha fundido completamente durante la soldadura.

weld metal area: The area of weld metal as measured on the cross-section of a weld.
área de metal de soldadura: El área de la sección de la soldadura

weld pass: A single progression of welding along a joint. The result of a pass is a weld bead or layer.
pase de soldadura: Una sola progresión de soldadura a lo largo de una juntura. El resultado es un cordón, o una capa, de soldadura.

weld pass sequence: The order in which the weld passes are made.
secuencia de pases: El número y el orden en que se deben hacer los pases en una soldadura.

weld penetration: A nonstandard term for joint penetration and root penetration.
penetración de la soldadura: Un término, no aprobado, para penetración de juntura, o de raíz.

weld pool: The localized volume of molten metal in a weld prior to its solidification as a weld metal.
balsa de soldadura: El volumen de metal fundido , resultado de una soldadura, antes de que se solidifique, cuando se llama metal de soldadura.

weld puddle: A nonstandard term for weld pool.
charco de soldadura: Término, no aprobado, para balsa de soldadura.

weld reinforcement: Weld metal in excess of the quantity required to fill a joint.
refuerzo de soldadura: La cantidad de metal que excede la necesaria para llenar la juntura.

weld root: The points, shown in a cross-section, at which the root surface intersects the base metal surfaces.
soldadura de raíz: Los puntos, vistos en una sección, en que la superficie de la raíz toca las superficies del metal de base.

weld symbol: A graphical character connected to the welding symbol indicating the type of weld.
símbolo de resultado de la soldadura: Un gráfico que se conecta con el símbolo de la soldadura para indicar que tipo de resultado se ha obtenido.

weld tab: Additional material that extends beyond either end of the joint, on which the weld is started or terminated.
pestaña: Material adicional que aparece a los extremos de la juntura, donde la soldadura comenzó y terminó.

weld toe: The junction of the weld face and the base material.
orilla de la soldadura: La zona donde la cara del cordón toca el material de base.

wetting: The phenomenon whereby a liquid filler metal or flux spreads and adheres in a thin continuous layer on a solid base metal.

mojar: *El fenómeno que se muestra en la adherencia de metal de aportación en forma líquida o el fundente al metal de base, en una capa delgada y continua a todo lo largo de la soldadura.*

wiped joint: A joint made with solder having a wide melting range and with the heat supplied by the molten solder poured onto the joint. The solder is manipulated with a hand-held cloth or paddle to obtain the required size and contour.

juntura limpiada: *Una juntura unida con soldadura que tiene una gama muy amplia de temperaturas de fusión, y con el calor proveniente de soldadura fundida que se vierte sobre la juntura. La cantidad de soldadura puede ser ajustada pasando un trapo o una espátula por las orillas hasta obtener el tamaño y la forma deseada.*

wire feed speed: The rate at which wire is consumed in arc cutting, thermal spraying, or welding.

velocidad de alimentación del alambre: *La rapidez con que se consume el alambre en soldadura con arco, corte térmico, y otras soldaduras.*

work angle: The angle less than 90° between a line perpendicular to the major work-piece surface and a plane determined by the electrode axis and the weld axis. In a T-joint or a corner joint the line is perpendicular to the non-butting member. This angle can also be used to partially define the positions of guns, torches, rods, and beams.

ángulo de trabajo: *El ángulo (menos de 90°) entre una línea perpendicular a la superficie mayor de la pieza y el plano determinado por el eje del electrodo y el eje del cordón. En una juntura a T o en una juntura de esquina, la línea es perpendicular al miembro que no es de tope. Este plano puede usarse para definir, parcialmente, la posición de pistolas de soldar, sopletes, varas y vigas.*

work hardening: Also called cold working; the process of forming, bending, or hammering a metal well below the melting point to improve strength and hardness.

endurecimiento por trabajo: *También llamado endurecimiento por trabajo en frío; es el proceso de martillear el metal a una temperatura muy baja con respecto a su tempera-tura de fusión, con el propósito de aumentar su dureza y su resistencia.*

workpiece: The part that is welded, brazed, soldered, thermal cut, or thermal sprayed.

pieza de trabajo: *La pieza que está siendo soldada, cortada o rociada térmicamente.*

work-piece lead: The electrical conductor between the arc welding current source and work-piece connection.

cable de la pieza de trabajo: *El cable aislado que conecta la pieza al suministrador de corriente de arco.*

wrought iron: A material composed almost entirely of iron, with very little or no carbon.

hierro forjado: *Material compuesto casi exclusivamente de hierro, y si acaso, con algunas trazas de carbono.*

Y

yield strength: The load at which a material will begin to yield, or permanently deform. Also referred to as yield point.

resistencia al relajamiento: *También llamado punto de relajamiento, es la magnitud de carga en tensión en la cual el material deja de ser elástico y comienza a deformarse permanentemente.*

Young's modulus: A ratio between the stress applied and the resulting elastic strain; the slope of a metal's elastic limit curve; a relative measure of a material's stiffness.

módulo de Young: *En un metal sometido a carga de tensión, el cociente entre el esfuerzo resistido por la muestra y la deformación unitaria elástica. En un gráfico de esfuerzo vs. deformación unitaria, este valor es representado por la inclinación de la porción linear, y es una constante para cada metal, por lo que sirve como una indicación de la dureza relativa de los metales. Su sigla es E.*

Index

Picture Credits

American Welding Society Publications
Figures 1-12, 1-13, 1-14,1-20,2-2, 2-3, 2-7, 2-17, 2-18, 2-19,
2-25, 2-26, 3-1, 3-3, 3-4, 3-5, 3-6, 3-10b, 3-11, 3-12, 3-13,
3-14, 3-16, 3-17, 3-18, 4-1, 4-2, 4-4, 4-9, 4-12, 4-13, 4-15,
4-16, 4-19, 4-23, 5-2, 5-6, 6-2, 6-3, 6-4, 6-5, 6-6, 6-7, 6-8, 6-9,
6-17, 6-18, 6-19, 6-20, 6-21, 7-2, 7-3, 7-5, 7-7, 7-8, 7-9, 7-10,
7-12, 7-13, 7-14, 7-15, 7-16, 7-17, 7-18, 7-19, 7-20, 8-1, 8-2,
8-3, 8-4, 8-5, 8-6, 8-7, 8-8, 8-9, 8-11, 9-7, 9-8, 9-8, 9-9, 9-10,
9-12, 9-13, 10-1, 10-2, 10-3, 10-4, 10-5, 10-6, 10-7, 10-8, 10-9,
10-10, 10-11, 10-12, 10-13, 10-14, 10-15, 10-16, 10-17, 10-18,
10-19, 10-20, 10-21, 10-22, 10-23, 10-24, 10-26, 10-27, 11-7,
11-11, 11-19, 11-23, 12-8, 12-9, 12-10, 12-11, 12-12, 12-14,
12-15, 12-17, 12-22, 12-23, 12-24, 12-25, 12-26, 12-27, 12-28,
12-29, 12-30, 13-16, 13-17, 13-18, 13-25, 13-26,13-27, 13-40,
13-41, 13-42, 13-43, 13-48, 14-1, 14-2, 14-3, 14-4, 14-5, 14-6;
Tables 1-1, 3-1, 5-2, 5-4, 5-5, 6-1, 6-5, 6-6, 6-7ab, 6-8, 6-10,
6-11, 7-1, 7-2, 7-3, 7-4, 7-5, 11-1, 11-2, 11-3, 12-1, 12-2.

ESAB Welding and Cutting Products
Figures 1-15,1-6,1-20, 2-20, 2-21, 2-22, 2-24, 2-30, 9-5, 12-16.

The Lincoln Electric Company
Figures 5-8, 5-9, 5-10, 5-11; Table 5-1.

Thermadyne Industries, Inc.
Figures 1-18,2-11,2-12, 2-13, 2-14.